FUNDAMENTAL PROBLEMS IN STATISTICAL MECHANICS

II

Edited by

E. G. D. COHEN

The Rockefeller University, New York

1968

NORTH-HOLLAND PUBLISHING COMPANY - AMSTERDAM

Library of Congress Catalog Card Number 68-18112

Publishers:
NORTH-HOLLAND PUBLISHING CO. – AMSTERDAM
Sole distributors for U.S.A. and Canada:
Interscience Publishers, a division of
JOHN WILEY & SONS, INC. – NEW YORK

PRINTED IN THE NETHERLANDS

PREFACE

This book contains the lectures and some seminars of the 1967 Summerschool on Fundamental Problems in Statistical Mechanics, the purpose of which was to review critically some of the developments in Statistical Mechanics which have come to the foreground since 1961, when the previous summerschool of this kind was held. As then, emphasis was in general placed more on clarity of presentation and an understanding of the problems as a whole than on specialized methods.

The present volume, is however, not merely a sequel to the book "Fundamental Problems in Statistical Mechanics", which contained the lectures of the 1961 summerschool. It is also intended to be the second in a series, which reviews fundamental problems in statistical mechanics about every five years. It is hoped that by doing this not only a useful survey of problems of interest is obtained at regular but not too frequent intervals, but also the *development* of these problems over the years will become apparent.

As far as the subjects are concerned, only some aspects of quantum statistical mechanics were discussed, in particular of interacting Bose- and Fermi-systems and their application to the problems of liquid helium and superconductors. The reason for this is that at the time of this summerschool at least two other schools, entirely devoted to this subject, were held.

The summerschool took place from June 20-July 8 in Hotel Noordzee, Noordwijk, The Netherlands, and was organized by the NUFFIC (Foundation for International Cooperation) in cooperation with the ROCKEFELLER UNIVERSITY, New York, while financial support was given by NATO and by ZWO, the Netherlands Organization for Pure and Scientific Research. The school was attended by about 75 students from 20 different countries. The scientific organizing committee was composed of:

E. G. D. Cohen, Rockefeller University, New York
N. M. Hugenholtz, University of Groningen
N. G. van Kampen, University of Utrecht.

The program of the school consisted of 36 regular lectures given during the morning by 8 lecturers and a number of seminars on related topics, given during the afternoon.

All the regular lectures of the school are published in this volume together with 3 seminar talks, viz. those of H. B. G. Casimir, B. Jancovici and L. Waldmann.

It is hoped that this book will prove useful both to the newcomer and to the specialist in the field of statistical mechanics.

E. G. D. Cohen

CONTENTS

Preface v

An outline of statistical mechanics
G. E. Uhlenbeck 1

Phase transitions
P. W. Kasteleyn 30

Toward a unified view on mathematical theories of phase transitions
M. Kac 71

A survey of the theories of freezing
B. Jancovici 106

On the Gibbs phase rule
D. Ruelle 113

Photon statistics
R. J. Glauber 140

On Bose-Einstein condensation
H. B. G. Casimir 188

Quantum mechanics of infinitely large systems
N. M. Hugenholtz 197

The kinetic theory of dense gases
E. G. D. Cohen 228

Kinetic theory of dilute gases with internal molecular degrees of freedom
L. Waldmann 276

Statistical mechanics of classical plasmas
N. G. van Kampen 306

AN OUTLINE OF STATISTICAL MECHANICS

G. E. UHLENBECK
The Rockefeller University

1. INTRODUCTION

In these lectures I will try to present a kind of rough outline of statistical mechanics. It will be very incomplete and one-sided, but I hope it may serve as a background for the more detailed expositions of specific problems which will follow [1] *.

In the first place one always must emphasize that the basic concern of statistical mechanics is the study and elucidation of the relation between the macroscopic description of the physical phenomena and the underlying atomic theory. Depending on whether one describes the molecular motion and the intermolecular interactions by the classical or by the quantum mechanics, one can then make the following conceptual scheme:

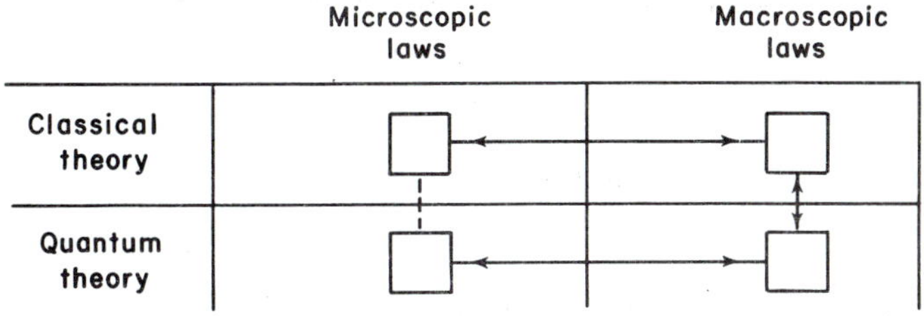

Fig. 1.

The microscopic laws are supposed to be known and also the relation between the classical and the quantum theory is supposed to be completely understood with regard to the atomic description. The general problem of statistical mechanics is to elucidate the other three relations indicated in the scheme.

Here one should say that the distinction between a microscopic and a macroscopic description is, of course, in principle not sharp since it depends on the number of particles and also on the zeal of

* Numbers refer to the notes at the end of these lectures.

the macroscopic observer and on his technical equipment. Although one must keep this in mind, it is usually clear in practice what one means by a macroscopic theory. So I will not try to elaborate the distinction; in the future it may be necessary to distinguish various levels of macroscopic description, especially when one thinks of the biological phenomena. It is perhaps also good to point out that the classical macroscopic laws, like the laws of thermodynamics or hydrodynamics, form the most familiar and well established part of physics. As a result, statistical mechanics has here more the character of an a-posteriori theory; it tries to explain phenomena which are already well known. This does not mean that there are not many unsolved problems, but it implies that there are, so to say, very few surprises, especially from the experimental point of view [2]. This academic flavor of classical statistical mechanics explains why the facts of plasma physics and especially the macroscopic quantum phenomena are of such great interest for statistical physics. In these fields the macroscopic theories and especially the limitations of these theories are *not* very well established; and they offer, therefore, a great challenge for the worker in statistical mechanics. In these fields the theory has sometimes recaptured the predictive power [3] which is so often absent in the classical theory. Unfortunately, lack of time and lack of competence will prevent me from including these fields in my outline, and I will therefore be exclusively concerned with only one of the three conceptual relations in the scheme of fig. 1, namely the relation between classical mechanics and classical macroscopic physics. And in order to exclude plasma physics, I will, in addition, always assume that the forces between the molecules have a short range and are of the familiar van der Waals-London type.

2. THE ZEROTH LAW OF THERMODYNAMICS

The most famous and well established macroscopic laws are the laws of thermodynamics and statistical mechanics in fact started from the attempt by Boltzmann and Gibbs to interpret these laws in terms of classical mechanics. You know that this seemed impossible because of the apparently irreconcilable conflict between mechanics and thermodynamics with regard to the role of time. According to thermodynamics, an isolated system (that is from the molecular point of view a conservative and bounded mechanical system consisting of a large number of molecules) will, in the course of time, approach a state of thermal equilibrium in which all macroscopic variables have reached steady values. This is the so-called *zeroth law of thermodynamics*, and it is in conflict with *the*

Poincaré theorem [4] according to which *any* bounded and conservative mechanical system will execute a quasi-periodic motion. Representing the state of the mechanical system by a point in the phase space (Γ-space), this point will move on the energy surface in such a way that it always will return to any given region around the initial position.

The resolution of this dilemma just lies in the realization that the two sides of the conflict refer to two different levels of description of the physical system. Thermal equilibrium is a *macroscopic* notion described in terms of a small number of macroscopic variables, while the Poincaré theorem is a property of the *microscopic* description, which requires a very large number of variables, say $6N$ if the system consists of N monoatomic molecules. More specifically, let us suppose that macroscopically the system is described by a small number of variables y_i, which are functions of the coordinates and momenta of all the particles. So:

$$y_i = f_i(x_1, x_2, \ldots, x_N) \,, \tag{1}$$

where $x_k \equiv (\boldsymbol{r}_k, \boldsymbol{p}_k)$ stands for the phase of the kth particle and where $i = 1, 2, \ldots, M$ with $M \ll N$. If all particles are identical, the functions f_i should, in addition, be symmetric in their arguments. Clearly, a macroscopic description is a contracted description, which means that for given values of the y_i there are very many values of the $x_1, \ldots, x_N$. In fact, we will assume that a given set of the y_i corresponds to a *region* on the energy surface in Γ-space. In addition, we require that for any sensible macroscopic description and for large N there is one region which is overwhelmingly the largest. The various values of the y_i will therefore divide the energy surface as shown very schematically in fig. 2.

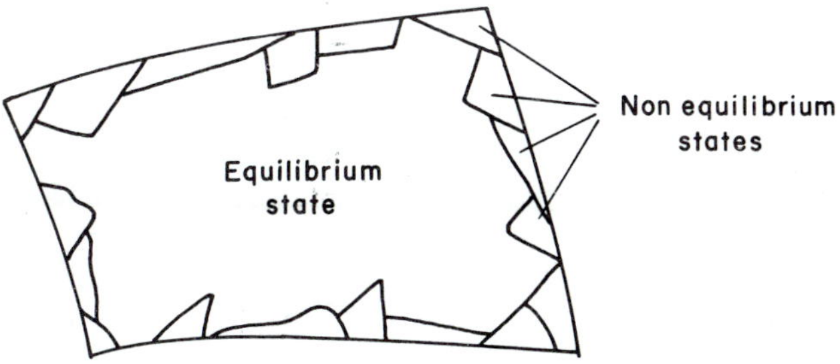

Fig. 2.

One now can argue in various ways. Following Boltzmann one can say that if the system is not in thermal equilibrium it should be represented by a point in a region which is not the maximum region. If one then assumes that in the motion of the Γ-point over the energy surface there is no special preference for any portion of the surface, then it is very plausible that the time $t(A)$ during which the Γ-point is in a region A is roughly proportional to the magnitude of the area of A. Identifying the maximum region with the macroscopic state of thermal equilibrium, one then can conclude that if the system is *not* in thermal equilibrium, it almost always will go into that state; and if the system is in thermal equilibrium, it almost always will stay in that state although fluctuations away from equilibrium will and must occur because of the quasi-periodic character of the motion of the Γ-point. This is the Boltzmann picture; it clearly reconciles the reversibility of the mechanical motion as expressed by the Poincaré theorem with the approach to equilibrium as required by the zeroth law of thermodynamics.

Following Gibbs one can make the argument more precise by introducing the notion of a probability distribution or of an ensemble as Gibbs calls it. Suppose one only knows that at $t = 0$ the system is in a definite non-equilibrium macroscopic state. Then the Γ-point can be anywhere in the corresponding region of the energy surface and this lack of knowledge can be represented by an initial probability distribution which is uniform in the region and zero everywhere else [5]. In the course of time the distribution or ensemble fluid will flow out of the region into other portions of the energy surface. One knows from the Liouville theorem that the area will always remain the same [6] but, of course, the shape varies and it is again plausible to assume that the fluid will be drawn out more and more in a thin filament which will wind itself more and more uniformly over the whole energy surface. After a long time the probability that the system is in any macroscopic state will then be proportional to the area of the corresponding region, and hence the equilibrium state will be overwhelmingly the most probable state. This is the so-called *phase mixing process* of Gibbs and besides interpreting the zeroth law, it shows in principle how at any time the probabilities of the various macroscopic states can be found from the knowledge of the initial macroscopic state, since they are determined by the amounts of the ensemble fluid in the corresponding regions at time t. Note also that the approach to the uniform distribution over the energy surface (which Gibbs calls the *microcanonical ensemble*) will continue when t increases *although* each Γ-point will move quasi-periodically as the Poincaré theorem demands.

Let me conclude with a number of remarks.

1. The notion of the macroscopic variables y_i is very abstract and schematic, so it may be helpful to have an example. For an ideal gas the distribution n_i of the N points over the cells ω_i in the phase space (μ-space) of a molecule can be considered as a kind of macroscopic description of the state of the gas. The sizes of the cells must be finite and their number $M \ll N$. The macroscopic variables are the occupation numbers n_i of the cells which can be written as function of the Γ-phase as in equation (1) by putting

$$n_i = \sum_{k=1}^{N} \Delta_i(x_k) \,, \tag{2}$$

where

$$\Delta_i(x) = \begin{cases} 1 \text{ if } x \text{ is in } \omega_i \\ 0 \text{ otherwise} \end{cases} .$$

Here it is clear that a set n_i corresponds to a volume in Γ-space, namely

$$W(n_1, n_2, \ldots, n_M) = \frac{N}{n_1!\, n_2! \ldots n_M!}\, \omega_1^{n_1}\, \omega_2^{n_2} \ldots \omega_M^{n_M}$$

and one proves [7] that the Maxwell-Boltzmann distribution corresponds to the overwhelmingly largest volume and hence describes the thermal equilibrium state.

2. Especially artificial are the sharp boundaries of the regions corresponding to the y_i or the n_i. An exchange of one molecule between two cells changes the distribution in μ-space but can hardly be said to be macroscopically observable. One should, of course, take into account the accuracy of a macroscopic measurement. This will deepen the description and make it more realistic, but I believe it will not change the qualitative features.

3. There is no indication how one should choose the y_i, except that they should fulfill the two conditions which we have imposed. Other requirements will be discussed later when we come to the non-equilibrium macroscopic theories. Of course the equilibrium state as described by the microcanonical ensemble is the *same* whatever macroscopic variables y_i one has chosen, although the approach may be different. Equilibrium states only differ in the overall *thermodynamic variables,* which are for our simple case the total energy E, the volume V and the number of particles N.

4. The role of the *ergodic theorems.* The Boltzmann-Gibbs ar-

gument depends besides on the notion of macroscopic variables, on an assumption about the motion of the Γ-point over the energy surface, which I expressed vaguely by saying that there should not be any preference for any portion of the energy surface. The precise mathematical formulation of this property is that one must assume that the system is *metrically transitive*. By this one means that the energy surface can *not* be divided in two regions of finite measure such that the orbits starting from points in one of the regions always remain in that region. For a metrical transitive system, one can show strictly from the Birkhoff ergodic theorems [8] that

$$\lim_{T \to \infty} \frac{t(A)}{T} = \frac{V(A)}{V}, \tag{3}$$

where $t(A)$ is the time the phase point spends in the region A and $V(A)$ is the volume of the part of the energy shell (whose total volume is V), which is based on A. Clearly (3) expresses precisely what Boltzmann had in mind. Also, the stronger phase mixing property required for the Gibbs argument can be shown to hold for a metrically transitive system.

So far one can only say that the gap in the Boltzmann-Gibbs argument has been formulated more precisely but that it has not been filled. How does one know whether a system of N particles with short range forces *is* metrically transitive? It is clearly a property of the mechanical motion which may be plausible but which should be proved from the equations of motion. This question, often called loosely the ergodic problem, remained open for a long time and only recently Sinai [9] has been able to show that a system of N hard spheres (with $N \geqslant 3$) moving in a volume with periodic boundary conditions is metrically transitive. When this remarkable result can be generalized to systems of particles with attractive *and* repulsive short range forces, then I think one can say that at least for systems of neutral molecules the relation between classical mechanics and thermodynamics has finally been completely understood.

3. ENSEMBLES AND CHARACTERISTIC THERMODYNAMIC FUNCTIONS

Once one has understood the zeroth law, the interpretation of the other laws of thermodynamics becomes more or less stardard [10]. Let me remind you of the introduction of the notion of temperature. To do this one must discuss the equilibrium of the given system A when it is weakly coupled to a very large heat reservoir B.

Under weakly coupled we mean that the total Hamiltonian can be taken to be $H_A + H_B$ although A and B interact and can interchange energy. We assume that the total system $A + B$ is in equilibrium in the sense of the microcanonical ensemble and one can then prove the theorem [11] that in the limit that system B is very large, the probability that system A is in a range $d\Gamma_A$ of its phase space is given by the *canonical distribution:*

$$D_A d\Gamma_A = C e^{-\beta H_A} d\Gamma_A \,. \tag{4}$$

Here β is a constant which depends on the system B, but which for large B is independent of the size and of the nature of $B_\perp$;β has all the required properties of temperature and it must be a universal function of the thermodynamic temperature T. The quantity C is a normalization constant; since

$$\int D_A d\Gamma_A = 1 \,,$$

one gets:

$$\frac{1}{C} = \int d\Gamma_A e^{-\beta H_A} = \int_0^\infty d\epsilon\, \mathcal{S}_A(\epsilon) e^{-\beta\epsilon} \,, \tag{5}$$

where $\mathcal{S}_A(\epsilon)$ is the area of the energy surface $H_A = \epsilon$ in the phase space of A.

The first law of thermodynamics becomes from the molecular point of view essentially a definition of the notion of quantity of heat Q. From the molecular interpretation of Q and T one then can *prove* the second law for reversible processes and find the expression for the entropy S. Comparison with the ideal gas laws then fixes β to be $1/kT$, and the whole discussion culminates in the formula:

$$e^{-A/kT} = \frac{1}{N!h^{3N}} \int \ldots \int d\boldsymbol{r}_1 \ldots d\boldsymbol{r}_N d\boldsymbol{p}_1 \ldots d\boldsymbol{p}_N e^{-H_N/kT} \,, \tag{6}$$

where H_N is the hamiltonian of the system of N particles in the volume V and $A(V, T, N)$ is the Helmholtz free energy of the system. This equation gives, therefore, a general recipe how to calculate the macroscopic thermodynamic quantities in terms of the inter-molecular forces.

I will omit the details of the derivation [12], but let me make some remarks about this basic final result.

1. From $A(V, T, N)$ one finds the entropy S, pressure p, and chemical potential μ, since:

$$dA = -SdT - pdV + \mu dN \tag{7}$$

and from these one finds all other thermodynamic quantities. They can therefore all be found once one has calculated A, which is why one calls A a *characteristic thermodynamic function.*

2. The right hand side of (6) is called the *canonical partition function*. The factors h^{3N} and $N!$ are conventional but they are based on very sound conventions! One needs the factor h^{3N} to make the volume element in Γ-space $d\Gamma = d\boldsymbol{r}_1 \ldots d\boldsymbol{r}_N d\boldsymbol{p}_1 \ldots d\boldsymbol{p}_N$ dimensionless and especially to insure that the quantum-mechanical expression [13] for the partition function goes over for $T \to \infty$ into the classical form in (6). These are, so to say, aesthetic reasons but they certainly justify the introduction of Planck's constant in a result based on classical mechanics, which may seem a bit artificial at first. The factor $N!$, although also conventional [14], is more important. One wants A to be an *extensive variable* in the so-called *thermodynamic limit*. By this we mean that for a large system, or more precisely in the limit $N \to \infty$, $V \to \infty$, $V/N = v$ is finite [15]:

$$\lim A(V, T, N) = Na(v, T) \, . \tag{8}$$

This must, of course, be *proved* [16]; it requires the factor $N!$, but the proof depends also on what one assumes about the intermolecular forces, although since (8) is quite generally valid, only some general features of the forces are required. Since the entropy S and internal energy U follow from A and are given by:

$$S = -\frac{\partial A}{\partial T} \, ,$$

$$U = A - T\frac{\partial A}{\partial T} \, ,$$

also S and U are extensive variables in the thermodynamic limit.

3. In (6) the Helmholtz function A seems to have a special significance. However, this is only apparent. Thermodynamically, the first and second law combined give for a homogeneous one-component system the basic relation:

$$U = TdS - pdV + \mu dN \, , \tag{9}$$

in which all independent variables are extensive. By a *Legendre transformation*, one then can transform one or more of the intensive variables T, p, and μ into independent variables. With $A =$

$U - TS$ this leads to (7), and going further one gets:

$$dG = -SdT + Vdp + \mu dN , \tag{10}$$

with $G = A + pV$ is the Gibbs free energy. Such transformations correspond to successive *Laplace transformations* of the partition functions of the corresponding ensembles. To see this, define the microcanonical partition function by:

$$Z_{\text{micro}}(U, V, N) = \frac{1}{N!h^{3N}} \mathcal{S}(U, V, N) , \tag{11}$$

where $\mathcal{S}$ is the area of the energy surface $H = U$. From (5) and (6), one then finds that

$$Z_{\text{can}}(V, T, N) = \int_0^\infty d\epsilon Z_{\text{micro}}(\epsilon, V, N) e^{-\epsilon/kT} . \tag{12}$$

To see that this Laplace transformation corresponds to the Legendre transformation of (9) into (7), one notes that $Z_{\text{micro}}(\epsilon)$ is for large N a very rapidly increasing function of ϵ, so that the integrand in (12) has a very sharp maximum which must be for $\epsilon = U$. Replacing the integral by this maximum term, one gets ($\beta = 1/kT$):

$$Z_{\text{can}} \cong \max_{(\epsilon)} Z_{\text{micro}}\, e^{-\beta\epsilon} ,$$

and:

$$\left(\frac{\partial Z_{\text{micro}}}{\partial \epsilon} - \beta Z_{\text{micro}}\right)_{\epsilon = U} = 0 ,$$

or:

$$\beta Z_{\text{micro}} = \frac{\partial Z_{\text{micro}}}{\partial U} \tag{13}$$

and:

$$Z_{\text{can}} = e^{-\beta U} Z_{\text{micro}}(U, V, N) . \tag{14}$$

So if we put:

$$S(U, V, N) = k \ln Z_{\text{micro}}(U, V, N) \;, \tag{15}$$

(which is, of course, the famous Boltzmann formula $S = k \ln W$), then (13) becomes:

$$\frac{\partial S}{\partial U} = \frac{1}{T}$$

and from (14) one gets:

$$A = -kT \ln Z_{\text{can}} = U - TS \;.$$

Similarly one can see that a Laplace transform in V:

$$Z_p = \int_0^\infty \mathrm{d}V \, Z_{\text{can}}(V, T, N) \, \mathrm{e}^{-pV/kT} \;,$$

which leads to the partition function of the so-called *pressure ensemble* corresponds to the Legendre transform of (7) into (9) with:

$$G = -kT \ln Z_p \;. \tag{16}$$

Since the different ensembles are therefore related to the different characteristic thermodynamic functions, that is to different choices of the independent variables, one should expect that in the thermodynamic limit, all ensembles are thermodynamically equivalent. Of course, this equivalence of the various ensembles should and can be proved quite rigorously [17]. It depends again only on quite general features of the molecular system.

Still a word about the so-called *grand ensemble*. Its partition function is formed from Z_{can} by a (discrete) Laplace transform in N:

$$Z_{\text{gr}}(T, V, \mu) = \sum_{N=0}^{\infty} Z_{\text{can}}(T, V, N) \, \mathrm{e}^{+\mu N/kT} \;.$$

Thermodynamically this corresponds to a Legendre transformation of (7) in which instead of the variables p, V one switches the variables μ, N around. One then gets from (7):

$$\mathrm{d}(A - \mu N) = -S\mathrm{d}T - p\mathrm{d}V - N\mathrm{d}\mu$$

and one finds:

$$A - \mu N = -kT \ln Z_{gr} \ . \tag{17}$$

Since in the thermodynamic limit μ is simply G/N, the meaning of the characteristic function $A - \mu N$ is just pV.

4. Finally one should expect that from the basic formula (6) it follows in the thermodynamic limit not only that A and the other characteristic functions become proportional to N, but also that the *thermodynamic stability conditions* are fulfilled. For our simple one-component system, these conditions are:

$$\begin{aligned} c_v &= -T\frac{\partial^2 a}{\partial T^2} > 0 \ , \\ \left(\frac{\partial p}{\partial v}\right)_T &= -\frac{\partial^2 a}{\partial v^2} \leqslant 0 \end{aligned} \tag{18}$$

and one has been able to prove them under quite general assumptions on the intermolecular forces [18].

4. THE PROBLEM OF THE PHASE TRANSITIONS

The most outstanding unsolved problem of equilibrium statistical mechanics is the problem of the phase transitions. Why do all substances occur in at least three phases, the solid, liquid, and vapor phase which can coexist in the triple point? Why is there, again for all substances, a critical point for the vapor-liquid equilibrium, while apparently there is no critical point for the fluid-solid transition. Note that since these are *general* phenomena, they must have a *general* explanation; the precise details of the molecular structure and of the intermolecular forces should not matter. Note also that thermodynamics do *not* give an answer to these questions. It can only say that *if* a substance can occur in more than one phase, then these phases are in equilibrium if, besides temperature and pressure, the chemical potentials of the phases are the same. The equation:

$$\mu_1(p,T) = \mu_2(p,T) \tag{19}$$

then determines the dependence of the equilibrium pressure on the temperature.

The starting point of the theory should, of course, be the basic formula (6). By calculating the partition function in the thermodynamic limit, one should find that for low enough temperatures and for quite general intermolecular forces, the pressure $p = -\partial a/\partial v$ as function of v has discontinuities in the first derivative and looks about as shown in fig. 3 when T is above the triple point and below the critical temperature [19].

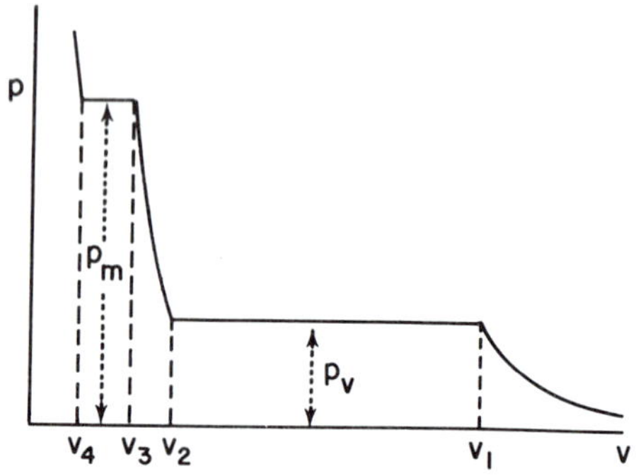

Fig. 3.

At $v = v_1$ the vapor starts to condense and at $v = v_3$ the liquid starts to solidify; p_v is the saturated vapor pressure and p_m is the melting pressure; p_v and p_m are functions of T alone which cross at the triple point, while $p_v(T)$ stops at the critical temperature (see fig. 4).

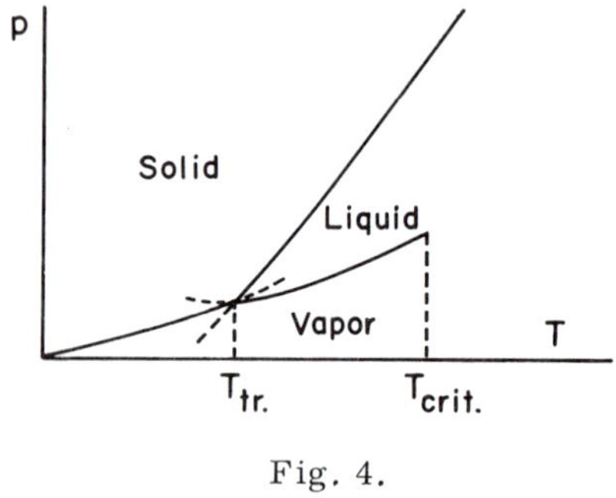

Fig. 4.

All this should follow from eq. (6) and I hardly need to say that this has not been accomplished so far!

The greatest progress has been made with respect to the liquid-vapor equilibrium mainly because of the clue provided by the van der Waals theory [20]. Let me remind you of the simple derivation of the van der Waals equation from eq. (6), which is due to Ornstein.

After integrating over the momenta the canonical partition function becomes:

$$Z(V,T,N) = \frac{1}{N!\,\lambda^{3N}} \int_V \ldots \int_V \mathrm{d}\boldsymbol{r}_1 \ldots \mathrm{d}\boldsymbol{r}_N \exp\Big(-\beta \sum_{i<j} \varphi(|\boldsymbol{r}_i - \boldsymbol{r}_j|)\Big) , \qquad (20)$$

where $\lambda = \hbar/(2\pi mkT)^{\frac{1}{2}}$ and where $\varphi(r)$ is the intermolecular potential assumed to be central and additive. Following van der Waals, suppose that $\varphi(r)$ consists of a hard core with diameter d and a weak but very long range attractive part. One can, for instance, put:

$$\varphi_{\text{attr}}(r) = -\gamma^3 F(\gamma r) ,$$

where F is a monotonic decreasing function and where γ is supposed to be very small; φ_{attr} is then very weak, has the range $1/\gamma$, and:

$$-\int \varphi_{\text{attr}}(r)\, \mathrm{d}\boldsymbol{r} = C = \text{constant} .$$

Each molecule interacts with the large number of molecules in its action sphere and the average potential energy will be proportional to the number density and is, in fact, equal to $(-CN/V)$. So for almost all configurations the total attractive potential energy can be taken to be $(-CN^2/2V)$ and one can approximate (20) by taking the attractive energy term outside the integral. One then gets:

$$Z = \frac{1}{N!\,\lambda^{3N}} \exp\Big(\frac{CN^2}{2kTV}\Big) \int_v \ldots \int_v \mathrm{d}\boldsymbol{r}_1 \ldots \mathrm{d}\boldsymbol{r}_N \prod_{i<j} S(|\boldsymbol{r}_i - \boldsymbol{r}_j|) , \qquad (21)$$

where:

$$S(x) = \begin{cases} 0 \text{ for } x < d \\ 1 \text{ for } x > d \end{cases} .$$

The integral is still difficult and cannot be done exactly. However, the result is surely a function of V and N only, and one can write:

$$Z = Z_{\text{h.c.}}(V,N) \exp\Big(\frac{CN^2}{2kTV}\Big) ,$$

where $Z_{\text{h.c.}}$ is the partition function for a gas of hard spheres. One concludes that:

$$p = p_{\text{h.c.}} - \frac{a}{v}, \tag{22}$$

where $a = C/2$ is the van der Waals constant a and $p_{\text{h.c.}}$ must be of the form:

$$p_{\text{h.c.}} = \frac{kT}{v} G\left(\frac{v}{b}\right), \tag{23}$$

where $b = (2\pi/3)d^3$ is the van der Waals constant b. For large v one can expand G in powers of $1/v$ (virial expansion) and one finds:

$$G = 1 + \frac{b}{v} + \frac{5}{8}\frac{b}{v^2} + \dots . \tag{24}$$

The van der Waals form for $p_{\text{h.c.}} = kT/(v - b)$ is therefore only a rough approximation. However, using the three terms is (24) one gets already a critical point. Below the critical temperature, the isotherm shows the van der Waals wiggle so that with the Maxwell rule one explains the vapor-liquid equilibrium. The argument is, of course, not kosher, but it can be fixed up [21]. One knows that the reason of the conflict of the van der Waals-like isotherm with the stability theorem (18) is the fact that there are *two* limiting processes, namely the thermodynamic limit and the limit $\gamma \to 0$, which should be done in the proper order. If one does that, then one gets theMaxwell rule automatically but otherwise the answer is the same.

The van der Waals theory gives an important clue, but, of course, it does *not* solve the condensation problem completely, because the assumption of very long range and weak attractive forces is artificial. We *know* that the attractive force has *not* a very long range and still all substances condense below a certain critical temperature. It is quite remarkable that the van der Waals theory explains at least qualitatively so many features of the vapor-liquid equilibrium, but there are also serious difficulties. I mention only:

1. The theory is *independent of the number of dimensions*. In the van der Waals limit $\gamma \to 0$ one even would get condensation for a one-dimensional gas, while it can be shown rigorously that for *finite* range (that is *real*!) forces there cannot be a phase transition in one dimension.

2. The singular behavior of probably all substances near the critical point is quite different from the predictions of the van der Waals theory [22]. For instance, the specific heat at constant critical volume should show as function of the temperature according to van der Waals a discontinuity at the critical point, while one

finds that the specific heat becomes logarithmically infinite near this point.

The basic question is therefore to find out what happens when the range of the forces is *finite*, and the only clue one has in this case is provided by the famous Onsager solution of the two dimensional Ising problem. Since this will be discussed in detail by Kasteleyn and Kac, let me only say that the Ising model is a model for the ferromagnetic and order-disorder phase transitions. The molecules are arranged on a lattice and they carry (scalar) spins between which there is a given interaction. One usually assumes that only nearest neighbors interact with each other, which is clearly the shortest possible range of the interaction. For this case, it is easy to show that in one dimension there is no phase transition. For a square lattice Onsager showed not only that there is a transition, but he was also able to discuss completely the nature of the singularity. Especially, he found the striking result that near the transition temperature the specific heat becomes logarithmically infinite. In three dimensions one can prove that there is a transition but its nature has not been completely elucidated.

One can solve the Ising problem also in the van der Waals limit, that is for weak and very long range interactions between the spins. This leads to the Weiss theory of ferromagnetism and to the Bragg-Williams theory of the order-disorder transitions. These theories are quite similar to the van der Waals theory; they also give a transition independent of the number of dimensions and with a discontinuity in the specific heat at the transition point.

Clearly the Onsager and the van der Waals-like solution of the Ising problem correspond to two opposite limits about the ratio of the range $1/\gamma$ of the attractive forces to the lattice distance d [23]. One may hope that in the not too distant future a complete theory will be developed which will combine the features of these two famous solutions. This theory should show the basic dependence on the dimensionality and also the basic independence on the details of the interaction mechanism [24].

5. THE PROBLEM OF THE SOLID PHASE

One might ask whether the Onsager and/or the van der Waals theories give any clue about the solid phase and the melting process. I think that this is *not* the case and as a result at least the theoretical investigation of the solid-fluid equilibrium seems completely at a standstill. The old ideas of Lindemann and others, which connect the melting process either with a critical size of the amplitudes of oscillation of the atoms in the crystal or with a critical weakening of the shear elasticity of the solid, have not led to any real theoretical insight. I believe that the solid-fluid transition is essentially different from the vapor-liquid transition and will require quite new methods for its understanding. To support this belief, let me make the following points.

1. Before trying to understand the melting process one might well ask first whether one can prove, assuming the usual type of intermolecular forces, that at low enough temperatures the molecules will be arranged in a regular lattice. Specifically, can one show that by taking the minimum of the potential energy:

$$E = \sum_{i<j}^{N} \varphi(|\boldsymbol{r}_i - \boldsymbol{r}_j|) , \tag{25}$$

where $\varphi(r)$ has the usual van der Waals-London form, one obtains for large N (strictly for $N \to \infty$) for the positions of the N points a discrete lattice? One can call this the problem of the *existence of the solid phase,* and so far as I know it has not been solved, although physically there is of course no doubt [25].

2. The vapor-liquid transition can be characterized by saying that at the condensation point there is an *onset of long range order*. I mean by this that in the two phase region the various distribution functions $n_2(\boldsymbol{r}_1, \boldsymbol{r}_2)$, $n_3(\boldsymbol{r}_1, \boldsymbol{r}_2, \boldsymbol{r}_3), \ldots$ do *not* fulfill the product property anymore, so that for instance if $|\boldsymbol{r}_1 - \boldsymbol{r}_2| \to \infty$, $n_2(\boldsymbol{r}_1, \boldsymbol{r}_2)$ will *not* go to $n_1(\boldsymbol{r}_1) n_1(\boldsymbol{r}_2) = 1/v^2$ or in other words even for large separations the particles will not become independent of each other. In fact, in the two phase region one expects (and for the one-dimensional van der Waals model one can prove) that:

$$n_s(\boldsymbol{r}_1, \ldots, \boldsymbol{r}_s, v) = \frac{1}{v}[x_1 v_1 n_s(\boldsymbol{r}_1, \ldots, \boldsymbol{r}_s, v_1) + x_2 v_2 n_s(\boldsymbol{r}_1, \ldots, \boldsymbol{r}_s, v_2)] , \tag{26}$$

where x_1, x_2 are the mole fractions and v_1, v_2 the specific volumes of the two phases so that $x_1 + x_2 = 1$, $x_1 v_1 + x_2 v_2 = v$. One can argue that (26) expresses the geometrical separation of the two phases [26] and it implies, say for $s = 2$, that:

$$n_2(\boldsymbol{r}_1, \boldsymbol{r}_2, v) \to \frac{1}{v}\left(\frac{x_1}{v_1} + \frac{x_2}{v_2}\right) > \frac{1}{v^2} \tag{27}$$

for $|\boldsymbol{r}_1 - \boldsymbol{r}_2| \to \infty$. Here one has used the fact that for each of the two phases separately, the product property holds for the distribution functions, which means that in *neither* of the two phases long range correlations are assumed to exist. Eq. (27) then shows that long range correlations begin to appear when x_1 or x_2 is different from zero or when a second phase begins to appear. I think that the onset of long range order or more precisely eq. (26) is a deeper characterization of the vapor-liquid transition than through the discontinuities in the isotherms, because these follow from (26) and the virial theorem. Furthermore, it expresses more precisely van der Waals' idea about the continuity of the vapor and liquid state and it makes the existence of a critical point at least in principle understandable.

3. Can one look upon the solid-fluid equilibrium in a similar way? I think this is not possible, because one cannot escape the fact (intuitively evident, although not proved!) that there is already long range order in the solid phase by itself. Therefore, in a solid all distribution functions cannot have the product property. This is not evident for $n_2(\boldsymbol{r}_1, \boldsymbol{r}_2)$ because from the definition of the functions $n_s(\boldsymbol{r}_1, \ldots, \boldsymbol{r}_s)$, it follows that they must be translationally and rotationally invariant, since the total potential energy has these properties. This means that one must average any configuration of the s molecules over all translations and rigid rotations. As a result, it is easy to see that *even* for a perfect lattice, $n_2(|\boldsymbol{r}_1 - \boldsymbol{r}_2|)$ will *still* become $n^2 = 1/v^2$ when $|\boldsymbol{r}_1 - \boldsymbol{r}_2| \to \infty$ and it will not look qualitatively different from the $n_2(|\boldsymbol{r}_1 - \boldsymbol{r}_2|)$ of a liquid (see fig. 5).

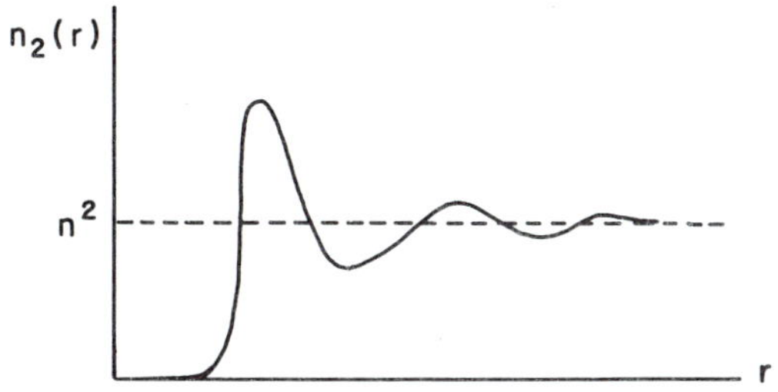

Fig. 5.

However the lattice structure will become more and more apparent in the higher distributions and in $n_5(\boldsymbol{r}_1, \ldots, \boldsymbol{r}_5)$ one will surely see long range order in $\boldsymbol{r}_5$ when $\boldsymbol{r}_1, \ldots, \boldsymbol{r}_4$ are fixed on lattice sites. Because the solid and the fluid are with respect to long range order qualitatively different, there cannot be a critical point, since by going around it this would imply that long range order would appear gradually, which is impossible. This is the argument of Landau and I find it completely convincing.

4. There are some empirical indications which suggest that the

solid-fluid transition can occur at *any* temperature when only the pressure is made high enough. Hydrogen and Helium can be solidified at temperatures many times the critical temperature T_c of the vapor-liquid transition and the melting line for He can be represented roughly by p_m = const. $T^{1.5}$. If it turns out that this behavior is a general property of at least all simple gases, then it should be explained by a general and common property of the intermolecular forces. Now one knows that the attractive part of these forces plays an essential role in the vapor-liquid transition and, in fact, kT_c is for all gases about equal to the minimum value of $\varphi(r)$. Since He can be solidified at a temperature of at least $10T_c$, this suggests that at high pressure the attractive force has little effect and that it is the sharp repulsive forces which are responsible for the solid-fluid transition. It is this train of thought which explains the great interest of the so-called *Kirkwood transition*. In 1941 Kirkwood posed the problem of whether a gas of hard spheres would show a phase transition, and he gave an argument which would give such a transition when the density of the gas was near the close-packed density limit. The argument was not convincing. Clearly, for instance, such a transition is impossible in one dimension, because then the equation of state has exactly the van der Waals form $p = kT/(v - b)$. However, in two and three dimensions a rigorous calculation of the partition function is beyond present day mathematical methods, and a phase transition, although rather paradoxical, is certainly *not* excluded. In fact there are numerical calculations for a gas of a few hundred hard spheres which seem to show that at high density there are two regimes of the motion of the spheres, one solid-like and the other fluid-like which sometimes can exist together. This is of course not a proof, but together with the empirical arguments mentioned above, it makes me believe that the Kirkwood transition *does* exist and is a kind of approximation of the behavior of the melting line of all substances at high pressures. Note that the Kirkwood transition would occur at *all* temperatures, since with hard spheres the two phases can differ only in their entropies and their densities which will be independent of the temperature. Since at fixed T, the energies of the two phases are the same, one obtains from the thermodynamic equilibrium condition $\mu_1(p,T) = \mu_2(p,T)$:

$$\frac{p}{T} = \frac{s(v_2) - s(v_1)}{v_1 - v_2} = \text{const.} \ ,$$

so $p \approx T$, which is not too much different from the observed $p \approx T^{1.5}$ for the He melting line. The difference must then, of course, be ascribed to the fact that the hard sphere repulsion is an idealization of the real short range repulsion.

6. THE MACROSCOPIC CAUSALITY PROBLEM

Let me return to the general question how to understand the macroscopic description of the physical phenomena from the molecular dynamics. We saw already that in order to understand equilibrium thermodynamics it is necessary that the macroscopic variables y_i fulfill the following two conditions:

I. The macroscopic description is a *contracted* description; the number of variables y_i is *much* smaller than N, so that a set of values of the y_i corresponds to a region in the phase space of the system.

II. The macroscopic variables must be *normal* variables, by which we mean that their fluctuations must be small, or more precisely that:

$$|\langle y_i^2\rangle - \langle y_i\rangle^2| \ll \langle y_i\rangle^2 \tag{28}$$

where $\langle\ \rangle$ indicates the averages over the microcanonical ensemble. Clearly if the system is metrically transitive this implies the existence of an equilibrium state, because then the sharpness of the distributions of the y_i over the phase space implies that also the distribution in time will be very sharp.

I also noted already that if these conditions are fulfilled and if the system is metrically transitive, the equilibrium state will be the same whatever the choice of the y_i and will always be described by the microcanonical ensemble. The various equilibrium states differ only by the values of the thermodynamic variables and these are therefore the only relevant macroscopic variables when the system is in thermodynamic equilibrium. The situation is quite different when one considers the temporal behavior, the *approach to equilibrium*. One then runs into the fundamental question, how is it possible that a contracted description can be *closed* and *causal*. In a bona-fide macroscopic theory it should of course *not* be necessary to go back to the microscopic, molecular picture (in this sense the theory must be closed), and it *should* be possible to make predictions, that is the theory must be causal. This is the macroscopic causality problem and although it is in my opinion still far from clarified, one begins to see some light thanks to the basic papers of Bogoliubov [27]. Bogoliubov pointed out that in any macroscopic theory the macroscopic variables must be in some sense *secular variables*, that is they must vary in time much slower than all the remaining variables needed to describe the molecular system. Bogoliubov maintained therefore that a closed contracted description is only possible if the system in its approach to equilibrium has relaxation times of quite different orders of magnitude, and he then tried to construct such a contracted description by expanding in an appropriate way in the ratios of these relaxation times, postulating that the macroscopic causality is fulfilled.

I will try to show by three examples how this works, and I can then discuss in a more concrete way the many puzzling features which remain.

Example I. *The van der Pol oscillator*. Consider the differential equation:

$$\frac{d^2x}{dt^2} - \epsilon(1 - x^2)\frac{dx}{dt} + \omega^2 x = 0 . \tag{29}$$

One knows that whatever the initial condition $x(t)$ will settle into a steady oscillation, which for small ϵ will be nearly harmonic with frequency ω. Physically this is almost obvious: in the beginning if x is small one has negative damping, so the amplitude will increase until on the average the damping is zero, while if the initial amplitude was large then there is positive damping and the amplitude decreases. There is therefore an equilibrium state and if $\epsilon \ll \omega$ there are two times in the problem of different order of magnitude. So one can ask whether a contracted description is possible in which one only wants to describe the slow variation of the amplitude, which is in this case the secular, macroscopic variable. It is well known that this can be done. Put:

$$x(t) = A(t)\sin(\omega t + \varphi) , \tag{30}$$

then

$$\ddot{x}(t) \cong 2\dot{A}\omega\cos(\omega t + \varphi) - \omega^2 A\sin(\omega t + \varphi) ,$$

$$\epsilon(1 - x^2)\dot{x} = \epsilon\frac{d}{dt}(x - \tfrac{1}{3}x^3) \cong \epsilon\frac{d}{dt}[(A - \tfrac{1}{4}A^3)\sin(\omega t + \varphi)]$$

$$\cong \epsilon\omega(A - \tfrac{1}{4}A^3)\cos(\omega t + \varphi) .$$

Here we have dropped the higher harmonic $\approx \sin 3\omega t$, and we have assumed that $\dot{A} \approx \epsilon A$. One sees therefore that up to order ϵ eq. (29) is fulfilled by (30), if $A(t)$ fulfills the equation:

$$2\frac{dA}{dt} - \epsilon(A - \tfrac{1}{4}A^3) = 0 . \tag{31}$$

This is a first order equation, and clearly constitutes a closed and causal "macroscopic" contraction of the original "microscopic" equation (29). Integrating (29) gives:

$$A^2 = \frac{4}{1 + \exp[-\epsilon(t - t_0)]}, \tag{32}$$

which show the approach to equilibrium. I think that this and similar examples with which Bogoliubov was quite familiar from his studies of non-linear mechanics, inspired his general ideas.

Example II. *The Kramers equation* [28]. This example is already from statistical mechanics. The distribution function $\rho(\boldsymbol{r}, \boldsymbol{v}, t)$ for a Brownian particle of mass m moving in a fluid, fulfills according to Kramers the equation:

$$\frac{\partial \rho}{\partial t} + \boldsymbol{v} \cdot \frac{\partial \rho}{\partial \boldsymbol{r}} + \boldsymbol{a} \cdot \frac{\partial \rho}{\partial \boldsymbol{v}} = \beta \left[\frac{\partial}{\partial \boldsymbol{v}} \cdot (\boldsymbol{v}\rho) + \frac{kT}{m} \Delta \rho \right], \tag{33}$$

where $\boldsymbol{a} = -(1/m)$ grad $U(\boldsymbol{r})$ is the acceleration due to the outside potential U and $m\beta$ is the friction coefficient for the motion of the particle in the fluid. Eq. (33) describes how the distribution ρ changes in time when the initial distribution $\rho(\boldsymbol{r}, \boldsymbol{v}, \mathrm{o})$ is given. For instance one might take $\rho(\boldsymbol{r}, \boldsymbol{v}, \mathrm{o}) = \delta(\boldsymbol{r} - \boldsymbol{r}_0)\delta(\boldsymbol{v} - \boldsymbol{v}_0)$ corresponding to a given initial phase $(\boldsymbol{r}_0, \boldsymbol{v}_0)$ of the particle. One can prove that *any* initial distribution will for $t \to \infty$ go to the Maxwell-Boltzmann (M. B.) distribution

$$\rho \to \rho_0 \approx \exp\left[-\frac{1}{kT} \left(\frac{mv^2}{2} + U \right) \right], \tag{34}$$

independent of the value of β. However, if β is large, or better if:

$$\beta \gg \langle v \rangle \left| \frac{\text{grad } U}{U} \right|, \tag{35}$$

then again one has two relaxation times of different orders of magnitude and one should expect that the approach to the equilibrium state (34) will go in two stages and that a contracted description for the final stage should be possible. In fact, putting:

$$\rho(\boldsymbol{r}, \boldsymbol{v}, t) = \sigma(\boldsymbol{r}, T) \exp\left[-\frac{mv^2}{2kT} \right], \tag{36}$$

Kramers shows that approximately $\sigma(\boldsymbol{r}, t)$ fulfills the Smoluchowski equation:

$$\frac{\partial \sigma}{\partial t} = -\frac{\partial}{\partial \boldsymbol{r}} \cdot \left(\frac{\boldsymbol{a}}{\beta}\sigma - \frac{kT}{m\beta}\frac{\partial \sigma}{\partial \boldsymbol{r}}\right) . \tag{37}$$

The approach to equilibrium would therefore first go quickly to a Maxwell distribution in the velocities and then the slow approach to the equilibrium distribution in $\boldsymbol{r}$ is described by the contracted "macroscopic" equation (37) which is clearly closed and causal.

The proof of (37) is as follows. Write (33) in the form:

$$\frac{\partial \rho}{\partial t} = \beta\left(\frac{\partial}{\partial \boldsymbol{v}} - \frac{1}{\beta}\frac{\partial}{\partial \boldsymbol{r}}\right) \cdot \left(\boldsymbol{v}\rho + \frac{kT}{m}\frac{\partial \rho}{\partial \boldsymbol{v}} - \frac{\boldsymbol{a}}{\beta}\rho + \frac{kT}{m\beta}\frac{\partial \rho}{\partial \boldsymbol{r}}\right) - \frac{\partial}{\partial \boldsymbol{r}} \cdot \left(\frac{\boldsymbol{a}}{\beta}\rho - \frac{kT}{m}\frac{\partial \rho}{\partial \boldsymbol{r}}\right) . \tag{38}$$

One verifies that this is identical with (33). Now integrate (38) along the straight line $\boldsymbol{r} + (\boldsymbol{v}/\beta) = \boldsymbol{r}_0$ from $\boldsymbol{v} = -\infty$ to $+\infty$ and call:

$$\sigma(\boldsymbol{r}_0, t) = \int_{-\infty}^{+\infty} \mathrm{d}\boldsymbol{v}\ \rho(\boldsymbol{r}_0 - \frac{\boldsymbol{v}}{\beta}, \boldsymbol{v}, t) .$$

Assuming that ρ vanishes sufficiently fast for $\boldsymbol{v} = \pm\infty$, the result from the first term in (38) vanishes and one gets:

$$\frac{\partial \sigma}{\partial t} = -\int_{-\infty}^{+\infty} \mathrm{d}\boldsymbol{v} \left[\frac{\partial}{\partial \boldsymbol{r}} \cdot \left(\frac{\boldsymbol{a}}{\beta}\rho - \frac{kT}{m\beta}\frac{\partial \rho}{\partial \boldsymbol{r}}\right)\right]_{\boldsymbol{r} = \boldsymbol{r}_0 - \frac{\boldsymbol{v}}{\beta}} .$$

This is still exact. Now if the variation in $\boldsymbol{r}$ is very slow compared to the variation in $(\boldsymbol{v}/\beta)$ which is the basic assumption (35), then one can integrate over ρ under the differentiation signs and one gets (37).

Example III. *Derivation of the hydrodynamical equations from the Boltzmann equation* [29]. This is of course one of the classical problems of the kinetic theory of gases. The famous Boltzmann equation:

$$\frac{\partial f}{\partial t} + \boldsymbol{v} \cdot \frac{\partial f}{\partial \boldsymbol{r}} + \boldsymbol{a} \cdot \frac{\partial f}{\partial \boldsymbol{v}} = \int \mathrm{d}\boldsymbol{v}_1 \int \mathrm{d}\Omega g I(g, \theta)(f' f_1' - f f_1) \tag{39}$$

for the distribution function $f(\boldsymbol{r}, \boldsymbol{v}, t)$ of the molecules of a dilute, mono-atomic gas describes how f changes in time due to the streaming of the molecules in phase space and due to the binary collisions $(\boldsymbol{v}, \boldsymbol{v}_1) \rightleftharpoons (\boldsymbol{v}', \boldsymbol{v}_1)$ between the molecules for which $I(g, \theta)$ is the differential cross section and $g = |\boldsymbol{v} - \boldsymbol{v}_1| = |\boldsymbol{v}' - \boldsymbol{v}_1'|$ the relative velocity. One knows from the H-theorem that *any* initial distribution will go for $t \to \infty$ to the M. B. distribution (34), and one also knows that this approach goes in two stages. First there is a quick approach in a time of order $\lambda/\langle v\rangle$ ($\lambda \approx$ mean free path) to a *local Maxwell distribution:*

$$f^{(0)} = n\left(\frac{m}{2\pi kT}\right)^{\frac{3}{2}} \exp\left[-\frac{m}{2kT}(\boldsymbol{v}-\boldsymbol{u})^2\right], \tag{40}$$

where the particle density n, average velocity $\boldsymbol{u}$ and temperature T are still slowly varying functions of $\boldsymbol{r}$ and t, and this is followed by the slow relaxation to complete equilibrium. The n, $\boldsymbol{u}$ and T are the secular, macroscopic variables, and one expects therefore that a closed contracted description for the temporal variation of these variables should be possible and the macroscopic equations must be of course the hydrodynamical equations. This can be accomplished by a well known successive approximation method which was developed by Chapman and Enskog. For a detailed account I refer to the literature [30], and I will here only remind you that in this method one seeks a solution of the Boltzmann equation (39) which can be written symbolically in the form:

$$f = f^{(0)}\left[1 + a_1(\boldsymbol{v})(\lambda\nabla) + a_2(\boldsymbol{v})(\lambda\nabla)^2 + \ldots\right], \tag{41}$$

where λ is again a mean free path and where the gradient operator ∇ acts on the macroscopic variables n, $\boldsymbol{u}$ and T. The expansion parameter in (40) is therefore the relative change of the macroscopic variables over a mean free path. This can also be interpreted as the ratio of the time between collisions to a macroscopic relaxation time, so that the expansion (40) agrees with the general ideas of Bogoliubov. At the same time as the successive corrections to $f^{(0)}$ one obtains equations for the macroscopic variables themselves. These are in zeroth order ($f = f^{(0)}$) the Euler or ideal fluid hydrodynamical equations, in first order the Navier-Stokes equations (with explicit expressions for the viscosity and heat conduction coefficients), and so on. Also the second order or Burnett equations have been completely worked out. All these successive order hydrodynamical equations are of the first order in the time derivatives of the macroscopic variables so that they fulfill the macroscopic causality requirement.

7. CONCLUDING REMARKS AND QUESTIONS

a. These examples have in common the existence of an equilibrium state and of two relaxation times of widely different orders of magnitude. The methods to derive the contracted description seem rather different, but I suspect that it must be possible to unify the methods, at least if from the abstract mathematical point of view these examples *are* really similar.

b. It is quite dark to me in which precise sense the contracted description approximates the exact solutions of the initial value problems, and I find it difficult to imagine how the successive orders of contractions in the ratios of the relaxation times can form a convergent sequence. In this respect it is, I think, also quite remarkable that in most cases it is difficult or it does not make much sense to try to go beyond the first approximation. This is not true for the van der Pol equation for which a successive approximation method has been worked out in detail, but even here I believe that the contraction aspect of the solution disappears after the first approximation. For the Kramers equation, I don't know of any attempt to find corrections to the Smoluchowski equation (37) [31]. For the Boltzmann equation I mentioned already that higher approximations beyond the well known Navier-Stokes equations have been worked out. Although no mathematical difficulties appear, it is unclear what the physical consequences of the Burnett and higher order equations are. This is due to the fact that these equations are of higher than the second order in the space derivatives of $\boldsymbol{u}$ and T, so that one needs more boundary conditions than for the Navier-Stokes equations. These additional conditions are *not* obtained from the Chapman-Enskog expansion and are therefore not known [32].

c. Bogoliubov tried to look upon the Boltzmann equation itself as a contraction from the Liouville equation after a time long compared to the collision time. The development parameter is in this case the ratio of the collision time to the time between collisions which is proportional to the density of the gas, so that the successive orders of contraction would be like the virial expansion of the equilibrium properties of the gas. Bogoliubov succeeded in deriving the Boltzmann equation in first approximation, but here again there are difficulties when one wants to go to higher approximations. These will be discussed in the lectures of Cohen.

d. One gets the impression that for a specific problem it is better to forget about the contraction procedure after the first approximation and to go back to the original equations. Let me illustrate this with the problem of finding the dispersion and absorption of high frequency sound in a dilute gas. One can discuss this by means of the hydrodynamical equations and one then finds from the Euler equations for the phase velocity:

$$U_0 = \left(\frac{5}{3}\frac{kT}{m}\right)^{\frac{1}{2}}$$

for a mono-atomic gas. At higher frequency, absorption sets in which was first calculated by Kirchhoff from the Navier-Stokes equations. In fact from these equations one finds for the velocity U and the absorption coefficient α as function of the frequency ω expansions of the form:

$$\frac{U}{U_o} = 1 + a_1\xi^2 + a_2\xi^4 + \dots ,$$

$$\frac{\alpha U_o}{\omega} = b_1\xi + b_2\xi^3 + \dots , \tag{42}$$

where:

$$\xi = \frac{\mu\omega}{\rho U_o^2} . \tag{43}$$

Here μ is the viscosity coefficient and ρ the density of the gas. The coefficients a_i and b_i are constants if one assumes that the heat conductivity K is related to μ by the relation

$$K = \tfrac{15}{4} \frac{k}{m} \mu ,$$

which is closely fulfilled for all mono-atomic gases. One gets from the Navier-Stokes equations:

$$b_1 = \tfrac{7}{6} ; \; a_1 = \tfrac{141}{72} ; \quad b_2 = -\tfrac{1559}{432} , \tag{44}$$

etc. Now note that the parameter ξ as given by (43) is of order λ/Λ, where λ is a mean free path and Λ the wavelength of the sound, since $\mu \approx \rho\lambda\langle v\rangle$ if $\langle v\rangle$ is a mean thermal velocity and $\langle v\rangle \approx U_o$. Therefore ξ is just the Bogoliubov expansion parameter used in the Chapman-Enskog development. Since from the point of view of this development the Navier-Stokes equations are only valid up to the first order in ξ, only the coefficient b_1 can be trusted, and to find a_1, b_2, etc., one should go to the Burnett and higher order hydrodynamical equations. This can and has been done, but it is quite clumsy and involved. It is *much* better to forget about the hydrodynamical equations and to go back to the original Boltzmann equation. As Foch [33] has shown in his dissertation, one can find the dispersion and the absorption of sound as function of the frequency ω by a direct perturbation calculation from the Boltzmann equation. His results can also be put in the form (42) and the coefficients a_i

and b_i turn out to be almost universal, that is almost independent of the intermolecular potential and almost independent of the temperature. He finds instead of (44):

$$b_1 = \frac{7}{6}\,; \qquad a_1 = \frac{215}{72}\,; \qquad b_2 = -\frac{5155}{432}\,. \tag{45}$$

So a_1 and b_2 are appreciably different from the Navier-Stokes values and the experiments of Greenspan fully confirm the values (45) [34].

e. Finally, let me conclude with the following general remark. It seems to me that just because on any level a macroscopic description is a contracted description, the macroscopic variables must be *stochastic variables*. The contracted equations, even when they are closed and causal, can only give the average behavior at the stage under consideration. The molecular background (which one has contracted away!) should show itself by fluctuations of the macroscopic variables, which is another way of saying that they must be stochastic variables. Of course, in a proper macroscopic description the fluctuations must be small, or in other words the macroscopic equations must somehow be stable. Perhaps the difficulties one has in trying to go to higher approximations are connected with this requirement. Also this stability may not hold near a critical or otherwise exceptional state of the macroscopic variables. The large density fluctuations near the vapor-liquid critical point are a familiar example of this kind of instability. It seems to me an interesting question to see whether macroscopic instability points (occurring for instance in the transition from laminar to turbulent flow) would also show large fluctuations.

It seems to me appealing to try at each stage of contraction to complete the macroscopic equations by adding terms representing *random sources*, which then would account for the fluctuations just as in the theory of Brownian motion. Landau and Lifshitz have tried to do this for the hydrodynamical equations [35]. However, the justification, say from the Boltzmann equation, and also the consequences are, I think, still far from clear. It may lead to an interesting extension of the classical fluctuation theory.

NOTES

[1] These lectures will partially be a re-hash of the "Lectures in Statistical Mechanics" by G. E. Uhlenbeck and G. W. Ford published by the American Mathematical Society (Providence, Rhode Island, 1963). Since it is awkward to repeat oneself, the reader should consult this book (to which I will refer as UF, Ch-) for some of the proofs and for detailed references.

[2] A recent surprise was the discovery of the logarithmic infinity of the specific heat at constant volume near the vapor-liquid critical point (Bagatskii, Voronel and Gusak, JETP 16 (1963) 517), which stimulated the current interest in critical phenomena.

[3] Examples are the prediction of the quantized magnetic flux in super conductors by London and of the quantized vorticity in liquid He II by Onsager.

[4] Compare UF. Ch. 1; a proof of the Poincaré theorem is given in Note 4 of that chapter.

[5] One can say with Poincaré and Smoluchowski that a lack of knowledge of the initial state is a prerequisite for the application of the probability calculus to closed and deterministic physical systems. In addition, the precise form of the distribution over the initial region should not matter when it is only sufficiently smooth. One should expect that after an appropriate chaotization or aging period, the evolution of the probabilities of the various macroscopic states (= amount of ensemble fluid in the corresponding regions) depends only on these probabilities themselves and not on the precise form of the distribution of the ensemble fluid in the regions. How this can come about is a difficult question related to the macroscopic causality problem which will be discussed in section 5.

[6] It is better always to think of an energy *shell* between E and $E+\Delta E$; it is the volume of the portion of the shell based on the area of the energy surface, which remains the same.

[7] See for instance UF, Ch. 1, p. 7.

[8] See UF, Ch. 1, p.12 and notes 7, 8, 9.

[9] Sinai, Statistical Mechanics; Foundations and Applications, Proceedings of the I. U. P. A. P. Meeting, Copenhagen 1966 (W. A. Benjamin, New York, 1967) p. 559.

[10] The words more or less refer especially to the degree of mathematical rigor and more important to the assumptions one makes about the molecular structure and the molecular interactions. Clearly, the validity of the laws of thermodynamics should be almost completely independent of these assumptions and should only require that N is large (so that fluctuations can be neglected) and that the molecular system is in some sense *stable*. The ideal treatment which makes transparent this independence is to my knowledge still lacking.

[11] In UF, Ch. 1, p. 18, a proof is given for the case that the heat reservoir B is an ideal gas. For a much more general derivation see Mazur and van der Linden, J. Math. Phys. 4 (1963) 271.

[12] Given in UF, Ch. 1.

[13] This is based on the very sensible *convention* that each non-degenerate energy level has the à priori statistical weight one.

[14] It is conventional because thermodynamically the energy, entropy, free energy, etc., are only determined up to an arbitrary constant, since one measures only differences. The dependence of the thermodynamic quantities on N can therefore only be agreed upon by convention.

[15] The term thermodynamic limit is perhaps unfortunate since it may suggest that thermodynamics is only applicable for very large systems. This is, of course, not true. There is for instance a thermodynamical theory of surface phenomena. Furthermore, if one defines a perpetuum mobile of the second kind as an *automatic* working machine which

transforms heat into mechanical energy, then according to Smoluchowski and Szilard the impossibility of such a machine (that is the second law of thermodynamics) remains valid even if one takes fluctuations into account so that it also holds for finite N. However, in most of the thermodynamic applications and especially in the theory of phase transitions one considers large systems for which the limit is appropriate.

[16] It was first attempted by van Hove, Physica 15 (1949) 951 and since then it has been generalized and rigorized considerably by the work of Ruelle, Penrose, Lebowitz, Fisher a. o.

[17] See van der Linden, ***Dissertation***, Leiden 1966 and also the lecture of Mazur in the Proceedings of the IUPAP Conference, l.c. note 9, p. 72.

[18] This was also first attempted by van Hove. See note [16].

[19] The figures are, of course, schematic and only indicate the behavior of normal substances, for which the solid phase has the highest density. Also Helium is excluded; it has no ordinary triple point and even at $T=0$ it can exist in two phases.

[20] Compare UF, Ch. 2, note 2 and Ch. 3. My view of the van der Waals theory has changed considerably by the work done in collaboration with Kac and Hemmer on the one-dimensional van der Waals-like models. See J. Math. Phys. 4 (1963) 216, 229; 5 (1964) 60, 75, where one finds also many more references.

[21] See van Kampen, Phys. Rev. 135 (1964) A364;
Lebowitz and Penrose, J. Math. Ph. 7 (1966) 98.

[22] See Fisher, J. Math. Ph. 5 (1964) 364 and also the paper by Uhlenbeck and Hemmer in the Proceedings of the IUPAP Conference, Aachen 1964 (North-Holland Publishing Co, Amsterdam 1965) p. 241.

[23] The Onsager solution is of course more realistic! Still one might think that for medium range forces and away from the transition point, the specific heat, for example, behaves (as function of the temperature) van der Waals-like, while close to the transition temperature it goes over into the logarithmic Onsager behavior. And the closeness should depend only on γd.

[24] All this is under active investigation and I refer to the lectures of Kac for some of the recent results. Here one also should mention that it is possible to translate the Ising model into the so-called lattice gas model (see Lee and Yang, Phys. Rev. 87 (1952) 410), which makes the analogy with the van der Waals model still closer. In this context the discovery of the logarithmic singularity of the specific heat at the vapor-liquid critical point (see note [2]) is, of course, very suggestive.

[25] In one dimension it is obvious that the energy will be minimum for a regular linear array. For this case there is an argument, which is due to Wigner, and which shows that for any small amount of thermal agitation the space correlation between distant points in the array tends to disappear with the distance. There is therefore no melting point which is in agreement with the fact that there cannot be a phase transition in one dimension. The same argument applies also for a two-dimensional lattice, so that it may be that the phenomena of melting can only occur in three dimensions. Finally, one should note that eq. (25) may have several minima corresponding to different regular arrangements, and this in fact must be the case when there are several solid phases.

[26] See Uhlenbeck, Hemmer and Kac, J. Math. Ph. 4 (1963) 229.

[27] The monograph of Bogoliubov from the year 1946 has been translated and has appeared in the "Studies in Statistical Mechanics" Vol. 1 (North Holland Publishing Co., Amsterdam, 1962). See also UF, Ch. VII.

[28] Kramers, Physica 7 (1940) 284 (= Collected Papers, North Holland Publishing Comp., Amsterdam 1956, p. 754).

[29] For a discussion of the Boltzmann equation see UF, Ch. IV.

[30] For a short account see UF, Ch. VI, and for a complete discussion the monograph of Chapman and Cowling (*The Mathematical Theory of Non-uniform Gases*, Cambridge University Press, 2nd edition, 1955).

[31] For special outside force fields like, for instance, a harmonic attraction to a fixed point, the exact solution of the initial value problem for the Kramers equation is known. For the overdamped case one can then find properties of the distribution ρ as for instance various mean values as function of β. However, it is not clear how to obtain the same results from Smoluchowski-like equations.

[32] As a result, one does not even known the first correction to the famous Stokes formula for the drag of a slowly moving sphere when the mean free path of the molecules is not very small compared to the radius of the sphere.

[33] Foch, *On the theory of sound propagation in monoatomic gases and binary mixtures.* Dissertation, Rockefeller University, 1967. See also Phys. Rev. Letters 19 (1967) 1025.

[34] Greenspan, J. Acous. Soc. 28 (1956) 644. About the convergence of the expansions (42) one knows that for hard spheres the series are convergent for small enough ξ. (See J. A. McLennan, Phys. Fluids 8 (1965) 1580). Because of the insensitivity of the coefficients on the molecular model, this makes it likely that the series always have a finite radius of convergence but a precise proof is lacking.

[35] Landau and Lifshitz, *Fluid Mechanics*, Ch. 17.

PHASE TRANSITIONS

P. W. KASTELEYN
Instituut-Lorentz voor Theoretische Natuurkunde, Leiden, The Netherlands

1. INTRODUCTION

One of the oldest problems in statistical mechanics is that of developing a theory of phase transitions. In recent years, this problem has received a great deal of attention. In particular the phenomena occurring very near critical points have been the subject of many new experimental and theoretical investigations. In these lectures we shall outline some of the results of this work. We shall successively deal with: (a) experimental data on the critical behaviour ("critical indices") of thermodynamic quantities; (b) theoretical models playing a role in the interpretation of these data; (c) relations between critical indices.

The limited size of this review does not allow us to go into many details or to discuss all points where important progress was made. We shall therefore restrict ourselves in (a) to one single group of closely related phase transitions, which includes the gas-liquid transition and magnetic transitions. In discussing models we shall mainly concentrate on the Ising model, about which we have the most detailed information. Special attention will be paid to the derivation of exact expressions for the partition function of 1- and 2-dimensional Ising systems, as an example of a method for obtaining exact information about critical phenomena. Where possible, the derivation is given in a recently developed form which is based on the application of well-known methods from many-particle theory.

Although we do not discuss ferroelectricity and antiferroelectricity, we have included a brief introduction to some models which are considered to be relevant to these phenomena, because they form the subject of a seminar lecture given by E. H. Lieb *.

For a more complete survey of both the experimental and the theoretical aspects of critical phenomena we refer to the excellent

* This lecture is not included in these proceedings. Cf., however, ref. 5.

reviews recently given by Fisher [1] and by Kadanoff et al. [2], from which we have borrowed many data, and to two still more recent review articles by Fisher [3] (theory) and Heller [4] (experimental techniques and data). These articles moreover contain an extensive and up-to-date bibliography. Therefore we shall not refer to the original literature except in a few special cases, but confine ourselves to referring to these four reviews.

2. EXPERIMENTAL DATA

What are the observed facts which a theory of phase transitions should be able to account for? Let us restrict ourselves to phase transitions for which there is a critical point (thus leaving aside phenomena like melting). Then first of all the theory should show that for a system of the type one is interested in there exists a critical temperature T_c below which the system can occur in two phases which become more and more alike as T_c is approached, and that when the system is cooled down under appropriate conditions, a temperature T is reached ($T \leqslant T_c$) at which the thermodynamic quantities change drastically. On the other hand, it should also explain why certain other systems do not exhibit such a phase transition. In other words, it has to answer the *existence question*.

Secondly, we would like the theory to tell us the value of the critical temperature and also the values of those quantities which remain finite at this temperature such as the energy and the entropy. For a long time most attention, both of experimental and theoretical physicists working in this field, has been devoted to the calculation of these *critical constants*. On the theoretical side, various approximative methods have successively been developed which provided a stepwise approach to the exact values of these constants. Unfortunately, the results obtained in this way showed an extremely slow convergence. In addition, numerical calculations (based on extrapolations from series expansions at high and low temperatures) produced results which were of a more rapidly increasing accuracy.

More recently, a third aim of a theory of critical phenomena has been stressed: it should also account for the way in which the thermodynamic quantities and their fluctuations behave as the critical point is approached (*critical* or *asymptotic behaviour*). This aspect of phase transitions has received particular attention because of two facts which gradually emerged. In the first place, it became apparent that there exist relations of a rather general nature between the parameters which describe the critical behaviour of the various physical quantities of a system (*critical indices*).

Secondly, it was found that even for systems of a vastly different physical nature the critical behaviour of certain quantities shows striking similarities.

Since the last-mentioned facts seem to indicate that the critical behaviour of physical quantities constitutes a more fundamental aspect of a phase transition than the critical constants, we concentrate in this review on the former.

We will not try to survey all the available experimental data but restrict ourselves to the data on a number of thermodynamic quantities. Other quantities whose behaviour near the critical point is of great importance for an understanding of phase transitions are the time-independent correlation functions. For a discussion of these we refer to the review articles mentioned in the Introduction. Furthermore we leave aside phase transitions in systems where quantum effects play a dominant role (superconductors, superfluids) and phenomena like ferroelectricity and antiferroelectricity. We will see that in this way we restrict ourselves to a family of closely related phase transitions.

In discussing the properties of a system near its critical point we will repeatedly describe the critical behaviour of a quantity Q by stating that

$$Q \sim (x - x_c)^p , \qquad (x \to x_c+); \tag{2.1}$$

in words: "when the variable x approaches its critical value x_c from above, Q behaves asymptotically as $(x - x_c)^p$". This means that, apart from a constant factor, Q is asymptotically equal (in the usual sense) to $(x - x_c)^p$, or

$$\lim_{x \to x_c+} \frac{Q(x)}{(x - x_c)^p} = \text{constant} . \tag{2.2}$$

The exponent p will be called the *critical exponent* of Q. When the exponent is negative (i.e. when Q diverges at the critical point) one usually writes $Q \sim (x - x_c)^{-p}$; in some other cases it is convenient to write $Q \sim (x - x_c)^{1/p}$; in each case we call the number p which is used to characterize the critical behaviour, the *critical index* of the quantity. It has become customary to denote the critical index for each quantity by a fixed (Greek) symbol, α for the specific heat, β for the spontaneous magnetization (both with $x = T$) etc.

If we encounter a case where Q diverges logarithmically:

$$Q \sim \ln(x - x_c) , \qquad (x \to x_c+) , \tag{2.3}$$

we will describe this by a critical exponent (and index) 0. That this is a useful extension of our definition follows from the fact that the logarithm can be considered as the limit of a very weak negative power:

$$\lim_{\alpha \to 0} \alpha^{-1}(x^{-\alpha} - 1) = -\log x \qquad \text{for } x > 0 \,. \tag{2.4}$$

(Notice that our definition of the symbol $\sim$ is not identical with that of refs. 1 and 3: Fisher uses $\sim$ to express asymptotic equality apart from a factor which varies slower than a power of $x - x_c$, such as $\ln(x - x_c)$ or $\ln \ln(x - x_c)$. For a logarithmically divergent quantity one can then simply write $Q \sim (x - x_c)^0$.)

2.1. *Gas-liquid transitions*

For the sake of completeness we begin by summarizing some elementary empirical facts. The isotherms of a fluid (gas-liquid system) fall into three classes. For temperatures higher than the critical temperature T_c, the pressure p is a monotonically increasing function of the density ρ with a finite, non-vanishing derivative, and without a trace of a discontinuity in any of the higher derivatives. When T_c is approached from above, the minimum value of $(\partial p/\partial \rho)_T$ along the isotherm decreases until, at $T = T_c$ (critical isotherm), $(\partial p/\partial \rho)_T = 0$ for a certain value ρ_c of ρ. For $T < T_c$, $(\partial p/\partial \rho)_T$ vanishes for a certain interval $\rho_G < \rho < \rho_L$, which increases with decreasing T and which represents the region where gas and liquid are coexisting; ρ_G is the maximum density of the pure gas phase, ρ_L the minimum density of the pure liquid phase.

Quantitatively, the behaviour of the isotherms can be described as follows. For most monoatomic gases and other simple gases, the top of the *coexistence curve* in the T versus ρ plane can be described by

$$\rho_L - \rho_G \sim (T_c - T)^{\frac{1}{3}} \,, \qquad (T \to T_c-) \,, \tag{2.5}$$

or, more precisely,

$$\rho_L - \rho_G \sim (T_c - T)^{\beta} \qquad \text{with } \beta = 0.346 \pm 0.01 \,, \qquad (T \to T_c-) \,. \tag{2.6}$$

For the same group of gases, the *critical isotherm* can in the neighbourhood of the critical point ($p = p_c$, $\rho = \rho_c$) be described by

$$|p - p_c| \sim |\rho - \rho_c|^{\delta} \qquad \text{with } \delta = 4.4 \pm 0.4 \,, \qquad (T = T_c, \rho \to \rho_c) \,. \tag{2.7}$$

The behaviour of the isotherms for $T > T_c$ can best be illustrated by specifying the critical behaviour of the *isothermal compressibility*

$$K_T = -\frac{1}{V}\left(\frac{\partial V}{\partial p}\right)_T = \frac{1}{\rho}\left(\frac{\partial \rho}{\partial p}\right)_T , \tag{2.8}$$

which, according to the above, goes to infinity for $T \to T_c+$. For a number of simple gases it is found that, if the critical point is approached at constant density $\rho = \rho_c$, K_T behaves asymptotically as

$$K_T \sim (T - T_c)^{-\gamma} \quad \text{with } \gamma = 1.37 \pm 0.2 , \quad (T \to T_c+, \rho = \rho_c) . \tag{2.9}$$

As the critical temperature is approached from below along the coexistence curve one finds

$$K_T(\rho_G) \sim K_T(\rho_L) \sim (T_c - T)^{-\gamma'} \quad \text{with } \gamma' = 1.0 \pm 0.3 , \quad (T \to T_c-) . \tag{2.10}$$

Finally, there is another important quantity which shows a singular behaviour at the critical point, viz. the *specific heat* at constant volume C_V. It is found that C_V increases sharply as T_c is approached from above or from below at constant density. It is hard to draw a definite conclusion about the critical index from the experimental data. They are in agreement both with a divergence in the form of a weak power of $|T - T_c|$ and with a logarithmic divergence:

$$C_V \sim (T - T_c)^{-\alpha} \quad \text{with } 0 \leqslant \alpha < 0.4 , \quad (T \to T_c+, \rho = \rho_c) , \tag{2.11}$$

$$C_V \sim (T_c - T)^{-\alpha'} \quad \text{with } 0 \leqslant \alpha' < 0.25 , \quad (T \to T_c-, \rho = \rho_c) . \tag{2.12}$$

In both cases the value 0 corresponds to a logarithmic divergence.

2.2. *Transitions in ferromagnets*

The isotherms of a ferromagnet fall again into three classes. For temperatures higher than the critical or Curie temperature T_c, the magnetization M is a monotonically increasing function of the magnetic field H along a given direction with a finite, non-vanishing derivative and with no trace of an anomaly near $H = 0$. The derivative at $H = 0$, which gives the zero-field susceptibility $\chi = (\partial M/\partial H)_{H=0}$, increases gradually as T_c is approached. For $T = T_c$ the isotherm has a vertical slope at $H = 0$. The isotherms for $T < T_c$ exhibit a discontinuity at $H = 0$. As H tends to zero from above or below, M tends to a non-zero limit $+M(0)$ and $-M(0)$ respectively; the spontaneous magnetization $M(0)$ increases with decreasing temperature.

This behaviour of the isotherms resembles strongly that of the isotherms of a fluid, M and H playing the roles of ρ and p, respectively. We will therefore denote the critical indices of the thermodynamic quantities again by the symbols $\alpha, \alpha', \beta, \gamma, \gamma', \delta$, using the same symbol for corresponding quantities. The basic reason for considering a thermodynamic quantity for a ferromagnet as corresponding to a quantity for a fluid will be explained in section 4.1.2.

Let us now look at the quantitative results. For a number of metallic and non-metallic ferromagnets the *spontaneous magnetization* behaves for $T \to T_c-$ as

$$M(0) \sim (T_c - T)^{\beta} \quad \text{with } \beta = 0.33 \pm 0.03\,, \quad (T \to T_c-,\, H = 0)\,. \tag{2.13}$$

Note the striking coincidence of this value with that found for fluids!

The *critical isotherm* has been investigated only for a small number of substances. For nickel and gadolinium the data are consistent with

$$|M| \sim |H|^{1/\delta} \quad \text{with } \delta = 4.1 \pm 0.1\,, \quad (T = T_c,\, H \to 0)\,. \tag{2.14}$$

Later measurements seem to indicate a somewhat higher value for δ, but it is clear that again the value of the critical index is close to the corresponding one for fluids.

The magnetic quantity analogous to the isothermal compressibility of fluids is the *isothermal susceptibility*. Most experiments in the critical region are described by

$$\chi \sim (T - T_c)^{-\gamma} \quad \text{with } \gamma = 1.33 \pm 0.03\,, \quad (T \to T_c+,\, H = 0)\,. \tag{2.15}$$

For $T \to T_c-$ we expect

$$\chi \sim (T_c - T)^{-\gamma'}\,, \quad (T \to T_c-,\, H = 0)\,, \tag{2.16}$$

but the experimental data are too scarce as yet to draw conclusions.

Finally, the *specific heat* at constant magnetic field C_H is, for $H = 0$, found to behave as

$$C_H \sim (T - T_c)^{-\alpha} \quad \text{with } 0 \leqslant \alpha < 0.16\,, \quad (T \to T_c+,\, H = 0)\,, \tag{2.17}$$

$$C_H \sim (T_c - T)^{-\alpha'} \quad \text{with } 0 \leqslant \alpha' < 0.2\,, \quad (T \to T_c-,\, H = 0)\,, \tag{2.18}$$

Again, a logarithmic singularity ($\alpha = 0$, $\alpha' = 0$) is among the possibilities.

2.3. *Transitions in antiferromagnets*

Whereas at sufficiently low temperatures the spins of a ferromagnet are to a large extent parallel, yielding a non-vanishing spontaneous magnetization, an antiferromagnet is characterized by the tendency of neighbouring spins to become antiparallel in a systematic way. In the simplest case (the only one we shall discuss) the lattice consists of two interpenetrating sublattices with opposite magnetizations M and $-M$ along a certain direction. The *sublattice magnetization* M is zero for temperatures higher than a certain temperature T_c, the Néel temperature, and non-zero for $T < T_c$, where it increases with decreasing temperature. The most accurate measurements on the critical behaviour of M are those for MnF_2. They show that

$$M \sim (T_c - T)^{\beta} \qquad \text{with } \beta = 0.335 \pm 0.01 \, , \qquad (T \rightarrow T_c-) \, . \qquad (2.19)$$

The critical behaviour of the *specific heat* C is again known less accurately. The available data are in agreement with

$$C \sim (T - T_c)^{-\alpha} \qquad \text{with } 0 \leqslant \alpha \lesssim 0.16 \, , \qquad (T \rightarrow T_c+) \, , \qquad (2.20)$$

$$C \sim (T_c - T)^{-\alpha'} \qquad \text{with } 0 \leqslant \alpha' \lesssim 0.16 \, , \qquad (T \rightarrow T_c-) \, . \qquad (2.21)$$

Whereas in ferromagnets the ordering of the spins can be enhanced by the application of an external magnetic field, there is no analogous procedure for antiferromagnets. The antiparallel ordering of the spins would be favoured by the application of a staggered magnetic field, pointing in opposite directions for neighbouring spins, but an experimental realization of such a field is not known. Therefore, there are no direct antiferromagnetic analogues to the critical isotherm and the susceptibility of a ferromagnet. An antiferromagnet can be placed in a homogeneous magnetic field, of course, but its properties are then completely different from those of a ferromagnet.

2.4. *Transitions in binary fluid mixtures*

Many mixtures of two fluids, consisting, say, of molecules A and B, exhibit at temperatures lower than a critical temperature T_c a separation into two phases, a phase 1 with a higher concentration $x_A^{(1)}$ of molecules A than the average concentration x_A, and a phase 2 with a lower concentration $x_A^{(2)}$ of A. We discuss only the *coexistence curve* in the T versus x_A plane which in some cases has been measured very carefully. For mixtures of CCl_4 and

n-C_7F_{16} it was found that the top of the coexistence curve is described by

$$x_A^{(1)} - x_A^{(2)} \sim (T_c - T)^\beta \quad \text{with } \beta = 0.33 \pm 0.02 \,, \qquad (T \to T_c-) \,. \tag{2.22}$$

2.5. *Transitions in binary alloys*

Alloys in which like atoms attract each other more strongly than unlike atoms undergo a phase separation at sufficiently low temperatures which is similar to that occurring in binary fluid mixtures. Alloys in which *unlike* atoms attract each other more strongly, on the other hand, tend to show the formation of superlattices at low T. These two types of alloys are to a large extent analogous to ferromagnets and antiferromagnets, respectively. Although qualitatively the behaviour of the various thermodynamic quantities is similar to that of the corresponding quantities in magnetic systems and fluids, the available data do as yet not permit to draw definite conclusions about their critical behaviour *.

3. BASIC FEATURES OF A THEORY OF PHASE TRANSITIONS

After having reviewed a number of the most characteristic experimental facts about phase transitions we may ask what kind of a theory will be able to account for these facts. It is obvious that the basis of such a theory will have to be sought in statistical mechanics. Having observed this, however, we are faced with a number of questions:

a. Is it possible to explain the occurrence of phase transitions by the methods of statistical mechanics alone (starting from the microscopic properties of the system) or are there other, special arguments required? By the study of explicit examples, the most famous of which is the two-dimensional Ising model, this question has been answered in the affirmative.

b. How can a sharp transition come out of a statistical mechanical theory? At first sight one might expect all thermodynamic quantities to be smooth functions of the primary variables p, T, H etc. Indeed, the N-particle canonical partition function Z_N is known to be an analytic function of the primary variables. Taking the logarithm in order to get the free energy does not introduce a singu-

* Only quite recently, accurate measurements on β-brass have been made (cf. ref. 3, section 2.7), which indicate that $\beta = 0.305 \pm 0.010$ and $\gamma = 1.25 \pm 0.02$.

larity either, because Z_N is always strictly positive. The answer to the question is that discontinuities and other singularities can occur if one considers the various quantities *per particle* and then goes to the limit of an infinite system ($N \to \infty$, $V \to \infty$), while keeping the local properties such as the density constant (and letting the fraction of the volume lying close to the surface go to zero). The necessity of taking this *thermodynamic limit* leads to another question:

c. *Do* the thermodynamic quantities of a given system tend to a well-defined limit when the thermodynamic limit is taken? For the model systems discussed in these lectures this turns out to be the case.

d. One might ask whether it is not possible to start with an infinite system right from the beginning rather than first considering a finite system and letting the system become infinite afterwards. This question is dealt with in the lectures by Hugenholtz and Ruelle.

e. Does a phase transition correspond to a *mathematical singularity* of the thermodynamic quantities (the partition function in the first place) or not? Two examples, which are entirely artificial but nevertheless not too far from more realistic examples, may serve to illustrate the difference.

1. Suppose that the N-particle partition function is given by

$$Z_N = \{f_1(x)\}^N + \{f_2(x)\}^N , \tag{3.1}$$

where x is some primary variable and f_1 and f_2 are positive functions of x. Then

$$Z = \lim_{N\to\infty} Z_N^{1/N}$$

exists and is given by

$$Z = f_1(x) \lim_{N\to\infty} [1 + \{f_2(x)/f_1(x)\}^N]^{1/N} = f_1(x) \qquad \text{for } x \text{ such that } f_1(x) \geq f_2(x) ,$$

$$Z = f_2(x) \qquad \text{for } x \text{ such that } f_1(x) \leq f_2(x) ,$$

or

$$Z = \max\{f_1(x), f_2(x)\} . \tag{3.2}$$

So if the two curves $y = f_1(x)$ and $y = f_2(x)$ cross at a certain value x_c of x, $Z = Z(x)$ jumps from one curve to the other. This is equivalent to saying that for each x there are two possible states, of which the one with the lowest free energy is the equilibrium state:

$$F = -kT \ln Z = \min\{-kT \ln f_1(x), -kT \ln f_2(x)\} . \tag{3.3}$$

The point $x = x_c$ is a point of *discontinuity*, a "crossing point", but not a point of singularity: the "branch" for $x \leqslant x_c$ can be continued analytically into the region $x \geqslant x_c$ and vice versa.

2. Suppose that

$$Z_N = \prod_{n=0}^{N-1} \left(x^2 + 1 - \frac{2nx}{N}\right) = \exp\left[\sum_{n=0}^{N-1} \ln\left(x^2 + 1 - \frac{2nx}{N}\right)\right] . \tag{3.4}$$

Since $x^2 + 1 - 2nx/N > x^2 + 1 - 2x \geqslant 0$, we have $Z_N > 0$. It is not difficult to show that

$$Z = \lim_{N \to \infty} Z_N^{1/N}$$

exists and is given by

$$Z = \exp[\int_0^1 d\lambda \ln(x^2 + 1 - 2\lambda x)] , \tag{3.5}$$

whence

$$Z = \exp[-\frac{1}{2x}\{(x-1)^2 \ln(x-1)^2 - (x^2+1)\ln(x^2+1) + 2x\}] . \tag{3.6}$$

This function is seen to have a *singularity* at $x = 1$. The function $\varphi(x) \equiv -(x-1)^2 \ln(x-1)^2$ goes to zero for $x \to 1$, but its second derivative (and hence the second derivative of $F = -kT \ln Z$) contains a term going to $+\infty$ for $x \to 1-$ and for $x \to 1+$:

$$\frac{d^2\varphi}{dx^2} \sim \ln|x-1| , \qquad (x \to 1) . \tag{3.7}$$

Up to now, all evidence indicates that it is the second type of transition which is realized in most physical systems. Only one rather artificial model exhibits a phase transition of the first type.

4. MODELS

In order to understand the most characteristic features of phase transitions without having to make ad hoc assumptions during the calculations one has introduced, as in other branches of physics, various idealized systems or models. The models which have been investigated most extensively are those where the particles occupy the sites of a regular crystal lattice. The advantages of these *lattice models* over models where the particles are allowed to vary their position continuously are twofold. Firstly, there are no configurational problems to be solved such as one encounters e.g. in the hard-sphere gas; the positions of all particles are known in advance. Secondly, the periodicity of the lattice simplifies the calculations considerably. The lack of spatial isotropy, however, is sometimes regarded as a disadvantage. On the other hand, the resulting multitude of possible systems enables one to investigate families of closely related systems and thus to study the relative importance of the dimensionality and other global properties as compared with the local structure.

4.1. *The Ising model*

The model which up to now has been most thoroughly investigated by rigorous, approximative and numerical methods, is the Ising model. This model was proposed by Lenz in 1920 as a model for ferromagnetism and worked out for the case of a one-dimensional lattice by Ising in 1925. Later on it was found that it can also be used as a model for a lattice gas and for a binary alloy; we shall briefly discuss the relation between the various interpretations.

4.1.1. *The Ising ferromagnet*

Consider a crystal where each atom has a magnetic moment (to be called "spin" for convenience, cf. section 4.1.4) which can point in two directions, "up" and "down". The two states of the jth spin are described by a variable σ_j (in the literature also denoted by μ_j or s_j) which is +1 for an "up" spin and -1 for a "down" spin. The energy of the spin system is assumed to be

$$E = -J \sum_{\langle j,k \rangle} \sigma_j \sigma_k - mH \sum_j \sigma_j \,, \tag{4.1}$$

where J is a positive constant, m the magnitude of the atomic magnetic moments and H the external magnetic field; the first sum is over all pairs of nearest neighbour spins.

The canonical partition function is

$$Z_N^{\mathrm{can}} = \sum_{\sigma_1 \ldots \sigma_N} \exp[K \sum_{\langle j,k \rangle} \sigma_j \sigma_k + L \sum_j \sigma_j] , \tag{4.2}$$

where $K \equiv J/kT$, $L \equiv mH/kT$. The total free energy F_N is given by

$$F_N = -kT \ln Z_N . \tag{4.3}$$

We are interested in the partition function per spin Z and the free energy per spin F in the thermodynamic limit ($N \to \infty$, J constant):

$$Z = \lim_{N \to \infty} Z_N^{1/N} ; \qquad F = -kT \ln Z , \tag{4.4}$$

from which the energy U, the specific heat at constant field C_H, the magnetization M and the susceptibility χ are derived in the standard way:

$$U = F - T\left(\frac{\partial F}{\partial T}\right)_H ; \quad C_H = \left(\frac{\partial U}{\partial T}\right)_H ; \quad M = -\left(\frac{\partial F}{\partial H}\right)_T ; \quad \chi = \left(\frac{\partial M}{\partial H}\right)_T . \tag{4.5}$$

The problem of calculating Z and the quantities derived from it is known as the "Ising problem". Up to now, only partial solutions to this problem are known.

An obvious extension of the Ising model is obtained by removing the restriction to nearest-neighbour interactions, putting:

$$E = -\sum_{(j,k)} J_{jk} \sigma_j \sigma_k - mH \sum_j \sigma_j , \tag{4.6}$$

where the first sum is over *all* pairs of spins and where J_{jk} depends only on the relative position r_{jk} of the spins j and k.

The mathematical analysis of the Ising model will form the subject of section 6.

4.1.2. *The lattice gas*

As was shown by Lee and Yang, the Ising model can be reinterpreted as a model for a lattice gas. This is a gas contained in a volume which is divided into cells so small that they cannot contain more than one molecule, and where the interaction between two molecules depends only on the cells in which they are, not on their position within the cells. Alternatively, it may be described as a system where the molecules sit on the sites of a lattice. If we now use the variable σ_j to describe the state of occupation of the jth cell (site) rather than the state of the jth spin, letting $\sigma_j = +1$ denote an occupied cell and $\sigma_j = -1$ an empty cell, we have established a one-to-one correspondence between the states of the gas and the spin states of a ferromagnet. The transition, at sufficiently low temperatures, of a spin system from the disordered, paramagnetic state to the ordered, ferromagnetic state has then its parallel in the condensation of the lattice gas.

Mathematically, this translation is effected by introducing occupation numbers for the cells of a lattice gas:

$$\nu_j = \begin{cases} 1 \\ 0 \end{cases} \text{if the } j\text{th cell is } \begin{cases} \text{occupied} \\ \text{empty} \end{cases},$$

which are related to the σ_j as follows:

$$\nu_j = \tfrac{1}{2}(1 + \sigma_j) \,. \tag{4.7}$$

The total number of particles is given by $N_1 = \sum_j \nu_j = \frac{1}{2}N + \frac{1}{2}\sum_j \sigma_j$; the total number of cells N plays the role of the available volume.

In the simplest case, where the only interactions are between molecules in adjacent cells (interaction strength V), the energy of the lattice gas is

$$E = V \sum_{\langle j,k\rangle} \nu_j \nu_k = \tfrac{1}{4}V \sum_{\langle j,k\rangle} \sigma_j\sigma_k + \tfrac{1}{4}qV \sum_j \sigma_j + \tfrac{1}{8}NqV \,, \tag{4.8}$$

where q is the coordination number of the lattice. The last term is a constant which does not influence the thermodynamic properties and may therefore be omitted.

Let us consider the canonical partition function of the lattice gas, neglecting the kinetic energy of the molecules; taking the kinetic energy into account would change the equations only in a trivial way. By definition, the canonical partition function is a sum over all states with a *fixed* number of particles $\sum_j \nu_j = N_1$, i.e. with $\sum_j \sigma_j = 2N_1 - N$:

$$Z_{N_1}^{\text{can}} = \sum_{\substack{\sigma_1 \ldots \sigma_N \\ (\sum_j \sigma_j = 2N_1 - N)}} \exp\Big[-\tfrac{1}{4}\beta V \sum_{\langle j,k\rangle} \sigma_j\sigma_k - \tfrac{1}{4}\beta q V \sum_j \sigma_j\Big] \,; \tag{4.9}$$

($\beta \equiv 1/kT$). Therefore it is not this function which is the strict lattice-gas analogue of the canonical partition function of an Ising ferromagnet, but the corresponding *grand-canonical partition function:*

$$Z^{\mathrm{gr}} = \sum_{N_1} \mathrm{e}^{\beta\mu N_1}\, Z^{\mathrm{can}}_{N_1}$$

$$= \sum_{\sigma_1 \dots \sigma_N} \exp[-\tfrac{1}{4}\beta V \sum_{\langle j,k\rangle} \sigma_j \sigma_k + (\tfrac{1}{2}\beta\mu - \tfrac{1}{4}\beta q V) \sum_j \sigma_j + \tfrac{1}{2}\beta\mu N] \;. \quad (4.10)$$

The comparison of (4.10) with (4.2) shows that we have the following correspondence:

$$V \Longleftrightarrow -4J \;, \quad (4.11)$$

$$\mu \Longleftrightarrow 2mH - 2qJ \;. \quad (4.12)$$

Replacing the left-hand quantities by the right-hand expressions changes (4.10) into (4.2), if we omit the irrelevant constant factor $\exp(\frac{1}{2}\beta\mu N)$, and vice versa. Using these "translation rules", we can associate with each thermodynamic quantity of the lattice gas a quantity of the ferromagnetic system. We will mention only a few examples. If v_0 is the volume of a cell of the lattice gas, and $\langle \dots \rangle$ denotes the thermal average of the quantity between the brackets, the density of the lattice gas is

$$\rho = \langle N_1 \rangle / N v_0 = [1 + N^{-1} \langle \sum_j \sigma_j \rangle] / 2v_0 \;. \quad (4.13)$$

Since the magnetization per spin of a magnetic system is $M = m\langle \sum_j \sigma_j \rangle / N$, the magnetic counterpart of $2v_0\rho$ is $1 + M/M_0$, where $M_0 = m$ is the saturation magnetization. To the *coexistence curve* of the lattice gas corresponds the *spontaneous-magnetization curve* of the ferromagnet. The critical value of M is 0, that of ρ is $\rho_c = 1/2v_0$. Eliminating v_0 we find the following correspondence

$$\rho/\rho_c \Longleftrightarrow 1 + M/M_0 \;. \quad (4.14)$$

The *configurational specific heat at constant volume* per unit cell of the lattice gas C_V corresponds to the *specific heat at constant magnetization* per spin of the magnetic system C_M:

$$C_V \Longleftrightarrow C_M \;. \quad (4.15)$$

For $H = 0$, $T > T_c$, C_M is identical to the specific heat at constant field C_H because M is a constant ($M \equiv 0$); in general, however, the two quantities are different.

Finally, the *isothermal compressibility* of the lattice gas $K_T = \rho^{-1}(\partial\rho/\partial p)_T$ finds its counterpart in the *isothermal magnetic susceptibility* $\chi = (\partial M/\partial H)_T$. To see this, consider the thermodynamic relation $d\mu = -s\,dT + v\,dp$, where s and v are the specific entropy and volume of the gas. For isothermal changes this relation reduces to $d\mu = v\,dp = \rho^{-1}\,dp$, from which we derive $K_T = \rho^{-2}(\partial\rho/\partial\mu)_T$, or

$$\rho^2 K_T/\rho_c = (\partial\rho/\partial\mu)_T/\rho_c \,. \tag{4.16}$$

From (4.12), (4.14) and (4.16) it follows that

$$\frac{\rho^2 K_T}{\rho_c} \Longleftrightarrow \frac{\chi}{2m^2} \,. \tag{4.17}$$

The above correspondences remain valid if in both systems interactions between more distant molecules are introduced, provided (4.11) is properly generalized.

Although the lattice gas model in its simplest form cannot be more than a crude approximation to a real gas, its relationship to the Ising ferromagnet nevertheless gives an insight into the origin of the near equality found experimentally for the critical indices of various quantities.

4.1.3. *Binary mixtures*

Another useful translation of the Ising model is that in terms of a binary mixture (alloy or fluid). It is obtained by letting $\sigma_j = +1$ describe the presence on the jth lattice site (or, if we consider a cell model of a binary fluid mixture, in the jth cell) of an atom of type A, and $\sigma_j = -1$ the presence of an atom of type B. This interpretation is very similar to the previous one, and will therefore not be discussed in these lectures.

4.1.4. *Quantum mechanical formulation of the Ising problem*

The Ising problem was formulated originally as a problem in classical physics, in which the restriction of the number of states of the magnetic moment to two constituted an ad hoc assumption. Later on this restriction was shown to find a natural explanation in the quantum mechanical theory of the spin, a fact which enables us to reformulate the Ising problem within the framework of quantum mechanics. This reformulation, trivial though it is, forms a useful preparation for the discussion of another, more strictly quantum mechanical problem, that of Heisenberg (cf. section 4.2) and a stepping stone towards the algebraic solution of the Ising problem in two dimensions (cf. section 6).

If we associate with each atom three spin component operators σ_j^x, σ_j^y, σ_j^z and we identify the two states of the jth atom ($\sigma_j = +1$ and $\sigma_j = -1$) with the two eigenstates of the operator σ_j^z, we can characterize the Ising model as a spin system described by the Hamiltonian

$$\mathcal{H} = -J \sum_{\langle j,k \rangle} \sigma_j^z \sigma_k^z - mH \sum_j \sigma_j^z . \tag{4.18}$$

The first term is now said to represent a strongly anisotropic interaction between neighbouring spins, because it acts only on the z components of the spins and not on the x and y components. The canonical partition function is

$$Z_N = \mathrm{Tr}\, e^{-\beta\mathcal{H}} , \tag{4.19}$$

with the trace taken over the states of all spins. Since all operators σ_j^z commute, the quantum mechanical aspect is a trivial one: the eigenvalues of $\mathcal{H}$ are immediately known (they are given by (4.1)) and only the statistical part of the problem remains.

The situation is entirely different if we consider a system of spins with Ising interaction which is placed in a magnetic field whose direction is not that of the interacting spin components, but e.g. the x direction. The Hamiltonian is then

$$\mathcal{H} = -J \sum_{\langle j,k \rangle} \sigma_j^z \sigma_k^z - mH \sum_j \sigma_j^x . \tag{4.20}$$

Since σ_j^x and σ_j^z do not commute this problem is a quantum mechanical one in the full sense (i.e., it has no classical interpretation). The properties of the Ising model in a perpendicular field are therefore much harder to derive than those of the Ising model in a parallel field.

Obviously, the choice of the z axis as the direction of the interacting spin components in (4.18) and (4.20) is completely arbitrary. We can just as well interpret the two states of the jth atom as the eigenstates of the operator σ_j^x. This fact will be made use of in section 6.3, where we investigate a system of the type (4.20).

4.2. *The Heisenberg model*

This model, essentially proposed by Heisenberg in 1928, is characterized by a Hamiltonian in which the spin-spin interaction term is completely isotropic; the direction of the magnetic field is chosen as the z direction:

$$\mathcal{H} = -J \sum_{\langle j,k \rangle} \boldsymbol{\sigma}_j \cdot \boldsymbol{\sigma}_k - mH \sum_j \sigma_j^z . \tag{4.21}$$

The second sum commutes with the first sum, but the various terms of the latter do not commute with each other. This problem turns out to be still far more difficult than that of the Ising model in a perpendicular field. Nevertheless it has been the subject of extensive investigations, and many interesting results have been obtained.

A generalization of (4.21) which at the same time contains the Ising model as a special case is obtained by allowing the terms $\sigma_j^x\sigma_k^x$, $\sigma_j^y\sigma_k^y$ and $\sigma_j^z\sigma_k^z$ to have arbitrary coefficients:

$$\mathcal{H} = -J \sum_{\langle j,k \rangle} (\sigma_j^x\sigma_k^x + \Gamma\sigma_j^y\sigma_k^y + \Delta\sigma_j^z\sigma_k^z) - mH \sum_j \sigma_j^z . \tag{4.22}$$

This model appears to play an important role in the analysis of the problems discussed in the next section (cf. ref. 5).

4.3. *The ice model and related models*

A class of models which very recently has received a great deal of attention is that which arose from three classic problems in solid state physics, viz. the problems of the residual entropy of ice, of ferroelectricity and of antiferroelectricity. Although the physical aspects of these problems are rather different, they have much in common from the mathematical point of view. We shall only sketch their formulation. For their solution, which forms the subject of a seminar lecture by E. H. Lieb, and for further details we refer to the literature [5].

4.3.1. *The ice model*

From X-ray diffraction experiments it is known that the oxygen atoms in ice form a wurtzite hexagonal lattice in which each oxygen has four nearest neighbours. In order to account for the residual entropy of 0.82 ± 0.05 cal/(deg mole) which ice is found to have at low temperatures, Pauling in 1935 proposed the following model for the distribution of the hydrogen atoms (which are not observed in the X-ray analysis): (a) there is one hydrogen atom between every two nearest neighbour oxygen atoms; (b) the hydrogen atoms are not at the centre of the oxygen-oxygen bonds, but in one of two equivalent positions closer to one end then to the other; (c) each oxygen is surrounded by two hydrogens close to it (as in water vapour) and two at a larger distance. One easily convinces oneself that there are many configurations satisfying the above rules (the latter of which is usually called the *ice condition*). To complete the model it is assumed that (d) all allowed configurations have the same energy. It is this degeneracy which is held responsible for the residual entropy of ice. If W_N is the number of allowed configurations of a crystal containing N oxygens, the corresponding entropy per H_2O group is $S = k \ln W$, where

$$W = \lim_{N\to\infty} W_N^{1/N} .$$

The various configurations of ice can be pictured geometrically by connecting each pair of neighbouring sites of the oxygen lattice by a line and drawing an arrow from the jth site to the kth site, if the hydrogen between the two sits closer to the jth site; this amounts to replacing O. . H. . . . O by o—>—o. The ice condition then requires that there are two arrows pointing into each lattice point. Fig. 1 shows an allowed configuration of the two-dimensional analogue of ice, based on a square lattice of oxygen atoms ("*square ice*"), and its representation by arrows.

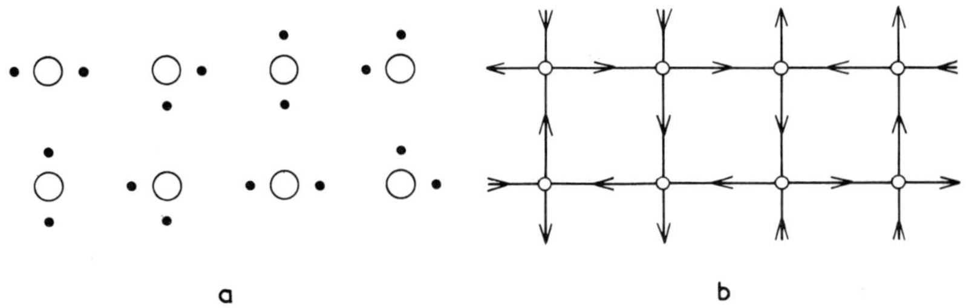

Fig. 1. Part of an allowed configuration of ice, represented (a) as an arrangement of atoms (circles = oxygen atoms, dots = hydrogen atoms) and (b) as a configuration of arrows.

For the calculation of W for real ice various approximation methods have been developed, but up to now the exact answer is unknown. The corresponding problem for square ice, however, has recently been solved by Lieb [5]. The calculation is lengthy and far from trivial, but the result is surprisingly simple: $W = (\frac{4}{3})^{\frac{3}{2}}$.

An interesting translation of the problem of square ice has been given by Lenard [6]. Consider the squares of an infinite chessboard to be coloured not by two, but by three colours, so that no two adjacent squares have the same colour. There is an infinite number of such colourings, and each of them can be represented in a simple way by a certain orientation of the boundaries between the squares. Call the three colours A, B and C. Consider an arbitrary square S_1 and a square S_2 adjacent to it. Draw an arrow on the line segment separating S_1 and S_2 according to the following rule: if in the cyclic triple A, B, C the colour of S_2 follows immediately after that of S_1, the arrow shall run to the right (clockwise) when viewed from the interior of S_1 (and hence to the left when viewed from the interior of S_2); if the colour of S_2 precedes that of S_1, it shall run the other way (fig. 2a). A simple enumeration of possible cases shows that the arrows satisfy the ice condition. Conversely, each set of arrows satisfying the ice condition determines an allowed colouring once the colour of one single square has been chosen (fig. 2b).

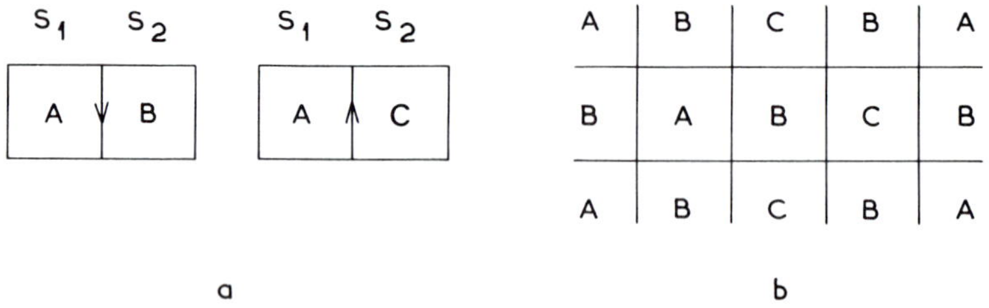

Fig. 2. (a) Two examples of the convention for representing colourings by arrows. (b) A colouring of the quadratic lattice corresponding to the ice configuration of fig. 1; the other compatible colourings are obtained by cyclically interchanging A, B and C.

A more physical interpretation considers A, B and C as representing three different kinds of atoms, which have to be so arranged on a quadratic lattice that no two like atoms are nearest neighbours. This model is likely to have interesting physical properties [6].

4.3.2. *The* F *model and the* KDP *model*

From the ice model we can construct other physically interesting models by attributing different energies to the various arrangements of arrows around a lattice point which are allowed by the ice condition. We consider two cases, which are usually referred to as the F model and the KDP model, and of those only the two-dimensional version. The models are defined by the following energy assignments ($\epsilon > 0$):

(1) (2) (3) (4) (5) (6)

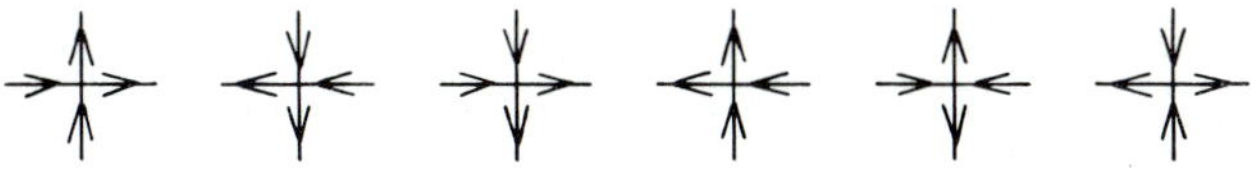

	(1)	(2)	(3)	(4)	(5)	(6)
F model	ϵ	ϵ	ϵ	ϵ	0	0
KDP model	0	0	ϵ	ϵ	ϵ	ϵ

In the applications of these models, the lattice points are not occupied by oxygen atoms but by certain molecular groups; on the bonds, on the other hand, there are again hydrogen atoms.

The ground state of the KDP model is doubly degenerate; it contains either only configurations (1) or only configurations (2). This implies parallel positions for all local groups, so that their dipole moments add up to give a net spontaneous electric polarization. Therefore the KDP model is a model for a *ferroelectric* solid; it was proposed as such by Slater in 1941. It derives its name from the ferroelectric compound KH_2PO_4 (potassium diphosphate, KDP).

In the ground state of the F model (which is also doubly degenerate), on the other hand, the configurations (5) and (6) alternate so that there is no net polarization. A closer investigation of the spatial arrangement of the groups of atoms shows that this model (which was proposed by Rys in 1963) is a suitable model for an *antiferroelectric* crystal.

Lieb has succeeded in also calculating the partition function of the "square" F model and KDP model, even in the presence of an electric field [5]. He found that both models show a phase transition, the nature of the transitions being completely different from that of the Ising model (and also different from each other)*.

5. THE "CLASSICAL" THEORIES

In the time when no exact results concerning phase transitions in model systems were known, various approximative theories have been developed, both for lattice models and for continuum models (Van der Waals' theory of condensation, Weiss' molecular-field theory of ferromagnetism, the so-called Bragg-Williams, Bethe-Peierls and Kramers-Wannier-Kikuchi approximations for the Ising model etc.).

* Recently, C.P.Yang, B.Sutherland and C.N.Yang have solved the most general case of the above-mentioned type of problem, in which each of the six configurations has a different energy [7].

Although these different theories gave widely varying results for the critical constants, they all agreed in their prediction of the critical behaviour of thermodynamic quantities. It was found e.g. that in ferromagnets M and χ should behave as

$$M \sim (T_c - T)^{\frac{1}{2}} , \qquad (T \to T_c-, \ H = 0) ,$$

$$\chi \sim (T - T_c)^{-1} , \qquad (T \to T_c+, \ H = 0) ,$$

$$|M| \sim |H|^{\frac{1}{3}} , \qquad (T = T_c, \ H \to 0) ,$$

for any number of dimensions, and that C_H should have a jump discontinuity. For the gas-liquid system analogous results were found:

$$\rho_L - \rho_G \sim (T_c - T)^{\frac{1}{2}} , \qquad (T \to T_c-) ,$$

$$K_T \sim (T - T_c)^{-1} , \qquad (T \to T_c+, \ \rho = \rho_c) ,$$

$$|p - p_c| \sim |\rho - \rho_c|^3 , \qquad (T = T_c, \ \rho \to \rho_c) .$$

Unfortunately, all these and other predictions are definitely at variance with the experimental data (cf. section 2). The origin of the failure of these "classical theories", as they are sometimes called, is now well understood (cf. ref. 3, section 5). We shall not go into it, however, but turn immediately to exact results.

6. THE ANALYSIS OF THE ISING MODEL

The mathematical analysis of the Ising model (and also that of other models) has led to results which can roughly be divided into three groups, to each of which important contributions have been made during the last few years. The first group consists of results of a general nature. It includes, among other things, the proofs of: the existence (or non-existence) of a phase transition, inequalities for the critical temperature and for certain thermodynamic quantities, symmetry properties, and relations between properties of models of different lattice structure.

The other groups consist of more specific results. Those of the second group concern series expansions of the thermodynamic quantities (mainly high and low temperature expansions). Those of the third group are the most detailed ones. They consist in closed-form expressions, which could be derived in a number of special cases. Since for the time being it is only from the last kind of results that we can draw rigorous conclusions about critical behaviour, we concentrate in this section on the problem of deriving these results. The main example we treat is the calculation of the partition function of the two-dimensional Ising model (quadratic lattice, no external field) by the algebraic method developed a few years ago by Schultz, Mattis and Lieb [8]. This method is, on one hand, closely related to the one originally used by Onsager in his classic paper. On the other hand, it is formulated in such a way as to fit into the general framework of many-particle theory. To this end the problem is cast into a form which makes it possible to apply well-known methods from the quantum theory of many-fermion systems.

As a preparation we first discuss the derivation of the partition functions of several one-dimensional systems. The section ends with a brief survey of critical indices for the Ising model which have been derived (for two-dimensional lattices) or estimated from series expansion data (for three-dimensional lattices).

6.1. *Linear chain in the absence of a magnetic field*

As we have seen in section 4.1 we can consider this problem as a classical one. If periodic boundary conditions are imposed ($\sigma_{N+1} \equiv \sigma_N$), the partition function is

$$Z_N = \sum_{\sigma_1 \ldots \sigma_N} \exp[K \sum_{j=1}^{N} \sigma_j \sigma_{j+1}]$$

$$= \sum_{\sigma_1 \ldots \sigma_N} \exp[K\sigma_1\sigma_2] \exp[K\sigma_2\sigma_3] \ldots \exp[K\sigma_{N-1}\sigma_N] \exp[K\sigma_N\sigma_1] \,. \quad (6.1)$$

If we use the abbreviation $V_{\sigma_j\sigma_{j+1}} \equiv \exp[K\sigma_j\sigma_{j+1}]$, and we introduce a 2×2 matrix $\boldsymbol{V}$ with elements $V_{\sigma_j\sigma_{j+1}}$:

$$\boldsymbol{V} = \begin{array}{c} \\ (\sigma_j = +1) \\ (\sigma_j = -1) \end{array} \begin{array}{cc} (\sigma_{j+1} = +1) & (\sigma_{j+1} = -1) \\ \left[\begin{array}{c} e^{K} \\ e^{-K} \end{array} \right. & \left. \begin{array}{c} e^{-K} \\ e^{K} \end{array} \right] \end{array} ,$$

we can write Z_N as follows

$$Z_N = \sum_{\sigma_1 \ldots \sigma_N} V_{\sigma_1\sigma_2} V_{\sigma_2\sigma_3} \cdots V_{\sigma_{N-1}\sigma_N} V_{\sigma_N\sigma_1}$$

$$= \sum_{\sigma_1} (\boldsymbol{V}^N)_{\sigma_1\sigma_1} = \mathrm{Tr}(\boldsymbol{V}^N) \,. \quad (6.2)$$

Since V connects the states of the jth atom with those of the $(j+1)$th atom, it is called the *transfer matrix*. If its eigenvalues are λ_1 and λ_2 (with $\lambda_1 \geqslant \lambda_2$) we have

$$Z_N = \lambda_1^N + \lambda_2^N , \tag{6.3}$$

and the partition function per spin in the thermodynamic limit is

$$Z = \lim_{N\to\infty} Z_N^{1/N} = \lambda_1 . \tag{6.4}$$

It is easily verified that $\lambda_1 = e^K + e^{-K}$, $\lambda_2 = e^K - e^{-K}$, so that finally we have

$$Z = 2 \cosh K . \tag{6.5}$$

The function $\cosh K$ is an analytic function of T for all $T > 0$, which shows that the linear Ising chain does not exhibit a phase transition.

6.2. *Linear chain in a parallel field*

The Hamiltonian of this system is given by (4.18). All terms of $\mathcal{H}$ commute and we can write down the partition function immediately (cf. 6.1)

$$Z_N = \sum_{\sigma_1 \ldots \sigma_N} \exp[K\sigma_1\sigma_2] \exp[L\sigma_2] \exp[K\sigma_2\sigma_3] \ldots \exp[K\sigma_N\sigma_1] \exp[L\sigma_1] . \tag{6.6}$$

Writing

$$V^{(1)}_{\sigma_j\sigma_{j+1}} \equiv \exp[K\sigma_j\sigma_{j+1}] , \qquad V^{(2)}_{\sigma_j\sigma_j} \equiv \exp[L\sigma_j] ,$$

and introducing the matrix $V^{(1)}$ with elements $V^{(1)}_{\sigma_j\sigma_{j+1}}$, the diagonal matrix $V^{(2)}$ with diagonal elements $V^{(2)}_{\sigma_j\sigma_j}$, and their product $V = V^{(1)}V^{(2)}$:

$$V^{(1)} = \begin{bmatrix} e^K & e^{-K} \\ e^{-K} & e^K \end{bmatrix}; \quad V^{(2)} = \begin{bmatrix} e^L & 0 \\ 0 & e^{-L} \end{bmatrix}; \quad V = \begin{bmatrix} e^{K+L} & e^{-K-L} \\ e^{-K+L} & e^{K-L} \end{bmatrix}, \tag{6.7}$$

we find that Z is again equal to the largest eigenvalue λ_1 of the transfer matrix

$$Z = \lambda_1 = e^K \cosh L + [e^{2K}(\sinh L)^2 + e^{-2K}]^{\frac{1}{2}} . \tag{6.8}$$

The results mentioned in 6.1 and 6.2 were first derived, along different lines, by Ising (1925).

6.3. *Linear chain in a perpendicular field*

According to (4.19) and (4.20) the partition function of this system is

$$Z_N = \mathrm{Tr}\,\exp\left[K\sum_{j=1}^{N} \sigma_j^z\sigma_{j+1}^z + L\sum_{j=1}^{N}\sigma_j^x\right]. \tag{6.9}$$

Since $[\sigma_j^z, \sigma_j^x] \neq 0$, the terms of the first sum in the exponent do not commute with those of the second. Therefore it is no longer possible to factorize the Boltzmann factor in (6.9) as we did in 6.1 and 6.2. The transfer-matrix approach is therefore not possible in this case, and we have to follow another method, which is based on the diagonalization of the exponent of (6.9). Although we could carry out this diagonalization on $\mathcal{H}$ alone, we prefer, with a view to the problem to be discussed in 6.4, to treat the partition function as a whole.

To simplify the discussion it is convenient to rotate first the coordinate system around the y axis until the field is in the z direction, i.e. to replace σ_j^x by σ_j^z; σ_j^z has then to be replaced by $-\sigma_j^x$:

$$\sigma_j^x \to \sigma_j^z\,; \qquad \sigma_j^z \to -\sigma_j^x\,. \tag{6.10}$$

We then introduce the spin raising and lowering operators

$$\sigma_j^{\pm} = \tfrac{1}{2}(\sigma_j^x \pm i\sigma_j^y) \tag{6.11}$$

(where the unconventional factor $\frac{1}{2}$ is inserted for convenience), and express σ_j^x and σ_j^z in terms of these:

$$\sigma_j^x = \sigma_j^+ + \sigma_j^-\,; \qquad \sigma_j^z = 2\sigma_j^+\sigma_j^- - 1\,. \tag{6.12}$$

Applying (6.10) to (6.9) and substituting (6.12) we get

$$Z_N = \mathrm{Tr}\,\exp\left[K\sum_{j=1}^{N}(\sigma_j^+ + \sigma_j^-)(\sigma_{j+1}^+ + \sigma_{j+1}^-) + L\sum_{j=1}^{N}(2\sigma_j^+\sigma_j^- - 1)\right], \tag{6.13}$$

which forms the starting point of our further discussion.

We observe that the exponent in (6.13), which is still nothing but $-\beta\mathcal{H}$, is quadratic in the operators σ_j^+ and σ_j^-, which can be considered as the creation and annihilation operators for "up" spins. Hence it resembles Hamiltonians occurring in certain problems of quantum field theory. This makes us hope that it can be diagonalized, and hence the Boltzmann factor factorized, by the application of well-known techniques.

To pursue the analogy farther we remark that the role of the vacuum is played by the state with all spins "down". (Notice that this is *not* the zero-field ground state of our Hamiltonian, because after the coordinate transformation this state has all spins oriented in the $+x$ or $-x$ direction). The analogy is particularly striking if we consider the lattice gas "translation" of the spin system. There the state corresponding to the state with all spins down is the real vacuum (all cells empty), and the operators σ_j^+ and σ_j^- create and annihilate a real particle in the jth cell.

Since a minus spin cannot be turned up more than once in succession, we expect the $\sigma_j^{\pm}$ to have the properties of fermion operators. However, this is only partially true as we see from their commutation properties. On one hand, we have for operators referring to the same spin the anticommutation relations of *fermion* operators:

$$\{\sigma_j^+, \sigma_j^-\} = 1 \; ; \qquad \{\sigma_j^+, \sigma_j^+\} = \{\sigma_j^-, \sigma_j^-\} = 0 \; . \tag{6.14}$$

The spin operators for different spins, on the other hand, commute:

$$[\sigma_j^+, \sigma_k^+] = [\sigma_j^+, \sigma_k^-] = [\sigma_j^-, \sigma_k^-] = 0 \; , \qquad (j \neq k) \; , \tag{6.15}$$

so that in this respect the $\sigma_j^{\pm}$ behave as *boson* operators. Obviously, if they are considered to describe the creation and annihilation of particles, these particles are neither fermions nor bosons, but a kind of intermediate entity, sometimes called *paulions*. As a consequence, it is not possible to find a unitary transformation diagonalizing $\mathcal{H}$ and at the same time leaving the commutation character of the operators invariant.

Fortunately, Jordan and Wigner have shown as early as 1928 how to get rid of the mixed commutation character of the spin operators. We introduce new quantities

$$a_j = (-1)^{\sum_{k=1}^{j-1} \sigma_k^+ \sigma_k^-} \sigma_j^- \; , \qquad a_j^* = (-1)^{\sum_{k=1}^{j-1} \sigma_k^+ \sigma_k^-} \sigma_j^+ \; . \tag{6.16}$$

It is easily verified that the a_j and a_j^* satisfy the anticommutation relations

$$\{a_j, a_k^*\} = \delta_{jk} \; ; \qquad \{a_j, a_k\} = \{a_j^*, a_k^*\} = 0 \; , \tag{6.17}$$

and are, therefore, genuine fermion operators. Inverting (6.16) we find

$$\sigma_j^- = (-1)^{\sum_{k=1}^{j-1} a_k^* a_k} a_j \,, \qquad \sigma_j^+ = (-1)^{\sum_{k=1}^{j-1} a_k^* a_k} a_j^* \,. \tag{6.18}$$

For $j = 1, 2, \ldots, N\text{-}1$ it follows from (6.18) and (6.17) that

$$\sigma_j^+ \sigma_{j+1}^+ = (-1)^{\sum_{k=1}^{j-1} a_k^* a_k} a_j^* \, (-1)^{\sum_{k=1}^{j} a_k^* a_k} a_{j+1}^*$$

$$= a_j^* \, (-1)^{a_j^* a_j} \, a_{j+1}^* = a_j^* a_{j+1}^* \,. \tag{6.19}$$

In the last step we have used the fact that $a_j^* (-1)^{a_j^* a_j} = a_j^*$, which is most readily verified by proving that both members of the equation have the same effect on the two eigenstates of $a_j^* a_j$. Similarly we find

$$\sigma_j^+ \sigma_{j+1}^- = a_j^* a_{j+1} \,, \tag{6.20}$$

but, on the other hand,

$$\sigma_j^- \sigma_{j+1}^+ = - a_j a_{j+1}^* \,; \qquad \sigma_j^- \sigma_{j+1}^- = - a_j a_{j+1} \,. \tag{6.21}$$

If we neglect a boundary effect resulting from the different character of the term $(\sigma_N^+ + \sigma_N^-)(\sigma_1^+ + \sigma_1^-)$, where the first factor bears a higher index than the second, we find the following expression for the partition function

$$Z_N = \mathrm{Tr} \exp[K \sum_j (a_j^* - a_j)(a_{j+1}^* + a_{j+1}) + L \sum_j (2a_j^* a_j - 1)] \,. \tag{6.22}$$

The minus sign in the first term of the exponent is the only visible trace of the minus signs introduced in (6.16).

We can now safely introduce a unitary transformation diagonalizing $\mathcal{H}$; it will leave the commutation relations (6.17) invariant. Naturally, the transformation is one to running-wave operators:

$$A_q = \frac{1}{\sqrt{N}} \, e^{-\frac{1}{4}\pi i} \sum_{j=1}^{N} e^{iqj} \, a_j \,, \tag{6.23}$$

where $q = 2\pi l/N$, $l = -\frac{1}{2}N, -\frac{1}{2}N+1, \ldots, \frac{1}{2}N-1$; for convenience we have assumed N to be even. The only unusual feature in (6.23) is the occurrence of the factor $e^{-\frac{1}{4}\pi i}$. It is inserted in order to get real coefficients in the following expression for Z_N, which is obtained by combining terms with opposite wave number:

$$Z_N = \operatorname{Tr} \exp[2K \sum_{q>0} \{(\cos q)(A_q^* A_q + A_{-q}^* A_{-q}) + (\sin q)(A_q A_{-q} + A_{-q}^* A_q^*)\}$$
$$+ 2L \sum_{q>0} \{A_q^* A_q + A_{-q}^* A_{-q} - 1\}] . \tag{6.24}$$

Strictly speaking, we should treat the terms with $q = -\pi$ and $q = 0$, which have no counterpart with opposite wave number, separately, but in the limit $N \to \infty$ their contribution to Z goes to zero. The exponent of (6.24) still contains non-diagonal terms, which can be removed by going over to new variables B_q defined by

$$B_q = (\cos \varphi_q)A_q + (\sin \varphi_q) A_{-q}^* , \quad B_{-q}^* = (-\sin \varphi_q)A_q + (\cos \varphi_q) A_{-q}^* , \tag{6.25}$$

where φ_q is chosen so as to make the coefficient of $B_q B_{-q} + B_{-q}^* B_q^*$ vanish. This transformation is similar to the Valatin-Bogoliubov transformation known from the theory of superconductivity. For Z_N we get

$$Z_N = \operatorname{Tr} \exp \sum_{q>0} [2(K^2 + 2KL \cos q + L^2)^{\frac{1}{2}}(B_q^* B_q + B_{-q}^* B_{-q} - 1) + 2K \cos q] . \tag{6.26}$$

Since the B_q and B_q^* are also fermion operators, terms in the exponent of (6.26) which correspond to different q values commute (being bilinear in the B_q and B_q^*). The exponent is thus split into mutually commuting parts, the Boltzmann factor can be factorized, and the trace can be taken for each q independently (notation Tr_q). We thus find

$$\begin{aligned} Z_N &= \prod_{q>0} \operatorname{Tr}_q \exp[2(K^2 + 2KL \cos q + L^2)^{\frac{1}{2}}(B_q^* B_q + B_{-q}^* B_{-q} - 1) + 2K \cos q] \\ &= \prod_{\text{all} q} \sum_{n_q = 0,1} \exp[(K^2 + 2KL \cos q + L^2)^{\frac{1}{2}}(2n_q - 1) + K \cos q] \\ &= \prod_q \Big\{(\exp[(K^2 + 2KL \cos q + L^2)^{\frac{1}{2}}] + \exp[-(K^2 + 2KL \cos q + L^2)^{\frac{1}{2}}]) \\ &\quad \times \exp[K \cos q]\Big\} = \exp \sum_q [\ln 2 \cosh(K^2 + 2KL \cos q + L^2)^{\frac{1}{2}} + K \cos q] , \end{aligned} \tag{6.27}$$

Extension of X-Y model theory to anisotropic case.

and, in the thermodynamic limit,

$$Z = \exp \frac{1}{2\pi} \int_{-\pi}^{+\pi} dq \ln 2 \cosh[(K^2 + 2KL \cos q + L^2)^{\frac{1}{2}}] . \tag{6.28}$$

For $L = 0$ this reduces to (6.5) as it should.

6.4. *Quadratic lattice in zero field*

We consider only the case of a symmetric quadratic lattice where all nearest-neighbour interactions have the same coupling constant J. All variables commute again so that we can factorize the Boltzmann factor and follow the transfer-matrix approach.

Let the lattice have n rows of m atoms each. The 2^m states of the rth row we indicate by $S_r \equiv \{\sigma_{r1}, \sigma_{r2}, \dots, \sigma_{rm}\}$. We write $Z_N \equiv Z_{m,n}$ as follows

$$\begin{aligned} Z_{m,n} &= \sum_{S_1 \dots S_N} \exp[K \sum_j \sigma_{1,j}\sigma_{1,j+1}] \exp[K \sum_j \sigma_{1,j}\sigma_{2,j}] \\ &\quad \times \exp[K \sum_j \sigma_{2,j}\sigma_{2,j+1}] \dots \exp[K \sum_j \sigma_{n,j}\sigma_{1,j}] \\ &= \sum_{S_1 \dots S_n} V^{(\mathrm{i})}_{S_1 S_1} V^{(\mathrm{m})}_{S_1 S_2} V^{(\mathrm{i})}_{S_2 S_2} \dots V^{(\mathrm{m})}_{S_n S_1} . \end{aligned} \tag{6.29}$$

The 2^m numbers

$$V^{(\mathrm{i})}_{S_r S_r} = \exp [K \sum_j \sigma_{r,j}\sigma_{r,j+1}] ,$$

describing the internal interactions within a row, we consider as the diagonal elements of a $2^m \times 2^m$ matrix $\boldsymbol{V}^{(\mathrm{i})}$, the $2^m . 2^m$ numbers

$$V^{(\mathrm{m})}_{S_r S_{r+1}} = \exp[K \sum_j \sigma_{r,j}\sigma_{r+1,j}] ,$$

corresponding to the mutual interactions between adjacent rows, as the elements of a matrix $\boldsymbol{V}^{(\mathrm{m})}$. The matrix $\boldsymbol{V}^{(\mathrm{i})}$ operating on the states S of a row of m spins can in an obvious way be expressed in terms of m simpler matrices σ_j^z acting effectively only on one single spin:

$$\boldsymbol{V}^{(\mathrm{i})} = \exp[K \sum_{j=1}^{m} \sigma_j^z \sigma_{j+1}^z] , \tag{6.30}$$

$$\sigma_j^z = \mathbf{1} \times \mathbf{1} \times \ldots \times \mathbf{1} \times \sigma^z \times \mathbf{1} \times \ldots \times \mathbf{1} \, . \tag{6.31}$$

In the direct product (6.31), the Pauli matrix σ^z is the jth factor, and $\mathbf{1}$ is the 2×2 unit matrix; for the rest we use the same symbols for matrices as for the corresponding operators.

Next we observe that $V^{(\mathrm{m})}$ is a direct product of m identical 2×2 matrices

$$\begin{bmatrix} e^K & e^{-K} \\ e^{-K} & e^K \end{bmatrix} = e^K \mathbf{1} + e^{-K} \sigma^x \, . \tag{6.32}$$

The matrix (6.32) can, by virtue of the relation $(\sigma^x)^2 = \mathbf{1}$, be written in an exponential form:

$$e^K \mathbf{1} + e^{-K} \sigma^x = A \exp[B\sigma^x]$$

with

$$A = (2 \sinh 2K)^{\frac{1}{2}} \, , \qquad B = -\tfrac{1}{2} \ln \tanh K \, . \tag{6.33}$$

To prove this, write out $\exp[B\sigma^x]$ as a power series in σ^x, put $(\sigma^x)^2 = \mathbf{1}$, and collect even and odd powers. This yields

$$A \exp[B\sigma^x] = (A \cosh B)\,\mathbf{1} + (A \sinh B)\sigma^x \, ; \tag{6.34}$$

putting $A \cosh B = e^K$, $A \sinh B = e^{-K}$ immediately yields (6.33). In the literature one usually writes K^* for B; we follow this convention. If we write, as before,

$$\exp[K^*\sigma^x] \times \exp[K^*\sigma^x] \times \ldots \times \exp[K^*\sigma^x] = \exp[K^* \sum_j \sigma_j^x] \, , \tag{6.35}$$

$$\sigma_j^x = \mathbf{1} \times \ldots \times \mathbf{1} \times \sigma^x \times \mathbf{1} \times \ldots \times \mathbf{1} \, , \tag{6.36}$$

we find for the matrix $V = V^{(\mathrm{i})} V^{(\mathrm{m})}$:

$$V = A^m \exp[K \sum_j \sigma_j^z \sigma_{j+1}^z] \exp[K^* \sum_j \sigma_j^x] \, . \tag{6.37}$$

$Z_{m,n}$ is equal to $\mathrm{Tr}\, V^n$, and its asymptotic behaviour for large n is

$$Z_{m,n} \sim (\lambda^{\max})^n \, , \tag{6.38}$$

where $\lambda^{\max}$ is the largest eigenvalue of V. For Z we thus find

$$Z = \lim_{\substack{m\to\infty\\ n\to\infty}} Z_{m,n}^{1/mn} = \lim_{m\to\infty} \left(\lambda^{\max}\right)^{1/m} . \tag{6.39}$$

The great similarity of (6.37) with (6.9) (which is a formula for the partition function of a one-dimensional system!) is striking. There are two points in which the present case differs from that of the preceding section: (1) Z is calculated from a product of two exponentials instead of from a single exponential. (2) It is derived from the *largest eigenvalue* of this product, and not from the trace (i.e. the sum over all eigenvalues); therefore its calculation is more analogous to the calculation of the *ground state energy* of the one-dimensional system (4.20) than to that of the *partition function*. In spite of these differences, the calculation runs largely parallel to that in the preceding section.

As a first step, we make again the substitutions:

$$\sigma_j^x \to \sigma_j^z = 2(\sigma_j^+\sigma_j^- - 1) ; \qquad \sigma_j^z \to -\sigma_j^x = -(\sigma_j^+ + \sigma_j^-) ; \tag{6.40}$$

notice that, in order to stress the similarity with the previous case, we write 1 instead of **1**.

By the application of the Jordan-Wigner transformation (6.16) and the unitary transformation (6.23), eq. (6.37) changes into

$$V = A^m[\exp 2K \sum_{q>0} \{(\cos q)(A_q^*A_q + A_{-q}^*A_{-q}) + (\sin q)(A_qA_{-q} + A_{-q}^*A_q^*)\}]$$

$$\cdot [\exp 2K^* \sum_{q>0} \{A_q^*A_q + A_{-q}^*A_{-q} - 1\}] = A^m\Big(\prod_{q>0} V_q^{(\mathrm{i})}\Big)\Big(\prod_{q>0} V_q^{(\mathrm{m})}\Big), \tag{6.41}$$

in an obvious notation. Since the A_q and A_q^* are fermion operators, each factor $V_q^{(\mathrm{i})}$ commutes with the factors $V_{q'}^{(\mathrm{m})}$ with $q' \neq q$ (the exponents of both being bilinear in the A_q and A_q^*), but not with $V_q^{(\mathrm{m})}$. Hence we may write:

$$V = A^m \prod_{q>0} V_q^{(\mathrm{i})} V_q^{(\mathrm{m})} . \tag{6.42}$$

In the n_q representation the matrices $V_q^{(\mathrm{i})}$ and $V_q^{(\mathrm{m})}$ are direct products of a 4×4 matrix acting on the states $|n_q, n_{-q}\rangle$, where $n_q = 0,1$ and $n_{-q} = 0,1$, and 4×4 unit matrices acting on the states of other waves. As a consequence, the eigenvalues λ of V are products of the eigenvalues λ_q of $V_q^{(\mathrm{i})} V_q^{(\mathrm{m})}$, multiplied by A^m. For the largest eigenvalue $\lambda^{\max}$ we thus find

$$\lambda^{\max} = A^m \prod_{q>0} \lambda_q^{\max} , \tag{6.43}$$

where $\lambda_q^{\max}$ is the largest of the four eigenvalues λ_q.

Since the A_q occur in two exponentials instead of one, the Valatin-Bogoliubov transformation cannot be applied directly. The exponentials have first to be expanded with the aid of the anticommutation relations obeyed by A_q and A_q^*. This expansion is not difficult (observe e.g. that $A_q^n = A_q^{*n} = 0$ for $n > 1$) but still too lengthy to carry out here in detail. We therefore only state the result, referring to ref. 8 for further details. For the eigenvalues λ_q a quartic equation is derived which factorizes into two linear and one quadratic equation; $\lambda_q^{\max}$ is a root of the quadratic equation, given by

$$\lambda_q^{\max} = [(\cosh 2K \cosh 2K^* - \cos q \sinh 2K \sinh 2K^*) \\ + \{(\cosh 2K \cosh 2K^* - \cos q \sinh 2K \sinh 2K^*)^2 - 1\}^{\frac{1}{2}}] \exp[2K \cos q] . \tag{6.44}$$

Taking into account the values which q may take and expressing $K^*(=B)$ in terms of K with the aid of eq. (6.33), we derive from the equations (6.39), (6.43) and (6.44) for Z, $\lambda^{\max}$ and $\lambda_q^{\max}$, respectively, the following expression for the partition function per spin of the quadratic lattice:

$$Z = \exp \frac{1}{4\pi} \int_{-\pi}^{+\pi} \mathrm{d}q \ln 2[(\cosh 2K)^2 - \cos q \sinh 2K \\ + \{((\cosh 2K)^2 - \cos q \sinh 2K)^2 - (\sinh 2K)^2\}^{\frac{1}{2}}] . \tag{6.45}$$

This result has also been derived by a variety of other methods but we do not discuss these.

An investigation of the right-hand member of (6.45) shows that it has a weak singularity at the temperature $T = T_c$ $(K = K_c)$ defined by $\sinh 2K_c = 1$. More precisely, it is found that near T_c, Z behaves as follows:

$$\ln Z(T) = \ln Z(T_c) + a\left(\frac{T}{T_c} - 1\right) + b\left(\frac{T}{T_c} - 1\right)^2 \ln\left(\frac{T}{T_c} - 1\right)^2 + \dots , \tag{6.46}$$

just as in the second example discussed in section 3. This analogy is not accidental: for finite m and n, $Z_{m,n}$ appears to have the form of a product (or rather a sum of 4 products) of a structure similar to that of Z_N in the above-mentioned example. The specific heat at constant field, which is proportional to the second derivative of ln Z with respect to T, has the following critical behaviour:

$$C_H \sim \ln|T - T_c| , \qquad (T \to T_c, H = 0) , \tag{6.47}$$

which implies that for this two-dimensional Ising model the critical indices α and α' for the specific heat (cf. section 2.2) are both zero. For other two-dimensional lattices the same result has been found.

6.5. *Quadratic lattice in a parallel field*

The next thermodynamic quantities which we would like to know for the quadratic-lattice Ising model involve all in some way or other the magnetic field. Up to now, however, nobody has been able to calculate the partition function for $H \neq 0$. All methods developed for the solution of the Ising problem have proved to fail in this case.

Fortunately, it has been possible to calculate the spontaneous magnetization along other lines. The result is

$$M = m[1 - (\sinh 2K)^{-2}]^{\frac{1}{8}} , \tag{6.48}$$

so that near T_c

$$M \sim (T_c - T)^{\frac{1}{8}} , \qquad (T \to T_c-, H = 0) .$$

A closed form for the zero-field susceptibility χ is not known, but from the relation between χ and the two-spin correlation functions Fisher was able to conclude that

$$\chi \sim |T - T_c|^{-\frac{7}{4}} , \qquad (T \to T_c, H = 0) . \tag{6.49}$$

Finally, there is no exact information on the form of the critical isotherm near T_c. The only available data are derived by numerical estimation from series expansions at high fields. They seem to indicate that

$$|M| \sim |H|^{1/\delta} \quad \text{with } \delta = 15.00 \pm 0.08 , \qquad (T = T_c, H \to 0) . \tag{6.50}$$

The value $\delta = 15$, if not exact, gives at least a good description of the known numerical data.

Again, the same results have been found for other two-dimensional lattices. Apparently, all two-dimensional Ising systems show the same critical behaviour.

6.6. *Three-dimensional lattices*

For three-dimentional lattices no thermodynamic quantity is

known exactly, and there are no exact data about the critical behaviour. The only available information is again from numerical estimates, based on series expansions. These data can be summarized as follows:

$$M \sim (T_c - T)^{\beta} \quad \text{with } \beta = 0.313 \pm 0.004\,, \quad (T \to T_c-,\ H=0), \quad (6.51)$$

$$|M| \sim |H|^{1/\delta} \quad \text{with } \delta = 5.2 \pm 0.15\,, \quad (T = T_c,\ H \to 0), \quad (6.52)$$

$$\chi \sim (T - T_c)^{-\gamma} \quad \text{with } \gamma = 1.250 \pm 0.001\,, \quad (T \to T_c+,\ H=0), \quad (6.53)$$

$$\chi \sim (T_c - T)^{-\gamma'} \quad \text{with } \gamma' = 1.31 \pm 0.05\,, \quad (T \to T_c-,\ H=0), \quad (6.54)$$

$$C_H \sim (T - T_c)^{-\alpha} \quad \text{with } \alpha \approx \tfrac{1}{8}\,, \quad (T \to T_c+,\ H=0), \quad (6.55)$$

$$C_H \sim (T_c - T)^{-\alpha'} \quad \text{with } \alpha' \approx \tfrac{1}{8}\,, \quad (T \to T_c-,\ H=0). \quad (6.56)$$

In estimating the value of α', use has been made of certain inequalities (cf. section 7) and the values found for other critical indices †.

Although no a priori reason is known why the critical indices should be simple fractions, the fact that they have this form in all cases where they are known exactly makes it worth while to ask with which simple fractions the empirical values listed above are compatible. The following set of values is often quoted:

$$\beta = \tfrac{5}{16}\,; \quad \delta = 5\,; \quad \gamma = \tfrac{5}{4}\,; \quad \gamma' = \tfrac{5}{4} \text{ (or } \tfrac{21}{16}\text{ ?)}\,; \quad \alpha = \tfrac{1}{8}\,. \quad (6.57)$$

In the choice of these fractions a certain preference for multiples of powers of $\frac{1}{2}$ has played a role, which preference has its origin in the occurrence of such values in the two-dimensional Ising model.

7. RELATIONS BETWEEN CRITICAL INDICES

Having obtained a number of critical indices (from experiments and from calculations on models) it may be asked whether for a given system these critical indices (and other ones which have been measured or calculated, but which have not been discussed in these lectures) are all independent of each other, or whether there may exist certain relations between them. This question has been the source of a large number of investigations during the last few years.

† The most recent estimates are $\alpha = 0.125 \pm 0.015$, $\alpha' = 0.066^{+0.160}_{-0.040}$ (ref. 3, section 8.1).

7.1. *A conjectured identity*

Widom [9] and Essam and Fisher were the first to suggest that there exist certain relations between critical indices. Essam and Fisher based their argument on a heuristic microscopic model which is an elaborate version of the *droplet* or *cluster model* of condensation. They showed that one of the consequences of this model would be the existence of the relation

$$\alpha' + 2\beta + \gamma' = 2 \; . \tag{7.1}$$

For a ferromagnetic system, this relation can be interpreted as follows. Since we have $C_H \sim (T_c - T)^{-\alpha'}$ for $T \to T_c-$ and $H = 0$, and on the other hand $C_H = T(\partial S/\partial T)_H = -T(\partial^2 F/\partial T^2)_H$, the "singular part" of F (i.e. that part which gives rise to the singularity in the specific heat) behaves as $(T_c - T)^{2-\alpha'}$ for $T \to T_c-$, $H = 0$. Further, $(\partial F/\partial H)_T = -M$ behaves as $(T_c - T)^{\beta}$ in this region, and $(\partial^2 F/\partial H^2)_T = -\chi$ as $(T_c - T)^{-\gamma'}$. So the relation (7.1) expresses the fact that the critical *exponent* of $(\partial F/\partial H)_T$ is the arithmetic mean of those of F and $(\partial^2 F/\partial H^2)_T$.

It is interesting to test the relation (7.1) against the known theoretical and experimental results. The two-dimensional Ising model has $\alpha' = 0$, $\beta = \frac{1}{8}$, $\gamma' = \frac{7}{4}$, and thus satisfies the relation precisely! The three-dimensional model has $\alpha' \approx \frac{1}{8}$, $\beta \approx \frac{5}{16}$, $\gamma' \approx \frac{5}{4}$, and again eq. (7.1) seems to be valid. Even for the classical theories the relation appears to hold: $\alpha' = 0$ (in this case representing a jump discontinuity and not a logarithmic singularity), $\beta = \frac{1}{2}$, $\gamma' = 1$. In this connection it should be remarked that at least some of the classical theories have been shown to become exact theories in the limit where the range of the forces becomes infinite while their strength goes to zero in a special way (cf. ref. 3, section 5.4). There is also some evidence that they become correct in the limit of an infinite number of dimensions. Therefore the validity of the relation (7.1) in this case expresses more than an accidental property of some approximation methods.

Finally, the experimental data are reasonably consistent with the relation (7.1). For a real test of its validity, however, the accuracy of the measurements will have to be increased.

7.2. *The Rushbrooke inequality*

Let us now see what thermodynamics can say about a possible interdependence of the critical indices. This question was first investigated by Rushbrooke. He was able to show that for a ferromagnet the following *inequality* follows from thermodynamics:

$$\alpha' + 2\beta + \gamma' \geqslant 2 \; . \tag{7.2}$$

The proof is based on the relation between the specific heat at constant field C_H and the specific heat at constant magnetization C_M, and on the fact that C_M is non-negative. The relation between C_H and C_M is the analogue of the well-known relation between C_p and C_V:

$$C_H = C_M + T\left(\frac{\partial M}{\partial T}\right)_H^2 \left(\frac{\partial M}{\partial H}\right)_T^{-1} . \tag{7.3}$$

That C_M is non-negative follows from the stability condition for the free energy at constant M:

$$C_M = T\left(\frac{\partial S}{\partial T}\right)_M = -T\left(\frac{\partial^2 F}{\partial T^2}\right)_M \geqslant 0 . \tag{7.4}$$

Alternatively, it is a consequence of the proportionality of C_M to the energy fluctuations at constant M:

$$C_M = \left(\frac{\partial U}{\partial T}\right)_M = \frac{\partial}{\partial T} \frac{\Sigma_n' E_n \exp[-\beta E_n]}{N\Sigma_n' \exp[-\beta E_n]}$$

$$= \frac{(\Sigma_n' \exp[-\beta E_n])(1/kT^2)(\Sigma_n' E_n^2 \exp[-\beta E_n]) - (1/kT^2)(\Sigma_n' E_n \exp[-\beta E_n])^2}{N(\Sigma_n' \exp[-\beta E_n])^2}$$

$$= \frac{1}{NkT^2}[\langle E^2\rangle - \langle E\rangle^2] = \frac{1}{NkT^2}\langle(E - \langle E\rangle)^2\rangle \geqslant 0 , \tag{7.5}$$

where the sums and averages are taken over all states with a fixed value of the magnetization.

From (7.3) and (7.4) it follows that

$$C_H \geqslant T\left(\frac{\partial M}{\partial T}\right)_H^2 \left(\frac{\partial M}{\partial H}\right)_T^{-1} . \tag{7.6}$$

For $T \to T_c-$, $H = 0$, the asymptotic behaviour of the right-hand member is easily derived. We know that $M \sim (T_c - T)^\beta$, $(\partial M/\partial H)_T = \chi \sim (T_c - T)^{-\gamma'}$; so

$$T\left(\frac{\partial M}{\partial T}\right)_H^2 \left(\frac{\partial M}{\partial H}\right)_T^{-1} \sim (T_c - T)^{2(\beta-1)+\gamma'} . \tag{7.7}$$

By the inequality (7.6), the singularity of C_H cannot be weaker than this. Therefore, if $C_H \sim (T_c - T)^{-\alpha'}$, we conclude that $-\alpha' \leqslant 2(\beta-1)+\gamma'$ which is equivalent to the inequality to be proved. The proof of the corresponding inequality for a fluid is slightly more complicated; it has been given by Fisher.

7.3. *Other relations between critical indices*

Using the same inequality $C_M \geqslant 0$ (eq. (7.4)), Griffiths has derived another important inequality for ferromagnets (afterwards also for fluids), which involves the critical index δ:

$$\alpha' + \beta(1+\delta) \geqslant 2 \,. \tag{7.8}$$

If we consider the known values of the indices α', β and δ, we find again that for the two-dimensional Ising model (with the choice $\delta = 15$, cf. section 6.5) and for the classical theories the inequality becomes an equality

$$\alpha' + \beta(1+\delta) = 2 \,; \tag{7.9}$$

moreover, the preliminary numerical data for the simple cubic lattice are not inconsistent with this equality! It is interesting to note that eq. (7.9) can be shown to be another consequence of the droplet model of Essam and Fisher. Further consequences of this model are the symmetry relations

$$\alpha' = \alpha \,; \qquad \gamma' = \gamma \,. \tag{7.10}$$

For the two-dimensional Ising model, these relations are certainly correct, but for the three-dimensional model the numerical data justify some doubt as to their validity.

7.4. *Scaling laws*

Whereas the inequalities (7.2) and (7.8) between critical indices are well founded in thermodynamics and statistical mechanics, the equalities (7.1), (7.9) and (7.10) lack such a foundation. Several ideas have been developed which aim at deriving these relations, the first of which was the original derivation from a droplet model by Essam and Fisher. Each of these approaches, however, is based on some crucial assumptions, so that up to now they have only served to make the relations more plausible, not to yield a true proof.

For want of a good theory, we shall, as an example, briefly sketch how one can arrive at the equalities by the method of the "scaling laws", as applied to a ferromagnetic system (cf. ref. 2, section 3; ref. 3, section 9.4). Its starting point is the conjecture that close to the critical point, where the spin-spin correlations are known to extend over very large distances, the properties of the system are rather insensitive to the details of the interactions between the spins. They should depend on the general nature of the system (e.g. the number of dimensions) and of the interaction (isotropic or anisotropic, long-range or short-range), but not on the precise lattice structure or the relative strengths of the interactions of a given spin with the surrounding spins, except for appropriate *scaling factors*, which may have different values for different systems. The information we have on the critical behaviour of a number of Ising systems with different lattice structure is in agreement with this conjecture.

Consider a simple d-dimensional lattice with interactions only between nearest neighbour spins. Let the lattice be divided into cells of $L \times \ldots \times L$ points each ($L \gg 1$), labeled J, K, ... As long as we choose $L \ll l_0$, where l_0 is a very large number characterizing the range of the spin-spin correlations $\langle(\sigma_j - S)(\sigma_k - S)\rangle$ (where $S = \langle\sigma_j\rangle$), the spins within a cell will show a strong coherence. The state of the Jth cell can then be characterized by a single variable μ_J, which plays a role analogous to that of the variable σ_j in the original problem. The cells form again a simple d-dimensional lattice, with an interaction of the individual cells with the external magnetic field and an interaction between nearest neighbour cells through the cell interfaces. It is the latter one which is responsible for the existence of long-range correlations and long-range order. We assume (and it is a very strong assumption indeed) that its effect can be described by one single variable of the dimension of an energy (say $\widetilde{J}$), which is then the "natural" quantity in which to express the temperature T, or rather the difference $T - T_c$. So if we describe the temperature of the original system by the reduced quantity $\epsilon = (T - T_c)/T_c$, where T_c is of the order of J/k (say $T_c = aJk^{-1}$), the appropriate reduced variable to describe the temperature of the "cell system" is $\widetilde{\epsilon} = (T - T_c)/(a\widetilde{J}k^{-1})$. For $\epsilon \to 0$ we expect also $\widetilde{\epsilon} \to 0$, probably as $\widetilde{\epsilon} \sim \epsilon$.

The interaction of the cells with the field we describe in a similar way by a variable $\widetilde{H}$ different from H (turning on a field has a different effect on a single spin and on a group of interacting spins in a cell). Again we expect that $\widetilde{H} \to 0$ for $H \to 0$, probably as $\widetilde{H} \sim H$. More precisely, $\widetilde{H}$ should be such that the change of the total free energy due to a change δH of the field H

$$\delta F_N = -m \sum_j \langle\sigma_j\rangle\, \delta H \tag{7.11}$$

can also be written as

$$\delta F_N = -m \sum_J \langle \mu_J \rangle \, \delta \widetilde{H} \, . \tag{7.12}$$

By the translation symmetry, $\langle \sigma_j \rangle$ is the same for all N spins and $\langle \mu_J \rangle$ is the same for all N/L^d cells, so that (7.11) and (7.12) imply that

$$N \langle \sigma_j \rangle \, \delta H = (N/L^d) \langle \mu_J \rangle \, \delta \widetilde{H} \, . \tag{7.13}$$

The proportionality constants in $\widetilde{H} \sim H$, $\widetilde{\epsilon} \sim \epsilon$ may depend on L. It is now assumed that this dependence is through a simple power:

$$\widetilde{H} = L^x H \, , \qquad \widetilde{\epsilon} = L^y \epsilon \, , \qquad \text{with } x > 0 \, , \qquad y > 0 \, . \tag{7.14}$$

Now comes the most crucial assumption, viz. that the size L of the cells is one of those "details" of the system of which the critical behaviour is independent (apart from scaling factors). This implies that all relations between physical quantities, when expressed in terms of the appropriate reduced variables of the cell system, must have a form which is independent of L. One such relation is that between the magnetization M, or $\langle \sigma_j \rangle = M/m$, and T and H. Suppose that for the original system close to the critical point this relation is

$$\langle \sigma_j \rangle = \Phi(\epsilon, H) \, . \tag{7.15}$$

Then the corresponding relation between $\langle \mu_J \rangle$, $\widetilde{\epsilon}$ and $\widetilde{H}$ should be

$$\langle \mu_J \rangle = \Phi(\widetilde{\epsilon}, \widetilde{H}) \text{ for all } L \, . \tag{7.16}$$

By (7.13) and (7.14) we have

$$\langle \mu_J \rangle = L^{d-x} \langle \sigma_j \rangle \, . \tag{7.17}$$

Comparing (7.15) and (7.16) with the aid of (7.14) and (7.17) we find the following identity in L:

$$L^{d-x} \, \Phi(\epsilon, H) \equiv \Phi(L^y \epsilon, L^x H) \, . \tag{7.18}$$

This identity expresses a "scaling law" for the magnetization near the critical point, which can be valid only if Φ (and hence M) depends on ϵ and H in a special way, viz. as

$$\Phi(\epsilon, H) = \frac{H}{|H|} \, |\epsilon|^{(d-x)/y} \, \varphi\!\left(\epsilon \, |H|^{-y/x}\right) , \tag{7.19}$$

where φ is a function of one variable only; the factor $H/|H|$ ensures that $\langle\sigma_j\rangle$ changes sign as the field is reversed.

The form (7.19) is so restrictive that it is not surprising that some far-reaching conclusions can be drawn from it. To begin with, let us try to determine the critical indices.

1. By letting $H \rightarrow 0$ in (7.19) we find the *spontaneous magnetization* for $\epsilon < 0$ $(T < T_c)$:

$$\frac{M(0)}{m} = \lim_{H \to 0+} \Phi(\epsilon, H) = |\epsilon|^{(d-x)/y} \varphi(-\infty) . \tag{7.20}$$

If it is true that $M \sim (T_c - T)^\beta \sim |\epsilon|^\beta$, then $\varphi(-\infty)$ must be a constant, and

$$\beta = \frac{d-x}{y} . \tag{7.21}$$

2. To find the *critical isotherm* we consider the limit $\epsilon \rightarrow 0+$. In order that M does not vanish or diverge, the power of $|\epsilon|$ in front of φ must be compensated by a factor $\epsilon^{-(d-x)/y}$ in φ, which should therefore behave as

$$\varphi(\epsilon |H|^{-y/x}) \sim (\epsilon |H|^{-y/x})^{-(d-x)/y} . \tag{7.22}$$

Therefore

$$M \sim \epsilon^{(d-x)/y} \epsilon^{-(d-x)/y} |H|^{(d-x)/x} \sim |H|^{1/\delta} , \qquad (\epsilon = 0,\ H \rightarrow 0) , \tag{7.23}$$

with

$$\delta = \frac{x}{d-x} . \tag{7.24}$$

3. The *susceptibility* is found by differentiating M with respect to H

$$\frac{\chi}{m} \sim \frac{H}{|H|} |\epsilon|^{(d-x)/y} \varphi'(\epsilon |H|^{-y/x}) \, \epsilon |H|^{-y/x-1} \frac{H}{|H|} . \tag{7.25}$$

In order that $\lim_{H \to 0} \chi$ exists and is $\neq 0$, φ should be such that $\varphi'(\epsilon |H|^{-y/x})$ contains a factor $|H|^{y/x+1}$, i.e. that

$$\varphi'(\epsilon |H|^{-y/x}) \sim \left|\epsilon |H|^{-y/x}\right|^{-(y+x)/y} , \qquad (H \rightarrow 0) , \tag{7.26}$$

both for $\epsilon > 0$ and $\epsilon < 0$ (which correspond to large positive and large negative values of the argument $\epsilon|H|^{-y/x}$, respectively). It follows that

$$\chi \sim |\epsilon|^{(d-x)/y-(y+x)/y+1} = |\epsilon|^{(d-2x)/y}, \qquad (\epsilon \to 0,\ H=0)\ . \qquad (7.27)$$

This can be written as

$$\begin{aligned} \chi &\sim |\epsilon|^{-\gamma}, && (\epsilon \to 0+,\ H=0)\ , \\ \chi &\sim |\epsilon|^{-\gamma'}, && (\epsilon \to 0-,\ H=0)\ , \end{aligned} \qquad (7.28)$$

with

$$\gamma = \gamma' = \frac{2x-d}{y}. \qquad (7.29)$$

4. The *specific heat* is found by integrating M with respect to H and differentiating twice with respect to T. By the same type of argument as was used above we can derive

$$\begin{aligned} C_H &\sim |\epsilon|^{-\alpha}, && (\epsilon \to 0+,\ H=0)\ , \\ C_H &\sim |\epsilon|^{-\alpha'}, && (\epsilon \to 0-,\ H=0)\ , \end{aligned} \qquad (7.30)$$

with

$$\alpha = \alpha' = 2 - \frac{d}{y}\ . \qquad (7.31)$$

Only when d/y is an integer, the proof requires a bit more care. The case $d/y = 2$ appears to correspond to a logarithmic singularity in C_H.

In this way we have derived the six critical indices $\alpha, \alpha', \beta, \gamma, \gamma', \delta$ from only two parameters, x and y (together with the dimensionality d). So there must be four relations between the indices. Two of them we have mentioned already: $\alpha = \alpha'$, $\gamma = \gamma'$; they express a symmetry between the high- and low-temperature critical behaviour of C_H and χ. The other relations are found by eliminating x and y from (7.21), (7.24), (7.29) and (7.31). We get

$$\alpha' + 2\beta + \gamma' = 2\ , \qquad \alpha' + \beta(1+\delta) = 2\ ,$$

which are exactly the identities (7.1) and (7.9)!

From eq. (7.19) one can derive critical indices for other thermodynamic quantities (higher derivatives of the free energy F) and, consequently, further index relations, but we shall not discuss these.

REFERENCES

[1] M. E. Fisher, *Lectures in Theoretical Physics* VII C (University of Colorado Press, Boulder, 1965), 1-159.
[2] L. P. Kadanoff et al., Rev. Mod. Phys. 39 (1967) 395.
[3] M. E. Fisher, Rept. Progr. Phys. 30, (1967) 615.
[4] P. Heller, Rept. Progr. Phys. 30, (1967) 731.
[5] E. H. Lieb, Phys. Rev. 162 (1967) 162;
E. H. Lieb, Phys. Rev. Letters 18 (1967) 1046;
E. H. Lieb, Phys. Rev. Letters 19 (1967) 108.
[6] A. Lenard, private communication.
[7] C. P. Yang, Phys. Rev. Letters 19 (1967) 586;
B. Sutherland, C. N. Yang, C. P. Yang, Phys. Rev. Letters 19 (1967) 588.
[8] T. D. Schultz, D. C. Mattis, E. H. Lieb, Rev. Mod. Phys. 36 (1964) 856;
D. C. Mattis, *The Theory of Magnetism* (Harper and Row, New York, 1965), chapter 9.
[9] B. Widom, J. Chem. Phys. 37 (1962) 2703.

Have a look at Frankel's clustering theory and see if it can see anything strange happening in a random system,

TOWARD A UNIFIED VIEW ON MATHEMATICAL THEORIES OF PHASE TRANSITIONS

MARK KAC
The Rockefeller University, New York 10021, *USA*

1. INTRODUCTION

It is well known that classical theories of phase transitions based on the ideas of van der Waals and Weiss lead to predictions which are *qualitatively* at variance with results of measurements near the critical point.

On the other hand detailed and rigorous analyses of the two-dimensional Ising model with nearest neighbor interactions as well as extensive numerical work on its three-dimensional counterpart have brought us much closer to the understanding of the striking phenomena observed in the critical region.

It is therefore natural that one tends to emphasize the differences between the theories based on the concept of molecular field (van der Waals, Weiss) and those based on the work of Onsager and his followers.

It is the purpose of these lectures to do the opposite and to stress the many similarities (at least in the mathematical structure) of both approaches and although many important questions still remain unresolved, a vision of useful synthesis begins to emerge.

Much of what I shall have to say I have said (and in greater detail) in the lectures I gave in the summer of 1966 at Brandeis University. These lectures will soon appear in print, and I shall refer to them freely. However some material is new, and moreover after a year of reflection I think that I can say certain things better, or at least put them in a clearer perspective.

2. TRANSFER MATRIX AND THE PARTITION FUNCTION

My main purpose will be to elucidate the *mathematical* mechanism of certain phase transitions, and I shall consider only lattice systems and use exclusively magnetic terminology.

Since one of the most striking features of the whole subject is the role dimensionality plays I shall make it the central theme of the lectures.

Let me begin by considering a two-dimensional $N \times M$ lattice. On each point P of the lattice there is a little magnet described by a certain "spin" μ_{P} capable of assuming the values +1 and -1

$$\mu_{\mathrm{P}} = \begin{cases} +1 \; ; \\ -1 \; . \end{cases} \tag{2.1}$$

Let the energy E associated with a given configuration of spins be

$$E = -J \sum v_{\mathrm{PQ}}(\epsilon) \mu_{\mathrm{P}} \mu_{\mathrm{Q}} \; , \tag{2.2}$$

where

$$v_{\mathrm{PQ}}(\epsilon) = \begin{cases} 0 \text{ if P and Q are not nearest neighbors ;} \\ 1 \text{ if P and Q are horizontal nearest neighbors ;} \\ \epsilon \text{ if P and Q are vertical nearest neighbours .} \end{cases} \tag{2.3}$$

As usual it is convenient to use periodic boundary conditions, i.e., wind the lattice on a torus.

This is simply an unisotropic Ising model with nearest neighbor interactions.

If $\epsilon = 0$ the model reduces to M *uncoupled* one-dimensional Ising models and $\epsilon = 1$ corresponds to the familiar isotropic two-dimensional model.

As is well known the partition function

$$Q_{NM} = \sum \exp(-E/kT) \tag{2.4}$$

can be written as the trace of the Nth power of the transfer matrix

$$L_{\epsilon}(\boldsymbol{\mu}, \boldsymbol{\mu}) = \exp\left(\tfrac{1}{2}\nu\epsilon \sum_{1}^{M} \mu_k \mu_{k+1}\right) \exp\left(\nu \sum_{1}^{M} \mu_k \mu_k'\right) \exp\left(\tfrac{1}{2}\nu\epsilon \sum_{1}^{M} \mu_k' \mu_{k+1}'\right) , \tag{2.5}$$

where from now on ν will always be given by the formula

$$\nu = \frac{J}{kT} \; , \tag{2.6}$$

Can one do an Onsager or Kaufman solution of this random system

It should be clear that μ and μ' are M-dimensional vectors all of whose components are +1 or -1 and that because of periodic boundary conditions we have $\mu_{M+1} \equiv \mu_1$ and $\mu'_{M+1} = \mu_1$. The transfer matrix L is really a $2^M \times 2^M$ matrix as mentioned above

$$Q_{N,M} = \mathrm{Tr}\{L_\epsilon^N\} . \tag{2.7}$$

The thermodynamic limit is performed in two stages: first $N \to \infty$ then $M \to \infty$. Denoting by

$$\Lambda_1, \Lambda_2, \ldots, \Lambda_{2M}, \tag{2.8}$$

the eigenvalues (in decreasing order) of L we have

$$-\frac{\psi}{kT} = \lim_{\substack{N\to\infty \\ M\to\infty}} \frac{1}{NM} \log Q_{N,M} = \lim_{M\to\infty} \frac{1}{M} \log \Lambda_1 , \tag{2.9}$$

$e^{-\beta \mathcal{H}}$

where ψ is the free energy per lattice site.

3. THE PRINCIPAL EIGENVECTOR AND ITS INTERPRETATION

The principal eigenvector $\phi_1(\boldsymbol{\mu})$ of L has also a simple interpretation. In fact, $\phi_1^2(\boldsymbol{\mu})$ is the probability of the configuration $\boldsymbol{\mu}$ in a lattice column.

In particular,

$$\langle \mu_{k,l}\mu_{k,l+r} \rangle = \lim_{N\to\infty} \frac{\sum \mu_{k,l}\mu_{k,l+r} \exp(-E/kT)}{\sum \exp(-E/kT)}$$
$$= \sum_{\boldsymbol{\mu}} \mu_l \mu_{l+r}\, \phi_1^2(\boldsymbol{\mu}) \tag{3.1}$$

? this looks quite interesting

How do you define a correlation fn in a random system? Must be careful.

and because of periodic boundary conditions the result is independent of l.

The correlation coefficient in the horizontal direction is given by a more complicated formula

$$\langle \mu_{k,l}\mu_{k+r,l} \rangle = \lim_{N\to\infty} \frac{\sum \mu_{k,l}\mu_{k+r} \exp(-E/kT)}{\sum \exp(-E/kT)}$$
$$= \sum_{j=1}^{2^M} \left(\frac{\Lambda_j}{\Lambda_1}\right)^r (\mu_l\, \phi_j(\boldsymbol{\mu}), \phi_1(\boldsymbol{\mu}))^2 . \tag{3.2}$$

where $\phi_j(\boldsymbol{\mu})$ is the (normalized) eigenvector belonging to the eigenvalue Λ_j and (ϕ, ψ) denotes as usual the scalar product of ϕ and ψ so that e.g.

$$(\mu_l \ \phi_j(\boldsymbol{\mu}), \ \phi_1(\boldsymbol{\mu})) = \sum_{\boldsymbol{\mu}} \mu_l \ \ \phi_j(\boldsymbol{\mu}) \ \phi_1(\boldsymbol{\mu}) \ . \tag{3.3}$$

It is of interest to note that for $\epsilon = 1$ (isotropic Ising model) we have

$$(\mu_l \ \phi_1(\boldsymbol{\mu}), \ \mu_{l+r} \ \phi_1(\boldsymbol{\mu}) = \sum_{j=1}^{2^M} \left(\frac{\Lambda_j}{\Lambda_1}\right)^r (\mu_l \ \phi_j(\boldsymbol{\mu}), \ \phi_1(\boldsymbol{\mu})) \ , \tag{3.4}$$

a formula which does not seem to be easy to derive by direct manipulations.

Look at it again. This is very interesting.

4. PHASE TRANSITION REFLECTED IN THE CHANGE IN THE NATURE OF THE PRINCIPAL EIGENVECTOR

It is a vague truism that a phase transition occurs when an ordered state becomes more probable than a disordered one.

With the help of the transfer matrix formalism reviewed above we can make it more precise.

The vector

$$\phi_{\mathrm{d}}(\boldsymbol{\mu}) = \frac{1}{\sqrt{2^M}} \tag{4.1}$$

clearly corresponds to complete disorder inasmuch as *all* configurations $\boldsymbol{\mu}$ are equally probable.

On the other hand the vectors

$$\phi_0^+(\boldsymbol{\mu}) = \begin{cases} 1 \text{ if } \boldsymbol{\mu} = (1, 1, \ldots, 1) \\ 0 \text{ otherwise} \end{cases} \tag{4.2a}$$

and

$$\phi_0^-(\boldsymbol{\mu}) = \begin{cases} 1 \text{ if } \boldsymbol{\mu} = (-1, -1, \ldots, -1) \\ 0 \text{ otherwise} \end{cases} \tag{4.2b}$$

correspond to complete order.

If $\epsilon = 0$ the transfer matrix L becomes

$$L_0(\mu, \mu') = \exp\left(\nu \sum_1^M \mu_k \mu_k'\right) \tag{4.3}$$

and we have

$$(L_0 \phi_d, \phi_d) = \frac{1}{2^M} \sum_{\mu, \mu'} \exp\left(\nu \sum_1^M \mu_k \mu_k'\right) (2 \cosh \nu)^M > \exp(M\nu)$$

$$= (L_0 \phi_0^+, \phi_0^+) = (L_0 \phi_0^-, \phi_0^-) .$$

Actually ϕ_d is the principal eigenvector of L_0 and we see that the system always "prefers" to be found in a disordered state except at absolute zero ($\nu = \infty$) when the vectors corresponding to the ordered states also yield (asymptotically) Λ_1.

If $\epsilon > 0$ the situation is quite different and for sufficiently low temperatures

$$(L_\epsilon \phi_0^+, \phi_0^+) = (L_\epsilon \phi_0^-, \phi_0^-)$$

becomes greater than

$$(L_\epsilon \phi_d, \phi_d) .$$

To see this note that

$$(L_\epsilon \phi_0^+, \phi_0^+) = (L_\epsilon \phi_0^-, \phi_0^-) = \exp(M(1 + \epsilon)\nu) , \tag{4.4}$$

while

$$(L_\epsilon \phi_d, \phi_d) = \frac{1}{2^M} \sum_{\mu, \mu'} L_\epsilon(\mu, \mu') . \tag{4.5}$$

It is a simple and useful exercise to verify that

$$\sum_{\mu, \mu} L_\epsilon(\mu, \mu') = \mathrm{Tr}\{A^M\} , \tag{4.6}$$

where A is the 4×4 matrix

$$A = \begin{pmatrix} \exp((1+\epsilon)\nu) & 1 & 1 & \exp((1-\epsilon)\nu) \\ 1 & \exp(-(1-\epsilon)\nu) & \exp(-(1+\epsilon)\nu) & 1 \\ 1 & \exp(-(1-\epsilon)\nu) & \exp(-(1-\epsilon)\nu) & 1 \\ \exp((1-\epsilon)\nu) & 1 & 1 & \exp((1+\epsilon)\nu) \end{pmatrix} \tag{4.7}$$

By symmetry considerations the principal eigenvector (i.e. the eigenvector belonging to the largest eigenvalue) is of the form (x, y, y, x) and the equations from which the largest eigenvalue λ is determined become

$$e^{\nu} \cosh(\epsilon\nu)x + y = \frac{\lambda}{2}x , \qquad x + e^{-\nu} \cosh(\epsilon\nu)y = \frac{\lambda}{2}y .$$

Thus

$$\lambda_{max} = 2\left(\cosh(\nu) \cosh(\epsilon\nu) + \sqrt{\cosh^2(\nu) \cosh^2(\epsilon\nu) - \sinh^2(\epsilon\nu)}\right)$$

and (using (4.5) and (4.6))

$$(L_\epsilon \phi_d, \phi_d) \approx \left(\cosh(\nu) \cosh(\epsilon\nu) + \sqrt{\cosh^2(\nu) \cosh^2(\epsilon\nu) - \sinh^2(\epsilon\nu)}\right)^M . \tag{4.8}$$

To determine when $(L_\epsilon \phi_o^+, \phi_o^+)$ becomes greater than $(L_\epsilon \phi_d, \phi_d)$ we must investigate the inequality

$$e^{(1+\epsilon)\nu} > \cosh(\nu) \cosh(\epsilon\nu) + \sqrt{\cosh^2(\nu) \cosh^2(\epsilon\nu) - \sinh^2(\epsilon\nu)} ,$$

which after some elementary (and slightly tedious) transformations can be reduced to the form

$$\sin(\epsilon\nu)(e^{2\nu} e^{\epsilon\nu} - \cosh(\epsilon\nu) > 1 . \tag{4.9}$$

For small ϵ and hence large ν so that 1 can be neglected compare to $\epsilon^{2\nu}$ (4.9) becomes (to lowest order)

$$e^{2\nu} \nu > \frac{1}{\epsilon} . \tag{4.10}$$

The rigorous transition temperature (or equivalently $\nu = J/kT$) is given by the known formula

$$\sinh(2\nu) \sinh(2\epsilon\nu) = 1 , \tag{4.11}$$

which is seen to be equivalent to (4.10) to lowest order in ϵ!

5. ASYMPTOTIC DEGENERACY AND LONG RANGE ORDER

The heuristic argument of the preceding section shows clearly that phase transition is reflected mathematically in a *qualitative* change in the nature of the principal eigenvector. For high temperatures it more or less assigns the same weight to all configurations for low temperatures it tends to concentrate all the weight on two configurations $(+,+,\ldots,+)$ and $(-,-,\ldots,-)$ which correspond to perfect order.

Not unrelatedly there is also a qualitative change in the nature of the maximum eigenvalue.

As long as M is finite a theorem of Frobenius assures us that the maximum eigenvalue is non-degenerate. But while for high temperatures

$$\frac{\Lambda_2}{\Lambda_1} = 1 - \text{a quantity of order } 1 , \tag{5.1}$$

for low temperatures

$$\frac{\Lambda_2}{\Lambda_1} = 1 - \text{a quantity of order } \exp(-cM) , \tag{5.2}$$

where c is a temperature dependent constant.

Although rigorous proofs of (5.1) and (5.2) require the full power of the Onsager-Kaufmann analysis a heuristic argument can be easily provided.

We may restrict ourselves to the case of small ϵ since it seems reasonable that changing ϵ will produce only quantitative changes.

For small ϵ and sufficiently high temperatures

$$\Lambda_1 \approx (2 \cosh \nu)^M ,$$

while

$$\Lambda_2 \approx (2 \cosh \nu)^{M-1} (2 \sinh \nu) .$$

Thus

$$\frac{\Lambda_2}{\Lambda_1} \sim \tanh \nu . \tag{5.3}$$

For sufficiently low temperatures one is tempted to use

$$\frac{\phi_o^+ + \phi_o^-}{\sqrt{2}}$$

as an approximate principal eigenfunction while one is equally tempted to use

$$\frac{\phi_o^+ - \phi_o^-}{\sqrt{2}}$$

as an approximation to the eigenfunction which corresponds to Λ_2.

One may thus expect that

$$\frac{\Lambda_2}{\Lambda_1} \approx \frac{(\phi_o^+ - \phi_o^-,\ L_\epsilon(\phi_o^+ - \phi_o^-))}{(\phi_o^+ + \phi_o^-,\ L_\epsilon(\phi_o^+ + \phi_o^-))} \approx 1 - 2\frac{(\phi_o^-,\ L_\epsilon\ \phi_o^+)}{(\phi_o^+,\ L_\epsilon\ \phi_o^+)} = 1 - 2\ \exp(-M\nu)\ . \tag{5.4}$$

The near degeneracy of Λ_1 for low temperatures as expressed by formula (5.2) implies a fortiori.

$$\lim_{M\to\infty} \frac{\Lambda_2}{\Lambda_1} = 1\ . \tag{5.5}$$

In the case of the Ising model the limit (1) is approached with great rapidity (i.e. exponentially).

The property embodied in formula (5.5) is called *asymptotic degeneracy*.

As first observed many years ago by Ashkin and Lamb asymptotic degeneracy is closely related (in fact, almost equivalent) to the presence of long range order.

To see this we turn to formula (3.2) and note first of all that since $\phi_1(\boldsymbol{\mu})$ is symmetric

$$(\mu_l\, \phi_1(\boldsymbol{\mu}),\ \phi_1(\boldsymbol{\mu})) = 0$$

and therefore

$$\rho^{(M)}_{\text{horizontal}}(r) = \langle \mu_{k,1}\mu_{k+r,1}\rangle = \sum_{j=2}^{\infty} \left(\frac{\Lambda_j}{\Lambda_1}\right)^r (\mu_l\ \phi_j(\boldsymbol{\mu}),\ \phi(\boldsymbol{\mu}))^2\ . \tag{5.6}$$

It follows immediately that

$$\left(\frac{\Lambda_2}{\Lambda_1}\right)^r (\mu_l\, \phi_2(\boldsymbol{\mu}),\ \phi_1(\boldsymbol{\mu}))^2 < \rho_{\text{hor.}}^{(M)}(r) < \left(\frac{\Lambda_2}{\Lambda_1}\right)^r \tag{5.7}$$

and setting

$$\rho_{\text{hor.}}^{(r)} = \lim_{M\to\infty} \rho_{\text{hor.}}^{(M)}(r)\ , \tag{5.8}$$

we can draw the following conclusions:

a) If $\rho_{\text{hor.}}^{(\infty)} > 0$ (long range order), then

$$\lim_{M\to\infty} \frac{\Lambda_2}{\Lambda_1} = 1 \text{ (asymptotic degeneracy) .}$$

b) If

$$\lim_{M\to\infty} \frac{\Lambda_2}{\Lambda_1} = 1$$

and

$$\lim_{M\to\infty} (\mu_l\, \phi_2(\boldsymbol{\mu}),\ \phi,\ (\boldsymbol{\mu})) = 0\ ,$$

then

$$\rho_{\text{hor.}}(\infty) > 0\ .$$

6. THE WEISS THEORY OF THE MOLECULAR FIELD

The Weiss theory of the molecular field corresponds to a model in which N "spins" $\mu, \ldots, \mu_N$ interact with the energy

$$E = -\frac{2J}{N} \sum_{1 \leq i<j \leq N} \mu_i \mu_j\ . \tag{6.1}$$

The corresponding partition function can be readily evaluated (see e.g. § 3 of my Brandeis Lectures) and, in the thermodynamic limit $N \to \infty$, the free energy per site is given by the familiar formula.

$$-\frac{\psi}{kT} = \log \max_{\eta} \{\exp(-\tfrac{1}{2}\eta^2) \cosh(\sqrt{2\nu}\eta)\} + \log 2 . \tag{6.2}$$

The function

$$\exp(-\tfrac{1}{2}\eta^2) \cosh(\sqrt{2\nu}\eta)$$

has one maximum at $\eta = 0$ if $2\nu < 1$ and two equal maxima at the non-zero roots of the equation

$$\eta = 2\nu \tanh(\eta\ 2\nu) , \tag{6.3}$$

if $2\nu > 1$.

Thus $-\psi/kT$ is equal to log 2 for $2\nu < 1$ and equal to log 2 plus

$$\kappa(\nu) = -\tfrac{1}{2}\eta_0^2(\nu) + \log \cosh(\sqrt{2\nu}\eta_0(\nu)) \tag{6.4}$$

for $2\nu > 1$, where $\eta_0(\nu)$ is either of the non-zero roots of (6.3) (these roots are equal in magnitude but have opposite signs).

Actually, the Weiss model was not based on discrete "spins" (the concept did not exist at that time). Much closer in spirit (and in fact) is the model in which with each site i one associates a little magnet characterized by an angle θ_i which can vary *continuously* from 0 to 2π.

The interaction energy is assumed to be of the form

$$E = -\frac{2J}{N} \sum_{1 \leq i < j \leq N} \cos(\theta_i - \theta_j) \tag{6.5}$$

and the corresponding partition function is

$$Q_N = \int_0^{2\pi} \cdots \int_0^{2\pi} \exp \frac{2\nu}{N} \sum_{1 \leq i < j \leq N} \cos(\theta_i - \theta_j)\, d\theta_1 \ldots d\theta_N . \tag{6.6}$$

The calculation is just as easy as in the case of two-valued spins but I shall give it for the sake of completeness.

We note that

$$\exp\Big(\frac{\nu}{N}\sum_{i,j,=1}^{N}\cos(\theta_i-\theta_j)\Big) = \exp\Big(\frac{\nu}{N}\Big[\Big(\sum_{i=1}^{N}\sin\,\theta_i\Big)^2 + \Big(\sum_{i=1}^{N}\sin\,\theta_i\Big)^2\Big]$$

$$= \frac{1}{2\pi}\int_{-\infty}^{\infty}\int_{-\infty}^{\infty}\exp\Big\{-\frac{\xi^2+\eta^2}{2}\Big\}\exp\Big\{\frac{2\nu}{N}\sum_{i=1}^{N}(\xi\cos\theta_i+\eta\sin\,\theta_i)\Big\}\,d\xi d\eta$$

and consequently

$$Q_N = e^{-\nu}\frac{(2\pi)^N}{2\pi}\int_{-\infty}^{\infty}\int_{-\infty}^{\infty}\exp\Big\{-\frac{\xi^2+\eta^2}{2}\Big\}\,I_0^N\Big(\frac{\sqrt{2\nu}}{\sqrt{N}}\sqrt{\xi^2+\eta^2}\Big)d\xi d\eta\,. \quad (6.7)$$

Setting

$$\xi = \sqrt{N}\,r\,\cos\,\theta\;;$$

$$\eta = \sqrt{N}\,r\,\sin\,\theta\;\;,$$

we obtain

$$Q_N = e^{-\nu}N(2\pi)^{N-1}\int_0^{\infty}\{\exp(-\tfrac{1}{2}r^2)\,I_0(\sqrt{2\nu}\,r)\}^N\,dr\;,$$

and (using the familiar saddle point method)

$$\frac{-\psi}{kT} = \lim_{N\to\infty}\frac{1}{N}\log Q_N = \log\,\max\{\exp(-\tfrac{1}{2}r^2\,I_0(\sqrt{2\nu}\;r)\} + \log\,2\pi\;. \quad (6.8)$$

Comparison of this result with (6.2) reveals that the only difference is the replacement of cosh $\sqrt{2\nu}\eta$ by the Bessel function $I_0(\sqrt{2\nu}\eta)$. Note that they have very similar asymptotic forms.

The only *qualitative* difference that results from this is apparent only near absolute zero ($\nu \to \infty$).

Not surprisingly the "continuous" model based on (6.5) violates the Third Law since the specific heat

$$C = -k\nu^2\,\frac{d^2}{d\nu^2}\Big(-\frac{\psi}{kT}\Big) \quad (6.9)$$

does not approach zero as $\nu \to \infty$. In fact, for large ν and for the continuous model

$$-\frac{\psi}{kT} \approx -\nu + \log I_0(2\nu) \approx \nu - \tfrac{1}{2}\log \nu + \text{const} \tag{6.10}$$

and consequently

$$\lim_{\nu \to \infty} C = k\,. \tag{6.11}$$

For the discrete model

$$-\frac{\psi}{kT} \approx -\nu + \log \cosh 2\nu \approx \nu + \text{const}$$

and

$$\lim_{\nu \to \infty} C = 0\,.$$

Apart from this everything is qualitatively the same and even the Curie point remains the same, i.e., $2\nu = 1$. In particular, the specific heat has a jump discontinuity in both models.

This is somewhat surprising since one might expect the continuous model to be "less singular" than the discrete one.

For example, the recent proof of Mermin and Wagner that the two-dimensional Heisenberg model does not exhibit spontaneous magnetization can be adapted [1] to show that the continuous counterpart of (2.2) with interaction energy

$$E = -J \sum v_{\mathbf{PQ}}(\epsilon) \cos(\theta_{\mathbf{P}} - \theta_{\mathbf{Q}})\,, \tag{6.12}$$

(and angles $\theta_{\mathbf{P}}$ varying continuously between 0 and 2π) also fails to exhibit spontaneous magnetization.

As we shall see in the sequel a more subtle approach to the molecular field theory will, in a way, confirm that a continuous model is less singular than the corresponding discrete one.

7. MODELS WITH EXPONENTIAL INTERACTION

From the point of view of fundamental principles of statistical mechanics the most serious objection to theories based on (6.1) or (6.5) is that the assumed interaction depend on the *size of the system*. That this flagrant violation of a basic cannon does not lead to difficulties is more or less an accident because the same violation (in spirit and in letter) in the case of a gas does give rise to the van der Waals loops. It is easy however to remedy this fault at the expense of an additional limiting process.

Let χ be an even, non-negative, positive-definite function (i.e. the Fourier transform of χ is non-negative) and let also (for normalization purposes) $\chi(0) = 1$.

Consider a one-dimensional model with the interaction energy given by the expression

$$E = -J\gamma \sum_{1 \leq i < j \leq N} \chi(\gamma(i-j)) . \tag{7.1}$$

We shall also consider the corresponding continuous model with the interaction energy

$$E = -J\gamma \sum_{1 \leq i < j \leq N} \chi(\gamma(i-j)) \cos(\theta_i - \theta_j) . \tag{7.2}$$

From our rather strong assumptions on χ it follows first of all that

$$\int_{-\infty}^{\infty} \chi(t)\, \mathrm{d}t < \infty$$

and one can then show (using e.g. the method of Lebowitz and Penrose [2]), that if after taking the thermodynamic limit $N \to \infty$ one also lets $\gamma \to 0$ one obtains the Weiss theory (6.2) or (6.8) depending on whether one starts with (7.1) or (7.2).

Although this result establishes rigorously the Weiss theory as a limiting case (in the same way as Kac, Uhlenbeck and Hemmer and then Lebowitz and Penrose established the van der Waals theory as a limiting case) it does not e.g. distinguish in *qualitative* way between (7.1) and (7.2) (except, of course, near absolute zero) and as will be seen later on it is too weak to bring out features which depend on dimensionality of the model.

A deeper insight is obtained by not letting γ apprøve but by keeping it small. To obtain concrete results it becomes desirable (and perhaps even necessary) to further specialize the interaction and the simplest choice is

$$\chi(t) = \exp(-|t|) . \tag{7.3}$$

This case has been studied so extensively (see e.g. my Brandeis Lectures) that only a cursory review of pertinent results will suffice.

The reason why the choice (7.3) is so convenient is that $\exp(-|t|)$ is the *covariance function* of the so-called Ornstein-Uhlenbeck process $U(\tau)$ and one can write

$$\exp\{\nu\gamma \sum_{1\leq i<j\leq N} \exp(-\gamma|i-j|)\,\mu_i\mu_j\}$$

$$= \exp(-\tfrac{1}{2}N\nu\gamma)\,\exp\left\{\tfrac{1}{2}\,\nu\gamma \sum_{i,j=1}^{N} \exp(-\gamma|i-j|)\,\mu_i\mu_j\right\},$$

as $\exp(-\frac{1}{2}N\nu\gamma)$ times the mathematical expectation of

$$\exp\left\{\sqrt{\nu\gamma}\sum_{k=1}^{N} U(k\gamma)\,\mu_k\right\}.$$

Because of this the partition function Q_N corresponding to the energy (7.1) (with exponential $\chi(t)$ of course) can be written in the form

$$Q_N = \sum \exp(-E/kT) = 2^N \exp(-\tfrac{1}{2}N\nu\gamma)\,\langle \prod_{k=1}^{N} \cosh\sqrt{\nu\gamma}\,U(k\gamma)\rangle\,, \tag{7.4}$$

where I use now (and shall use in the sequel) $\langle\ \rangle$ to denote also mathematical expectations referring to stochastic processes.

Because the Ornstein-Uhlenbeck process is *Markoffian* (in addition to being Gaussian and stationary) the density $W(u_1, u_2, \ldots u_N)$ of the joint distribution of $U(\gamma), U(2\gamma), \ldots, U(N\gamma)$ *factorizes*, i.e.,

$$W(u_1, u_2, \ldots, u_N) = W(u_1)\,P_\gamma(u_1|u_2)\,P_\gamma(u_2|u_3)\cdots P_\gamma(u_{N-1}|u_N)\,, \tag{7.5}$$

where

$$W(x) = \frac{1}{\sqrt{2\pi}}\exp(-\tfrac{1}{2}x^2)\,; \tag{7.6}$$

$$P_\gamma(x|y) = \frac{1}{\sqrt{2\pi(1-\exp(-2\gamma))}}\exp\left\{-\frac{(y-x\exp(-\gamma))^2}{2(1-\exp(-2\gamma))}\right\}. \tag{7.7}$$

Thus we obtain that

$$Q_N = 2N \exp(-\tfrac{1}{2}N\nu\gamma) \int_{-\infty}^{\infty} \int_{-\infty}^{\infty} \ldots \int_{-\infty}^{\infty} \prod_{k=1}^{\infty} \cosh(\sqrt{\nu\gamma}\, u_k)\, W(u_1)$$
$$\times \prod_{k=1}^{N-1} P_\gamma(u_k | u_{k+1})\, du_1\, du_2 \ldots du_N \,. \tag{7.8}$$

Those unfamiliar with the rudiments of the theory of stochastic processes need merely verify the identity

$$\exp\left\{\tfrac{1}{2}\nu\gamma \sum_{i,j=1}^{N} \exp(-\gamma|i-j|)\, \mu_i \mu_j\right\}$$
$$= \int_{-\infty}^{\infty} \ldots \int_{-\infty}^{\infty} \exp\left(\sqrt{\nu\gamma} \sum_{1}^{N} u_k \mu_k\right) W(u_1) \prod_{k=1}^{N-1} P_\gamma(u_k | u_{k+1})\, du_1 \ldots du_N) \tag{7.9}$$

(which is not too difficult an exercise in multiple integration) when (7.8) follows almost at once.

To derive a formula similar to (7.8) for the model based on (7.2) one must introduce two *independent* Ornstein-Uhlenbeck processes $U(t)$ and $V(t)$ having the same covariance $\exp(-|t|)$ and note that

$$\exp\left\{\tfrac{1}{2}\nu\gamma \sum_{i,j=1}^{N} \exp(-\gamma|i-j|) \cos(\theta_i - \theta_j)\right\}$$
$$= \exp \sqrt{\nu\gamma} \sum_{k=1}^{N} (U(k\gamma) \cos \theta_k + V(k\gamma) \sin \theta_k) \,. \tag{7.10}$$

Since

$$\int_{0}^{2\pi} \exp\{\sqrt{\nu\gamma}\, (U \cos \theta + V \sin \theta)\, d\theta = 2\pi\, I_0\, (\sqrt{\nu\gamma}\, \sqrt{U^2 + V^2}) \tag{7.11}$$

the analogue of (7.8) is $\qquad Q_N =$

$$(2\pi)^N \exp(-\tfrac{1}{2}N\nu\gamma) \int_{-\infty}^{\infty} \cdots \int_{-\infty}^{\infty} \prod_{k=1}^{N} I_0 \left(\sqrt{\nu\gamma}\ \sqrt{u_k^2 + v_k^2}\right) W(u_1)\, W(v_1)$$

$$\times \prod_{k=1}^{N-1} P_\gamma(u_k | u_{k+1})\, P_\gamma(v_k | v_{k+1})\, du_1\, dv_1, \ldots du_N\, dv_N\,. \tag{7.12}$$

Returning to (7.8) we introduce the symmetric kernel

$$K(x, y) = \{\cosh(\sqrt{\nu\gamma}x)\}^{\frac{1}{2}} \frac{W(x)\, P_\gamma(x | y)}{(W(x)W(y))^{\frac{1}{2}}} \{\cosh(\sqrt{\nu\gamma}y\}^{\frac{1}{2}} \tag{7.13}$$

and note that

$$Q_N = 2^N \exp(-\tfrac{1}{2}N\nu\gamma) \int_{-\infty}^{\infty} \cdots \int_{-\infty}^{\infty} \{W(u_1) \cosh(\sqrt{\nu\gamma}\, u_1)\}^{\frac{1}{2}} K(u_1, u_2)$$

$$\times\ K(u_2, u_3) \ldots K(u_{N-1}, u_N)\ \{W(u_N) \cosh(\sqrt{\nu\gamma}\, u_N)\}^{\frac{1}{2}}\, du_1 \ldots du_N\,. \tag{7.14}$$

Denoting by $\lambda_1, \lambda_2, \ldots$ the eigenvalues (in decreasing order) and by $\phi_1 \phi_2 \ldots$ the corresponding normalized eigenfunctions of the Kernel $K(x, y)$ we see that

$$Q_N = 2^N \exp(-\tfrac{1}{2}N\nu\gamma) \sum_{j=1}^{\infty} \lambda_j^{N-1} \left(\int_{-\infty}^{\infty} \phi_j(x) \{W(x) \cosh(\sqrt{\nu\gamma}x)\}^{\frac{1}{2}} dx\right)^2 \tag{7.15}$$

and it follows at once that

$$-\frac{\psi}{kT} = \lim_{N\to\infty} \frac{1}{N} \log Q_N = \log 2 - \tfrac{1}{2}\nu\gamma + \log \lambda_1\,. \tag{7.16}$$

Similarly one can derive a formula for the correlation coefficient

$$\rho(r) = \lim_{N\to\infty} \langle \mu_k \mu_{k+r} \rangle\,, \tag{7.17}$$

where k must now approach ∞ with N to avoid boundary effects (it is no longer convenient to introduce periodic boundary conditions in this model).

The result is embodied in the formula (valid only for $r > 1$)

$$\rho(r) = \sum_{j=1}^{\infty} \left(\frac{\lambda_j}{\lambda_1}\right)^r \left\{ \int_{-\infty}^{\infty} \phi_j(x)\, \phi_1(x) \tanh(\sqrt{\nu\gamma}x)\, dx \right\}^2 \tag{7.18}$$

and the analogy with (3.2) is evident. Equally evident is the analogy between (7.16) and (2.9).

The reader should have no difficulty in deriving formulas for higher correlations, i.e.,

$$\rho(r_1, r_2, \ldots, r_l) = \lim_{N\to\infty} \langle \mu_k \mu_{k+r_1} \mu_{k+r_1+r_2} \cdots \mu_{k+r_1+\ldots r_l} \rangle, \tag{7.19}$$

where again k approaches ∞ with N and $\langle\ \rangle$ denotes the canonical average, i.e., average with respect to the weight $\exp(-E/kT)$. The formulas involve the kernels

$$K_r(x, y) = \sum_{j=1}^{\infty} \left(\frac{\lambda_j}{\lambda_1}\right)^r \phi_j(x)\, \phi_j(y) \tag{7.20}$$

and are as follows:

$$\rho(r_1, r_2 \ldots r_l)$$

$$= \int_{-\infty}^{\infty} \ldots \int_{-\infty}^{\infty} \phi_1(u_1) \tanh(\sqrt{\nu\gamma}\, u_1)\, K_{r_1}(u_1, u_2) \tanh(\sqrt{\nu\gamma}\, u_2)\, K_{r_2}(u_2, u_3)$$

$$\times \tanh(\sqrt{\nu\gamma}\, u_3)\, K_{r_3}(u_3, u_4) \ldots K_{r_l}(u_{l-1}, u_l)\, \phi_1(u_l)$$

$$\times \tanh(\sqrt{\nu\gamma}\, u_l)\, du_1 \ldots du_l\,. \tag{7.21}$$

In this connection there arises the following interesting and important problem:

Given the correlations $\rho(r_1, r_2, \ldots, r_l)$, $l = 1, 2, 3, \ldots$ (which presumably contain *all* the physically pertinent information about the system) is it possible to construct a kernel (the solution surely is not unique) which gives rise to them through the formula (7.21)? Analogous formulas can be derived from the "continuous spin" model based on (7.2) with χ given by (7.3).

Instead of $K(x, y)$ given by (7.13) we must now introduce the kernel $K(x_1, x_2; y_1, y_2)$ given by the formula

$K(x_1, x_2; y_1, y_2)$

$$= \left\{ I_0(\sqrt{\nu\gamma}\sqrt{x_1^2+x_2^2}) \right\}^{\frac{1}{2}} \frac{W(x_1)W(x_2)P_\gamma(x_1|y_1)P_\gamma(x_2|y_2)}{\{W(x_1)W(x_2)W(y_1)W(y_2)\}^{\frac{1}{2}}} \left\{ I_0(\sqrt{\nu\gamma}\sqrt{y_1^2+y_2^2}) \right\}^{\frac{1}{2}} \tag{7.22}$$

and the analogue of (7.16) is

$$-\frac{\psi}{kT} = \log(2\pi) - \tfrac{1}{2}\nu + \log \lambda_1 . \tag{7.23}$$

It is, of course, understood that λ_1 is now the largest eigenvalue of the kernel (7.22).

The analogue of (7.18) where $\rho(r)$ is now defined as

$$\lim_{N\to\infty} \langle \cos\theta_k \cos\theta_{k+r} \rangle$$

(with the usual proviso about k approaching ∞ with N) is

$$\rho(r) = \sum_{j=1}^{\infty} \left(\frac{\lambda_j}{\lambda_1}\right)^r \left\{ \int_{-\infty}^{\infty}\int_{-\infty}^{\infty} \phi_j(x,y) R(x,y) \phi_j(x,y) dx dy \right\}^2 , \tag{7.24}$$

where

$$R(x,y) = \frac{x}{\sqrt{x^2+y^2}} \frac{I_0'(\sqrt{x^2+y^2})}{I_0(\sqrt{x^2+y^2})} \tag{7.25}$$

is the obvious analogue of $\tanh x$ and $\phi_j(x, y)$ are the normalized eigenfunctions of the kernel (7.22) *.

8. DISCUSSION FOR SMALL γ

To study what happens for small γ we start with (7.13) and note that after a few simple algebraic transformations the kernel K can be written in the form

* It would be more consistent to use $\phi_j(x_1, x_2)$ or $\phi_j(y_1, y_2)$ but it is easier on the eye to use fewer subscripts and there is little chance of confusion.

$$K(x, y) = \exp(\tfrac{1}{2}\gamma)\exp(-\tfrac{1}{2}\gamma q(x)) \frac{\exp\left\{-\frac{(y-x)^2}{4\sinh\gamma}\right\}}{\sqrt{4\pi\sinh\gamma}} \exp(-\tfrac{1}{2}\gamma q(x)) , \qquad (8.1)$$

where

$$\gamma q(x) = \tfrac{1}{2}\tanh(\tfrac{1}{2}\gamma x^2) - \log\cosh(\sqrt{\nu\gamma}x) . \qquad (8.2)$$

It is easily verifiable that for a smooth function f

$$\int_{-\infty}^{\infty} K(x, y) f(y) \mathrm{d}y \qquad (8.3)$$

$$= \exp(\tfrac{1}{2}\gamma)\exp(-\tfrac{1}{2}\gamma q(x)) \exp\left\{(\sinh\gamma)\frac{\mathrm{d}^2}{\mathrm{d}x^2}\right\} \exp(-\tfrac{1}{2}\gamma q(x)) f(x)$$

and to *lowest order* in γ the operators

$$A = \gamma q(x) ,$$

and

$$B = (\sinh\gamma)\frac{\mathrm{d}^2}{\mathrm{d}x^2}$$

commute. (If one wants to go beyond the lowest order one must use the Baker-Hausdorf formula [3] or follow a perturbation technique similar to that used by Kac, Uhlenbeck and Hemmer [4,5]).

Thus, to lowest order, the integral operator K can be replaced by the operator

$$\exp(\tfrac{1}{2}\gamma)\exp\left[\sinh\left\{\gamma\frac{\mathrm{d}^2}{\mathrm{d}x^2} \gamma q(x)\right\}\right] ,$$

or (replacing $\sinh\gamma$ by γ and $\tanh\gamma/2$ in (8.2) by $\gamma/2$) by the operator

$$\exp(\tfrac{1}{2}\gamma)\exp\{\gamma\frac{\mathrm{d}^2}{\mathrm{d}x^2}[\tfrac{1}{4}\gamma x^2 - \log\cosh(\sqrt{\nu\gamma}x)]\} .$$

It follows that (again to order γ)

$$\log\lambda_1 \approx \tfrac{1}{2}\gamma - \gamma E_o , \qquad (8.4)$$

where E_o is the *lowest* eigenvalue of the Schrödinger equation

$$\frac{d^2\phi}{dx^2} - \{\tfrac{1}{4}x^2 - \frac{1}{\gamma}\log\cosh(\sqrt{\nu\gamma}x)\}\phi = -E\phi \tag{8.5}$$

and that the eigenfunctions of K can be taken to be those of (8.5).

To discuss the correlation function (7.18) we need, of course, the whole spectrum of the Schrödinger operators in (8.5) as well as its eigenfunctions. What about this spectrum?

Its character undergoes a dramatic change at $2\nu = 1$.

In fact, the potential

$$\tfrac{1}{4}x^2 - \frac{1}{\gamma}\log\cosh(\sqrt{\nu\gamma}\,x) \tag{8.6}$$

has a unique minimum at $x = 0$ if $2\nu < 1$ but has *two* (equal) minima if $2\nu > 1$.

These two minima occur at

$$x = \pm\frac{\eta_0(\nu)\sqrt{2}}{\sqrt{\gamma}}, \tag{8.7}$$

where $\eta_0(\nu)$ is the positive root of the transcendental equation (6.3).

The depth of each minimum is

$$-\frac{1}{\gamma}\kappa(\nu), \tag{8.8}$$

with $\kappa(\nu)$ as defined by (6.4).

If $2\nu < 1$ we can expand $\log\cosh(\sqrt{\nu\gamma}x)$ in a power series retaining only terms of order γ or lower (i.e. $\log\cosh\sqrt{\nu\gamma}x \approx \tfrac{1}{2}\nu\gamma x^2$) and replace the potential by

$$\tfrac{1}{4}(1-2\nu)x^2 .$$

Thus if $2\nu < 1$

$$E_n = (n+\tfrac{1}{2})\sqrt{1-2\nu} \tag{8.9}$$

and

$$\log\lambda_n \approx \tfrac{1}{2}\gamma\{1-(2n+1)\sqrt{1-2\nu}\}. \tag{8.10}$$

The $\log\lambda_n$ are therefore more or less evenly spaced, the spacing being of order γ.

For $2\nu > 1$ because the potential has two deep minima each eigenvalue and, in particular, the lowest one is almost degenerate.

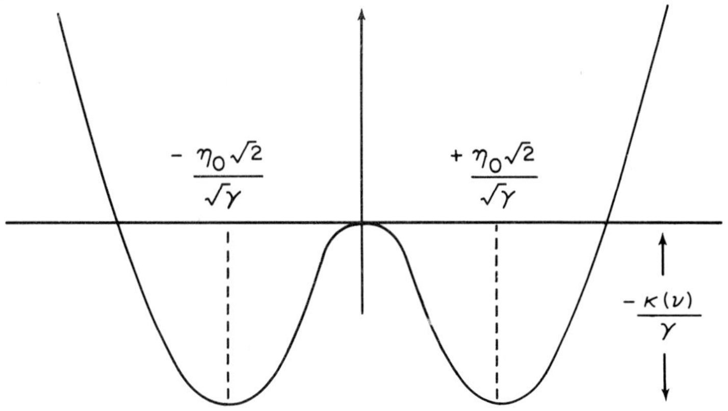

The degree of near-degeneracy can be calculated by a familiar application of the WKB method and it is not too difficult to prove rigorously (see end of §5 of my Brandeis Lectures) that

$$\frac{\lambda_2}{\lambda_1} = 1 - O\{\exp(-c/\gamma)\} . \tag{8.11}$$

A heuristic argument justifying (8.11) and strongly reminiscent of the heuristic argument by which we justified (5.4) is as follows.

We replace each of the potential wells by parabolic wells (it is easily checked that $1 - 2\nu \operatorname{sech}^2(\eta_0\sqrt{2\nu}) > 0$)

$$-\frac{1}{\gamma}\,\kappa(\nu) + \tfrac{1}{4}(1 - 2\nu \sinh^2(\eta_0\sqrt{2\nu}))\left(x \pm \frac{\eta_0\sqrt{2}}{\sqrt{\gamma}}\right)^2 \tag{8.12}$$

and (setting $\nu' = \nu \operatorname{sech}^2(\eta_0\sqrt{2\nu})$) and $\psi_0(x) = \exp\{-\tfrac{1}{4}\sqrt{1 - 2\nu'}\,x^2\}$ we can take

$$\psi^+ = \psi_0\left(x - \frac{\eta_0\sqrt{2}}{\sqrt{\gamma}}\right) + \psi_0\left(x + \frac{\eta_0\sqrt{2}}{\sqrt{\gamma}}\right)$$

and

$$\psi^- = \psi_0\left(x - \frac{\eta_0\sqrt{2}}{\sqrt{\gamma}}\right) - \psi_0\left(x - \frac{\eta_0\sqrt{2}}{\sqrt{\gamma}}\right)$$

as (un-normalized) trial wave functions corresponding to the lowest and next to the lowest eigenvalues.

A simple calculation based on the Raleigh-Ritz principle, then shows that $E_2 - E_1$ is of the order of the "overlap integral"

$$\int_{-\infty}^{\infty} \psi_0\left(x - \frac{\eta_0\sqrt{2}}{\sqrt{\gamma}}\right) \psi_0\left(x + \frac{\eta_0\sqrt{2}}{\sqrt{\gamma}}\right) dx\,, \tag{8.13}$$

(divided by the normalizing factor

$$\int_{-\infty}^{\infty} \psi_0^2\left(x - \frac{\eta_0\sqrt{2}}{\sqrt{\gamma}}\right) dx = \int_{-\infty}^{\infty} \psi_0^2\left(x + \frac{\eta_0\sqrt{2}}{\sqrt{\gamma}}\right) dx = \sqrt{2\pi}/\sqrt{1-2\nu'}\,)$$

which is clearly of the order

$$\exp(-c/\gamma)\,. \tag{8.14}$$

Moreover, for $2\nu > 1$ we also have

$$E_1 \approx -\frac{1}{\gamma}\kappa(\nu) + \tfrac{1}{2}\sqrt{1-2\nu'}\,, \tag{8.15}$$

or, equivalently

$$\log \lambda_1 \approx \kappa(\nu) + \tfrac{1}{2}\gamma(1 - \sqrt{1-2\nu'})\,. \tag{8.16}$$

Combining this with the result valid for $2\nu < 1$

$$\log \lambda_1 \approx \tfrac{1}{2}\gamma(1 - \sqrt{1-2\nu'})\,, \tag{8.17}$$

(see (8.10)) we have by (7.16) the following formula for free energy which is correct to order γ

$$-\frac{\psi}{kT} = \begin{cases} \log 2 - \tfrac{1}{2}\nu\gamma + \tfrac{1}{2}\gamma(1 - \sqrt{1-2\nu}), & 2\nu < 1\,, \\ \log 2 + \kappa(\nu) - \tfrac{1}{2}\nu\gamma + \tfrac{1}{2}\gamma(1 - \sqrt{1-2\nu'}), & 2\nu > 1\,. \end{cases} \tag{8.18}$$

We recognize at once that to order 1 (i.e. in the limit $\gamma \to 0$) we recover the molecular field theory and that this theory (at least as far as the analytic behavior of thermodynamic functions is concerned) is the result of replacing the lowest eigenvalue E_1 of the Schrödinger equation (8.5) by the *minimum of the potential*. In other words, the molecular field theory corresponds to the *classical approximation* which takes no account of the quantum effects caused by the presence of the *kinetic energy term* d^2/dx^2.

It is however the quantum effects that are responsible for the marked change in the behavior of the correlation function as one goes from high ($2\nu < 1$) to low ($2\nu > 1$) temperatures.

In fact, for $2\nu < 1$ since

$$\frac{\lambda_1}{\lambda_2} = 1 - O(\gamma) ,$$

the "correlation length" is of the order $1/\gamma$, i.e., of the same order as the range of the interaction.

For $2\nu > 1$, on the other hand, we have

$$\frac{\lambda_2}{\lambda_1} = 1 - O\{\exp(-c/\gamma)\}$$

and the correlation length becomes of the order $\exp(+c/\gamma)$ which is incomparably (in fact enormously!) larger than the interaction length $1/\gamma$.

The decay rate of the correlation function is governed by the rate λ_2/λ_1 which is approximately

$$\exp\{-\gamma(E_1 - E_o)\} .$$

and the quantity $\Delta E = E_2 - E_1$ is par excellence quantum mechanical. In particular, the fact that for $2\nu > 1$, ΔE is of the order $\exp(-c/\gamma)$ is intimately related to the phenomenon of penetration of the barrier.

9. DISCUSSION FOR SMALL γ CONTINUED

The discussion of the last section can be imitated step by step for the "continuous spin" model.

The analogue of (8.3) is

$$\int_{-\infty}^{\infty}\int_{-\infty}^{\infty} K(x,y;\xi,\eta) f(\xi,\eta) d\xi d\eta$$
$$= \exp(\gamma)\exp(-\tfrac{1}{2}\gamma q(x,y))\exp\left\{(\sinh\gamma)\left(\frac{d^2}{dx^2}+\frac{d^2}{dy^2}\right)\right\}\exp(-\tfrac{1}{2}\gamma q(x,y)) f(x,y) , \tag{9.1}$$

with

$$\gamma q(x,y) = \tfrac{1}{2}\tanh\{\tfrac{1}{2}\gamma(x^2+y^2)\} \log I_o(\sqrt{\nu\gamma}\sqrt{x^2+y^2}) \tag{9.2}$$

and, again as before, the integral operator can be replaced (to order γ) by the operator

$$\exp(\gamma)\ \exp\{\sinh\gamma\nabla^2 - \gamma q(x,y)\},$$

or, replacing $\sinh\gamma$ by γ and $\tanh\frac{1}{2}\gamma$ by $\frac{1}{2}\gamma$, by the operator

$$\exp(\gamma)\ \exp\{\gamma\nabla^2 - [\tfrac{1}{4}\gamma(x^2+y^2) - \log I_0(\sqrt{\nu\gamma}\ \sqrt{x^2+y^2})]\ . \qquad (9.3)$$

We thus have to order γ

$$\log\lambda_1 \approx \gamma - \gamma E_0\ , \qquad (9.4)$$

where E_0 is now the lowest eigenvalue of the Schrödinger equation

$$\frac{d^2\phi}{dx^2} + \frac{d^2\phi}{dy^2} - \left(\frac{x^2+y^2}{4} - \frac{1}{\gamma}\log I_0(\sqrt{\nu\gamma}\ \sqrt{x^2+y^2})\right)\phi = -E\phi\ , \qquad (9.5)$$

Also to lowest order the eigenfunctions of K can be taken to be these of (9.5).

Again in the limit $\gamma \to 0$ one obtains the molecular field theory (in the form (6.8) of course) and again this theory corresponds to using the classical approximation (i.e. the minimum of q) for E_0.

If $2\nu < 1$ one obtains, as before,

$$\frac{\lambda_2}{\lambda_1} = 1 - O(\gamma)\ , \qquad (9.6)$$

but for $2\nu > 1$ we are led to the first important qualitative difference (aside from the failure of the Third Law) from the case of ± 1 spins.

To see how this comes about we rewrite (9.5) in polar coordinates.

$$\frac{1}{r}\frac{\partial}{\partial r}\left(r\frac{\partial\phi}{\partial r}\right) + \frac{1}{r^2}\frac{\partial^2\phi}{\partial\theta^2} - \left(\frac{r^2}{4} - \frac{1}{\gamma}I_0(\sqrt{\nu\gamma}r)\right)\phi = -E\phi\ , \qquad (9.7)$$

and note that the lowest state in an S-state, i.e., ϕ is independent of θ.

The next lowest state is a (doubly degenerate) state whose wave function is of the form

$$\psi(r)e^{\pm i\theta} \qquad (9.8)$$

and the radial part $\psi(r)$ consequently satisfies the equation

$$\frac{1}{r}\frac{d}{dr}\left(r\frac{d\psi}{dr}\right) - \left(\frac{1}{r^2}+\frac{r^2}{4}-\frac{1}{\gamma}I_o(\sqrt{\nu\gamma}r)\right)\psi = -E\psi . \tag{9.9}$$

By the Raleigh-Ritz principle we have immediately that

$$E_o < \int_o^\infty \left\{\left(\frac{d\psi}{dr}\right)^2 + \left[\frac{r^2}{4}-\frac{1}{\gamma}I_o(\sqrt{\nu\gamma}\, r)\right]\psi^2\right\} rdr , \tag{9.10}$$

while, of course,

$$E_1 = \int_o^\infty \left\{\left(\frac{d\psi}{dr}\right)^2 + \left[\frac{1}{r^2}+\frac{r^2}{4}-\frac{1}{\gamma}I_o(\sqrt{\nu\gamma}\, r)\right]\psi^2\right\}2\ rdr . \tag{9.11}$$

It is understood that ψ is normalized, i.e.,

$$\int_o^\infty r\psi^2(r)dr = 1 . \tag{9.12}$$

It follows immediately that

$$E_1 - E_o > \int_o^\infty \frac{r}{r^2}\psi^2(r)dr \tag{9.13}$$

and, for small γ, $r\psi^2(r)$ is strongly concentrated around the value of r which minimizes

$$\frac{r^2}{4} - \frac{1}{\gamma}I_o(\sqrt{\nu\gamma}\, r)$$

and which is of the order $1/\sqrt{\gamma}$. Thus (9.13) implies that $E_1 - E_o$ is larger than a quantity of the order γ and it is not too difficult to show that actually $E_1 - E_o$ *is* of order γ. This in turn implies that

$$\frac{\lambda_2}{\lambda_1} = 1 - O(\gamma^2) ,$$

in marked contrast with (8.11).

All this should not be too surprising because the potential in (9.5) allows the particle to move freely (and classically) inside the potential barrier while in the two minimum case (8.5) $(2\nu > 1)$

for sufficiently low energies the classical passage from one well to the other is impossible and one needs the quantum mechanical penetration of the barrier to accomplish such a passage.

10. A TWO-DIMENSIONAL MODEL

I shall now sketch how effects of dimensionality can be studied on models with weak and long range interactions.

To do this I shall restrict myself to a particular two-dimensional model first introduced by Helfand and myself [3].

The model is a $N \times M$ (M rows, N columns) lattice of scalar spins $\mu_{k,l}$ $(1 \leq k \leq N,\ 1 \leq l \leq M)$ with the interaction energy

$$E = -\tfrac{1}{2} \sum_{(k,l)=(k',l')} v(k,l;k',l')\mu_{k,l}\mu_{k',l'}\ , \tag{10.1}$$

where

$$v(k,l;k',l') = J\gamma e^{-\gamma|k-k'|}\{\delta_{ll'} + \tau[\delta_{l',l+1} + \delta_{l',l-1}]\} \tag{10.2}$$

and $0 < \tau < 1$. The model is most conveniently placed on a cylinder, i.e., $M+1$ is identified with 1.

The thermodynamic limit is performed in two stages: first $N \to \infty$, then $M \to \infty$.

One obtains then (by a generalization of the procedure of section 7) that the free energy ψ per lattice site is given by the formula

$$-\frac{\psi}{kT} = \log 2 - \tfrac{1}{2}\nu\gamma + \lim_{M\to\infty} \Lambda_{max} \tag{10.3}$$

where $\Lambda_{max} = \Lambda_1$ is the largest eigenvalue of the M-dimensional integral equation

$$\int_{-\infty}^{\infty} \cdots \int_{-\infty}^{\infty} K(\mathbf{x},\mathbf{y})\phi(\mathbf{y})d\mathbf{y} = \Lambda\phi(\mathbf{x})\ , \tag{10.4}$$

where

$$\boldsymbol{x} = (x_1, \ldots, x_M), \quad \boldsymbol{y} = (y_1, \ldots, y_M), \quad \mathrm{d}\boldsymbol{y} = \mathrm{d}y_1 \ldots \mathrm{d}y_M ,$$

$$K(\boldsymbol{x}, \boldsymbol{y}) = \prod_{k=1}^{M} [\cosh\{\sqrt{\nu\gamma}\,(\sqrt{\xi}\, x_k + \sqrt{1-\xi}\, x_{k+1})\}]^{\frac{1}{2}}$$
$$\times \prod_{k=1}^{M} \left(\frac{W(x_k)}{W(y_k)}\right)^{\frac{1}{2}} P_\gamma(x_k | y_k) \prod_{k=1}^{M} [\cosh\{\sqrt{\nu\gamma}\,(\sqrt{\xi}\, y_k + \sqrt{1-\xi}\, y_{k+1})\}]^{\frac{1}{2}} \tag{10.5}$$

and

$$\xi = \sqrt{\tau(1-\tau)} \,. \tag{10.6}$$

Because of periodic boundary condition (recall that the model is placed on a cylinder) $x_{M+1} = x_1$ and $y_{M+1} = y_1$.

The definitions of W and $P_\gamma(x|y)$ are the same as in section 7 and the reader may try himself by deriving the analogue of (7.9); if pressed for time he may consult §7 of my Brandeis Lectures or the paper by Helfand and myself referred to above.

One derives also quite easily formulas for the correlations:

$$\rho(r) = \lim_{\substack{N\to\infty \\ M\to\infty}} \langle \mu_{k,l}\, \mu_{k+r,l} \rangle$$
$$= \lim_{M\to\infty} \sum_{j=1}^{\infty} \left(\frac{\Lambda_j}{\Lambda_1}\right)^{r} \left(\int \varphi_1(\boldsymbol{x}) \tanh\{\sqrt{\nu\gamma}\,(\sqrt{\xi}\, x_l + \sqrt{1-\xi}\, x_{l+1})\}\, \varphi_j(\boldsymbol{x}) \mathrm{d}\boldsymbol{x}\right)^2 \tag{10.7}$$

and

$$\sigma(r) = \lim_{\substack{N\to\infty \\ M\to\infty}} \langle \mu_{k,l}\, \mu_{k,l+r} \rangle = \lim_{M\to\infty} \int \varphi_1^2(\boldsymbol{x}) \tanh\{\sqrt{\nu\gamma}\,(\sqrt{\xi}\, x_l + \sqrt{1-\xi}\, x_{l+1})\}$$
$$\times \tanh\{\sqrt{\nu\gamma}\,(\sqrt{\xi}\, x_{l+r} + \sqrt{1-\xi}\, x_{l+r})\} \mathrm{d}\boldsymbol{x} \,. \tag{10.8}$$

In deriving (10.7) care must be taken that $k \to \infty$ as $N \to \infty$ so as to avoid boundary effects. Also, because of the periodic boundary condition, both ρ and σ are *independent of* l.

In analogy with (8.3) the integral equation (10.4) can be written in the symbolic form

$$\exp(-\tfrac{1}{2}\gamma q(\boldsymbol{x}))\exp\Big(\gamma\sum_{1}^{M}\frac{\partial^2}{\partial x_k^2}\Big)\exp(-\tfrac{1}{2}\gamma q(\boldsymbol{x}))\varphi(x) = \exp(-\tfrac{1}{2}M\gamma)\,\Lambda\varphi(x)\,, \tag{10.9}$$

where $q(\boldsymbol{x})$ is now given by the formula

$$\gamma q(\boldsymbol{x}) = \tfrac{1}{2}\tanh(\tfrac{1}{2}\gamma)\sum_{1}^{M}x_k^2 - \sum_{k=1}^{M}\log\cosh\{\sqrt{\nu\gamma}\,(\sqrt{\xi}\,x_k + \sqrt{1-\xi}\,x_{k+1})\}\,. \tag{10.10}$$

For the sake of definiteness I shall set $\tau = \frac{1}{2}$ so that

$$\xi = 1 - \xi = \tfrac{1}{2}\,. \tag{10.11}$$

As in the case of the one-dimensional model discussed in section 8 we can, for small γ, approximate the operator in (10.9) by

$$\exp\Big\{\gamma\sum_{1}^{M}\frac{\partial^2}{\partial x_k^2} - \gamma q(\boldsymbol{x})\Big\} \tag{10.12}$$

and moreover since to lowest order in γ, $\tanh\frac{1}{2}\gamma$ can be replaced by $\frac{1}{2}\gamma$ we can take $q(\boldsymbol{x})$ in (10.12) to be given by the formula

$$q(\boldsymbol{x}) = \tfrac{1}{4}\sum_{1}^{M}x_k^2 - \frac{1}{\gamma}\sum_{k=1}^{M}\log\cosh\{\sqrt{\tfrac{1}{2}\nu\gamma}\,(x_k + x_{k+1})\}\,. \tag{10.13}$$

(recall that by (10.11) we have set $\xi = \frac{1}{2}$).

It is now clear that for small γ and to lowest order (in γ) the analogue of (8.4) is

$$\log\Lambda_1 = \tfrac{1}{2}M\gamma - \gamma E_0\,, \tag{10.14}$$

where E_0 is the lowest eigenvalue of the Schrödinger equation

$$\sum_{1}^{M}\frac{\partial^2\phi}{\partial x_k^2} - \Big[\tfrac{1}{4}\sum_{1}^{M}x_k^2 - \frac{1}{\gamma}\sum_{k=1}^{M}\log\cosh\Big\{\sqrt{\tfrac{1}{2}\nu\gamma}\,(x_k + x_{k+1})\Big\}\Big]\phi = -E\phi\,. \tag{10.15}$$

The potential (10.13) undergoes a dramatic change at

$$\nu = \tfrac{1}{4}\,, \tag{10.16}$$

since for $\nu < \frac{1}{4}$ it has only one minimum at $(0, 0, \ldots, 0)$ while at $\nu > \frac{1}{4}$ it has two (equal) absolute minima at

$$\left(\frac{\eta_0\sqrt{2}}{\sqrt{\gamma}}, \frac{\eta_0\sqrt{2}}{\sqrt{\gamma}}, \ldots, \frac{\eta_0\sqrt{2}}{\sqrt{\gamma}}\right) . \tag{10.17}$$

and at

$$\left(-\frac{\eta_0\sqrt{2}}{\sqrt{\gamma}}, -\frac{\eta_0\sqrt{2}}{\sqrt{\gamma}}, \ldots, -\frac{\eta_0\sqrt{2}}{\sqrt{\gamma}}\right) . \tag{10.18}$$

Here η_0 is the (unique) positive root of the equation

$$\eta = 2\sqrt{\nu}\,\tanh(2\sqrt{\nu}\,\eta) . \tag{10.19}$$

So far the analogy with the discussion of section 8 is almost complete and indeed as long as M remains *finite* one can expect only *quantitative* differences between the model under discussion and the one-dimensional model of section 8 (which can be viewed as corresponding to the case $M = 1$).

It is only in the limit $M \to \infty$, which completes the taking of the thermodynamic limit, that one should see the deeper effects of dimensionality.

11. ASYMPTOTIC DEGENERACY

First of all, if one calculates the analogue of the "overlap integral" (8.13) one obtains instead of (8.14) the answer

$$\exp(-cM/\gamma) . \tag{11.1}$$

Thus one should expect the lowest eigenvalue of (10.15) (and therefore the highest eigenvalue of (10.4)) to become asymptotically degenerate in very much the same way as the highest eigenvalue of the transfer matrix becomes asymptotically degenerate (see (5.2)).

Unfortunately no rigorous proof of asymptotic degeneracy even in the weakest form, i.e.,

$$\lim_{M\to\infty} (E_1 - E_0) = 0 \tag{11.2}$$

has been found so far.

Some doubt on the correctness of the stronger statement i.e.

$$E_1 - E_0 \approx \exp(-cM/\gamma) , \qquad M\to\infty \tag{11.3}$$

is cast by the observation that for large M and *fixed* (though small) γ a particle of energy E_0 placed in one of the wells of the potential

$$\tfrac{1}{4}\sum_{1}^{M} x_k^2 - \frac{1}{\gamma}\sum_{1}^{M} \log\cosh\{\sqrt{\tfrac{1}{2}\nu\gamma}\,(x_k + x_{k+1})\} \tag{11.4}$$

can get to the other well by a *classically allowed motion*. Therefore there is no necessity for barrier penetration and the exponential nature of $E_1 - E_0$ becomes suspect since it is usually associated with this phenomenon.

On the other hand Griffiths [6] has recently shown that if a lattice model of ± spins exhibits long-range order then a model obtained from it by an addition of *ferromagnetic* interactions will also exhibit long-range order. Since our model of section 10 can be obtained from an unisotropic Ising model with nearest neighbor interactions by an addition of ferromagnetic interactions it follows that for sufficiently low temperatures our model will exhibit long-range order. This in turn implies that for *every* $\gamma > 0$

$$\lim_{M\to\infty} \frac{\Lambda_2}{\Lambda_1} = 1 \; . \tag{11.5}$$

and it is very likely (though again I have no rigorous proof) that this should imply (11.2), i.e.,

$$\lim_{M\to\infty} (E_1 - E_0) = 0 \; .$$

The fact that rigorous proofs are so difficult to come by is an indication that we are missing an essential link between our intuition and our formalism.

12. CORRELATION AT LARGE DISTANCES

The next instance where the effect of dimensionality (i.e. of the limit $M \to \infty$ since until this limit is taken the model is essentially one-dimensional) is in the study of the so-called critical region. Although I discussed this matter in some detail at the Summer School I shall not reproduce it here since it could only be a repetition of the latter part of §7 of my Brandeis Lectures.

Instead I shall discuss the behavior of the correlation function $\rho(r)$ (see (10.7)) for large r and for temperatures above the critical and compare the result with a corresponding result for the ordinary Ising model recently obtained by T. T. Wu [7].

For $\nu < \frac{1}{4}$ the potential $q(x)$ can be replaced (to lowest order in γ) by the quadratic potential

$$\tfrac{1}{4}\sum_{1}^{M} x_k^2 - \tfrac{1}{4}\nu \sum_{1}^{M} (x_k + x_{k+1})^2 \,. \tag{12.1}$$

The matrix of this quadratic form is cyclic and it can be easily diagonalized. Assuming, for the sake of simplicity, that M is odd the diagonalization is accomplished by the substitution

$$x = \sum_{r=1}^{M} a_{lr} y_{r-1} \,, \tag{12.2}$$

where

$$a_{lr} = \begin{cases} 1/\sqrt{M} \quad \text{if } r = 1 \,, \\ \sqrt{2/M}\, \sin \dfrac{\pi r(l-1)}{M} \quad \text{if } r \text{ is even} \,, \\ \sqrt{2/M}\, \cos \dfrac{\pi (r-1)(l-1)}{M} \quad \text{if } r \text{ is odd} \,. \end{cases} \tag{12.3}$$

The final result is

$$\sum_{1}^{M} x_k^2 - \nu \sum_{1}^{M} (x_k + x_{k+1})^2 = \sum_{s=0}^{M-1} \omega_{\bar{s}}\, y_s^2 \,, \tag{12.4}$$

where *

$$\bar{s} = \left[\frac{s+1}{2}\right] \tag{12.5}$$

and

$$\omega_l = 1 - 4\nu + 4\nu \sin \frac{2\pi l}{M} \,. \tag{12.6}$$

* $[x]$ denotes as usual the greatest integer not exceeding x, e.g., $[3] = 3$, $[4.2] = 4$, $[\sqrt{2}] = 1$ etc.

To the approximation considered here the lowest eigenvalue is given by the formula

$$E_o \approx \tfrac{1}{2} \sum_{s=0}^{M-1} \sqrt{\omega_{\bar{s}}} \approx \frac{M}{2\pi} \int_0^{\pi} \sqrt{1-4\nu+4\nu\sin^2\theta}\,d\theta\ , \tag{12.7}$$

while the higher eigenvalues are of the form

$$E_o \sum_{0}^{M-1} n_s \sqrt{\omega_{\bar{s}}}\ , \tag{12.8}$$

where the n_s are non-negative integers with at least one of them positive.

The next to the lowest eigenvalue is

$$E_1 = E_o + \sqrt{\omega_o} = E_o + \sqrt{1-4\nu}\ , \tag{12.9}$$

while the higher ones bunch up above E_1 spaced at intervals of order $1/M$.

Since

$$\log \Lambda_j \approx \tfrac{1}{2}M\gamma - \gamma E_{j-1}\ , \tag{12.10}$$

(an obvious extension of (10.14)) we see that $\log \Lambda_2$ is $\gamma\sqrt{1-4\nu}$ below $\log \Lambda_1$ while for $j > 2$ the $\log \Lambda_j$ bunch up (below $\log \Lambda_2$) with the spacing of order γ/M.

Qualitatively this description is the same as that of the spectrum of the transfer matrix *above* the transition temperature. The main quantitative difference is that in our model the gap between $\log \Lambda_1$ and $\log \Lambda_2$ as well as the spacings between the higher indices $\log \Lambda_j$ are by a factor γ smaller than the corresponding quantities in the case of the standard Ising model.

Let us now turn to (10.7) which I copy (with $\xi = \tfrac{1}{2}$ though) here for easier reference

$$\rho(r) = \lim_{M\to\infty} \sum_j \left(\frac{\Lambda_j}{\Lambda_1}\right)^r \left(\int \phi_1(\boldsymbol{x}) \tanh\{\sqrt{\tfrac{1}{2}\nu\gamma}\,(x_l + x_{l+1})\}\,\phi_j(\boldsymbol{x})\,d\boldsymbol{x}\right)^2 . \tag{12.11}$$

To lowest order in γ we have

$$\tanh\{\sqrt{\tfrac{1}{2}\nu\gamma}\,(x_l + x_{l+1})\} = \sqrt{\tfrac{1}{2}\nu\gamma}\,(x_l + x_{l+1})$$

and (by (12.10))

$$\frac{\Lambda_j}{\Lambda_1} = \exp\{-\gamma(E_{j-1} - E_o)\} .$$

Changing to the y's (see (12.2)) the square of the matrix element in (12.11) becomes

$$\tfrac{1}{2}\nu\gamma\left(\int \phi_1(y)\left(\sum_1^M b_{lp}y_{p-1}\right)\phi_j(y)\,dy\right)^2 ,$$

where

$$b_{lp} = a_{lp} + a_{l+1,p} .$$

It should be clear that (to lowest order in γ)

$$\phi_1(y) = \prod \frac{\sqrt[4]{\omega_{\bar{s}}}}{\sqrt[4]{2\pi}} \exp\left\{-\tfrac{1}{4}\sum_0^{M-1} \omega_{\bar{s}}\, y^2\right\}$$

and that the only ϕ_j which give non-vanishing contributions are

$$\phi_j(y) = \prod \frac{\sqrt[4]{\omega_{\bar{s}}}}{\sqrt[4]{2\pi}} \sqrt[4]{\omega_{\bar{j}}}\; y_j \exp\left\{-\tfrac{1}{4}\sum_0^{M-1} \omega_{\bar{s}}\, y_s^2\right\} .$$

It follows easily that the square of the matrix element (to lowest order in γ) is

$$\tfrac{1}{2}\nu\gamma \frac{1}{\sqrt{\omega_{\bar{j}}}}\, b_{l,j+1}^2$$

and therefore (again to lowest order in γ)

$$\rho(r) = \tfrac{1}{2}\nu\gamma \lim_{M\to\infty} \sum_0^{M-1} \frac{\exp(-\gamma\sqrt{\omega_{\bar{j}}}\,r)}{\sqrt{\omega_{\bar{j}}}}\, b_{l,j+1}^2$$

$$= \frac{2\nu\gamma}{\pi}\int_o^{\frac{1}{2}\pi} \frac{(1+\cos 2\theta)}{\sqrt{1-4\nu+4\nu\sin^2\theta}} \exp(-\gamma r\sqrt{1-4\nu+4\nu\sin^2\theta})\,d\theta . \qquad (12.12)$$

For large r (in fact $r \gg 1/\gamma$) we obtain, by a standard application of Laplace's method that

$$\rho(r) \approx \frac{2\nu\gamma}{\pi\sqrt[4]{1-4\nu}} \frac{\exp(-\gamma r\sqrt{1-4\nu})}{\sqrt{r}}, \quad r \gg 1/\gamma . \tag{12.13}$$

This result is of the *same form* as that derived by T. T. Wu [7] for the ordinary Ising model.

In fact, Wu proves that for large r

$$\rho(r) \approx A \frac{\exp(-\kappa r)}{\sqrt{r}} \tag{12.14}$$

and he determines explicitly A and κ.

It is the $\sqrt{r}$ in the denominator that is tied essentially to the models being two-dimensional. It is also directly traceable to the following property of the spectrum:

Let $n(\epsilon;M)$ denote the number of $\log \Lambda_j$ between $\log \Lambda_2 - \epsilon$ and $\log \Lambda_2$ then (for small ϵ and large M)

$$n(\epsilon;M) \approx M\sqrt{\epsilon} . \tag{12.15}$$

It should be also mentioned that for temperatures below the critical Wu proved that for large

$$\rho(r) - \rho(\infty) \approx A' \frac{\exp(-\kappa' r)}{r^2} . \tag{12.16}$$

The appearance of r^2 in the denominator is quite remarkable and I don't see how a similar result could hold for our model. Perhaps a more penetrating analysis will attribute the r^2 to some accidental features of the ordinary Ising models. This belief is based on the fact that, formally at least, the structure of the spectrum for our model in the case there are two wells is again quite similar to the corresponding structure of the spectrum of the transfer matrix. Some matrix elements may however accidentally vanish and (12.16) may have to be compared with γ^2 or even higher terms in the expansion of $\rho(r)$ for our model.

Finally it should be said that even the analysis of this section, compelling as it is, and leading to a result whose correctness can hardly be questioned is not easy to render rigorous. This is because we are compelled to use perturbation techniques (in γ) while the spectrum becomes continuous (since $M \to \infty$). Such situations are notoriously delicate and call for further investigations.

13. FINAL REMARKS

Let me conclude by posing a problem whose solution would go a long way in supporting the claim, implicit in these lectures, that models based (at least to some extent), on weak long range interactions can be used to "diagnose" many of the general features connected with phenomena of phase transitions.

The problem concerns the "continuous spin" model with interaction energy

$$E = -\tfrac{1}{2} \sum_{(k,l)} v(k,l;k',l') \cos(\theta_{k,l} - \theta_{k',l'}) , \tag{13.1}$$

where v is given by (10.2) with $\tau = \frac{1}{2}$.

Mermin [1] proves that at *no temperature* will magnetization at external field zero be different from zero.

This suggests (and almost implies) that at no temperature will there be a residual correlation at infinity and one might therefore expect that as long as γ is different from zero the maximum eigenvalue of the analogue of (10.4) remains *simple* as $M \to \infty$.

Proceeding formally we replace for small γ the integral operator by a differential one and the problem can be stated as follows:

Consider the Schrödinger equation

$$\sum_1^M \left(\frac{\partial^2\phi}{\partial x_k^2} + \frac{\partial^2\phi}{\partial y_k^2}\right) - \left[\tfrac{1}{4} \sum_1^M \left(x_k^2 + y_k^2\right)\right.$$
$$\left. - \sum_1^M \log I_0\{\sqrt{\tfrac{1}{2}\nu}\,\sqrt{(x_{k+1} - x_k)^2 + (y_{k+1} - y_k)^2}\}\right]\phi = -E\phi , \tag{13.2}$$

which is the analogue of equation (10.15).

Show, at least formally, that $E_1 - E_0$ does *not* approach zero as $M \to \infty$.

REFERENCES

[1] N.D. Mermin, J. Math. Phys. 8 (1967) 1061.
[2] Lebowitz and Penrose, J. Math. Phys. 7 (1966) 98.
[3] M. Kac and E. Helfand, J. Math. Phys. 4 (1963) 1078.
[4] M. Kac, G.E. Uhlenbeck and P.C. Hemmer, J. Math. Phys. 4 (1963) 216.
[5] M. Kac, *Statistical Mechanics of some one-dimensional systems*, Studies in Math. Anal. and Related Topics, Stanford University Press (1962) 165-169.
[6] R.B. Griffiths, J. Math. Phys. 8 (1967) 478, 484.
[7] T.T. Wu, Phys. Rev. 149 (1966) 380.

A SURVEY OF THE THEORIES OF FREEZING

B. JANCOVICI
Faculté des Sciences, Orsay, Essonne, France

1. INTRODUCTION

There does not exist at present any satisfactory quantitative theory of freezing, working from first principles, and furthermore the subject seems to have given rise to comparatively little research during the recent years.

Qualitatively, we understand what happens. Thermodynamics tell us that, for given temperature and pressure, freezing will occur if the liquid and the solid states have the same thermodynamic potential $U - TS + pV$. Since however the variation in volume between the two states is rather small, the free energies $U - TS$ are essentially equal. On freezing, the entropy S decreases, because of the transition to a more ordered state, and this is compensated for by a decrease in the internal energy U as a lattice is formed.

It must be noted however that in the special (academic but very interesting) case of a system of hard spheres, for which freezing is believed to occur, since there is no potential energy, the change in pV is indeed relevant, and freezing occurs as a new balance between $-TS$ and pV rather than between $-TS$ and U; a similar situation might prevail for real substances when under a very high pressure.

In the following sections, we shall review a few salient topics.

2. COMPUTER "EXPERIMENTS"

The results obtained with computers are interesting by themselves, and become even more helpful due to the inexistence of good theories. The computers deal with systems of a few hyndreds or thousands of particles. Two approaches may be used. In the molecular dynamics method, the equations of motion are actually integrated step by step. In the Monte Carlo method, configurations are generated at random with the probability they should have in a canonical ensemble. These studies do not replace a theory in the usual sense, but offer a nice insight in the behaviour of the systems, and rather replace or complete experiments.

2.1. *Hard spheres and discs* [1]

It is now rather generally believed that hard spheres, or hard discs on a plane, undergo a phase transition, called the Alder transition, when the volume is varied; for both spheres and discs, the transition occurs at a density which corresponds to an inter-neighbour average distance of about 1.16 times its value at close packing. A plot of the pressure versus the volume shows two branches, but in the transition region which lies between them, the fluctuations are too large to allow a plateau to be seen; the transition nevertheless seems to be a first-order one. This lack of clarity in the transition region is related to the finite size of the system; also, no phase separation in an external field has been achieved up to now.

For hard spheres, the transition is likely to be a fluid-solid one; the solid phase would be the high density one. For hard discs, it is possible to print or to display on a screen the positions of the particles at a given time (or, better, their average positions during, say, 100 moves [2]); the high-density phase looks as made of crystallites. There are however theoretical reasons which forbid the occurrence of a solid in two dimensions and the high-density phase for hard discs can at best be a pseudo-solid; we shall come back to that point in the last section.

2.2. *Argon*

Recent molecular dynamics studies [3] of argon reproduce very well, among other facts, its freezing. One starts with a non equilibrium configuration in which the particles sit on a perfect lattice and have random velocities with a Maxwell distribution at temperature T_0. As equilibrium is reached, part of the kinetic energy goes into raising the potential energy of the molecules which are no longer right on their sites, and therefore the equilibrium temperature T is lower than T_0. One tries increasing values of T_0, until the solid melts. The corresponding temperature T is in excellent agreement with experiment.

It is amusing that the computer can also generate a supercooled liquid. In a given liquid space configuration, one replaces the velocity distribution by a colder one, and one waits for equilibrium to be reestablished.

3. INSTABILITIES

Several approximate theories of freezing and melting are based on the search for an instability of one of the phases. One must remember however that an instability in one phase can at best signal the end of possible metastable states, and that the actual transition is likely to occur before the instability is reached.

3.1. *Instability of the solid*

The old theory of Lindemann simply states that a solid melts when the thermal vibration amplitude of an atom reaches a given fraction of the lattice parameter. The success of such a simple model is rather fair: the melting temperature is predicted by ±20%. This agreement is actually even a bit surprising, for such a simple theory does not make use at all of the properties of the liquid.

3.2. *Instability of the liquid*

There are various approximate integral equations for the pair correlation function in a liquid. In some cases, the solution becomes of a long range oscillatory character, and this may be considered as a freezing signal. For hard spheres, a long range appears in the solution of the Born-Green equation (at a specific volume of 1.48 times its value v_0 at close packing) or of the Kirkwood equation (at a specific volume of 1.24 v_0). Unfortunately, both these equations, which rely on the superposition approximation for the three-body density, are rather bad, as it can be seen by comparison with the computer results. On the contrary, another approximate integral equation, the Percus-Yevick one, although it is much better in describing the hard sphere liquid, predicts no instability!

Taking however for granted that a system of hard spheres shows a phase transition, or at least a rather abrupt ordering, one may think of an ordinary liquid as a fluid of hard spheres perturbed by the attractive part of the intermolecular potential. Using this philosophy, Brout [4] has derived an approximate relation between the density fluctuations of a plain liquid and the density fluctuations of the corresponding hard-sphere fluid. This relation is such that the attractive part of the potential has the effect of drastically amplifying the density fluctuations of the hard-sphere fluid, which makes the real liquid instable against freezing. This theory however does not seem to lead to a detailed agreement with experiment.

4. COEXISTENCE THEORIES

More elaborate theories must take into account the properties of both phases, since freezing is a first-order transition. One should obtain the free energies as functions of the volume for the liquid *and* the solid, and make a double-tangent construction. One can use different types of approximations for the description of each phase; it may however be considered as more elegant to have one set of equations, which would admit solutions of both the liquid and the solid types.

4.1. *A program* [5]

For building such a set of equations, one can first use an exact relation [6] between the one-body density $\rho(\boldsymbol{r})$ and the pair correlation function $g(\boldsymbol{r}, \boldsymbol{r}';\lambda)$ with a coupling parameter λ:

$$\rho(\boldsymbol{r}) = z \exp\{-\beta \int \mathrm{d}\boldsymbol{r}' \int_0^1 \mathrm{d}\lambda\, g(\boldsymbol{r}, \boldsymbol{r}';\lambda)\, u(|\boldsymbol{r} - \boldsymbol{r}'|)\, \rho(\boldsymbol{r}')\, \mathrm{d}\boldsymbol{r}'\} ,$$

where z is the activity, $\beta = 1/kT$, u the interparticle potential. If there were no correlations ($g \equiv 1$), this relation would express the molecular field theory. One expects possible solutions of the liquid type, with a uniform density $\rho(\boldsymbol{r})$, and also of the solid type, with a periodic density $\rho(\boldsymbol{r})$. For the actual determination of g, one needs another relation, which can be some approximate integral equation relating ρ and g; for instance, the Percus-Yevick equation may be chosen (it can be readily generalized to the case of uncomplete coupling λ and possibly non uniform density $\rho(\boldsymbol{r})$).

This set of equations has not been solved in realistic cases; one should not hope too much from it, since the Percus-Yevick equation is known to be only fair for real fluids, and has an unknown validity for solids.

4.2. *Actual calculations*

In the calculations which have been actually carried out, the liquid and the solid have been described with different approximations.

For instance, Tsuzuki [7] estimates the free energy in each phase, using a free volume approximation for the entropy and an energy derived from an extremely rough approximation for the pair correlation function. He gets unexpectedly fair results (for argon, he finds a melting temperature of 0.82, in reduced units, while the experimental value is 0.672).

Emtage [8] considers the liquid as made of mismatched bits of lattice; he finds the rather good melting temperature 0.736 for argon.

The Alder transition for hard spheres can be fairly well reproduced [9] by using a free volume approximation for the free energy of the solid, and computing the free energy of the liquid by integration of the Percus-Yevick equation of state starting from the low-density limit. The plateau is found for a specific volume to close packing specific volume ratio v/v_0 lying between 1.30 and

1.48, while the computer "experiments" locate the plateau between 1.35 and 1.50.

5. MODELS

Some simple lattice models, although not exactly soluble, can be very accurately studied. The behaviour of hard squares or cubes [10] on a lattice can be obtained by a clever extrapolation of the high and low density series. In the case of hard squares, it is also possible to consider systems of infinite length but finite width, and use the transfer matrix method [11]; one considers increasing widths and extrapolates the results. The model exhibits freezing in that sense that, at some transition density, one of the two interwoven sublattices of which the lattice is made becomes preferentially occupied. The transition is of the continuous type: the pressure versus density curve has only an inflexion point at the transition. The model may be used as a check on approximation methods: a calculation based on the Percus-Yevick approach described in section 4 predicts a transition density too low by 20 (cubes) or 30 (squares) %; this is a further reason for having not attempted to use the method in realistic cases.

The model of a system of hard cubes on a lattice, in any number of dimensions, can be modified into an exactly soluble "spherical" model [12]: the constraints of mutual exclusion on the same site or on adjacent sites are replaced by weaker ones (which look like the equation of an hypersphere, and are responsible for the adjective spherical). The model exhibits a second-order transition.

6. HELIUM AT ZERO TEMPERATURE

Helium at absolute zero is liquid at zero pressure but freezes when the pressure is increased enough. This transition has been studied [13] by the variational method of quantum mechanics. The trial wave functions which were used are a Jastrow wave function

$$\prod_{i>j} f(r_{ij})$$

for liquid ^{4}He, a Jastrow wave function times a Slater determinant for liquid ^{3}He, a Jastrow wave function times a product of Gaussians centered around lattice sites for both solid ^{4}He and ^{3}He. Although the problem belongs to pure quantum mechanics, there is a *formal* analogy between a Jastrow wave function and the statical mechanical distribution

$$\prod_{i>j} e^{-\beta u(r_{ij})}$$

in a classical fluid: therefore, the problem can be put on a computer with the use of the same methods as for the computer "experiments" on classical fluids. It is found that above a certain pressure the best wave function becomes of the solid type. The computed density of the solid phase at the transition is in agreement within 3% with the experimental value, for both ^{4}He and ^{3}He. The volume change, and the transition pressure in the case of ^{4}He, however, do not come out as well. Still, considering that a variational calculation is well adapted only for the computation of the energy, and also that there are uncertainties on the intermolecular potential, the results must be considered as good as it can be expected with this method.

7. TWO-DIMENSIONAL "SOLIDS"

The apparent occurrence of crystallites in a system of hard discs seems to violate a well-known argument according to which no solid can exist in less than 3 dimensions: the mean square displacement of a particle in a solid is given by a sum of phonon contributions proportional to $\int d\boldsymbol{k}/k^2$, and this sum diverges at low wave numbers in one and two dimensions. There can however exist in two dimensions pseudo-solids, an example of which is provided by the usual harmonic model (a lattice of particles interacting through harmonic forces). Although there is no long-range order, neighbouring particles vibrate in phase, and a finite sample would look much like a crystallite. Furthermore, even in an infinite system, the two-body correlation function is anisotropic, and this lower symmetry makes the pseudo-solid qualitatively different from a fluid. A transition between this pseudo-solid phase and a liquid phase is therefore perfectly conceivable (under conditions, of course, in which the harmonic approximation is no longer valid). The Alder transition for hard discs might very well be something of that kind.

One should be careful in two dimensions with a criterion which is commonly taken as a signal for a phase transition: the divergence of a properly defined susceptibility. The author has given an explicit example [14] (again that two-dimensional harmonic pseudo-solid) in which the divergence of the susceptibility is spurious and does not correspond to any phase transition.

REFERENCES

[1] W. W. Wood, F. R. Parker and J. D. Jacobson, Nuovo Cimento Suppl. 9 (1958) 133;
W. G. Hoover and B. J. Alder, J. Chem. Phys. 46 (1967) 686, and references quoted there.
[2] D. Schiff and M. Kalos, private communication.
[3] L. Verlet, Phys. Rev. 159 (1967) 98.
[4] R. Brout, *Phase transitions* (Benjamin, 1965).
[5] B. Jancovici, Physica 31 (1965) 1017.
[6] J. G. Kirkwood and E. Monroe, J. Chem. Phys. 9 (1941) 514.
[7] T. Tsuzuki, J. Phys. Soc. Japan 21 (1966) 25.
[8] P. R. Emtage, Physica 32 (1966) 1734.
[9] D. Levesque, private communication.
[10] D. S. Gaunt and M. E. Fisher, J. Chem. Phys. 43 (1965) 2840;
D. S. Gaunt, J. Chem. Phys. 46 (1967) 3237.
[11] F. H. Ree and D. A. Chestnut, J. Chem. Phys. 45 (1966) 3983; Phys. Rev. Letters 18 (1967) 5.
[12] J. L. Lebowitz and J. K. Percus, Phys. Rev. 144 (1966) 251;
B. Jancovici, Physica 32 (1966) 1234.
[13] W. L. McMillan, Phys. Rev. 138 (1965) A442;
D. Schiff and L. Verlet, Phys. Rev. 160 (1967) 208;
J. P. Hansen and D. Levesque, to be published in Phys. Rev.
[14] B. Jancovici, Phys. Rev. Letters 19 (1967) 20.

GENERAL REFERENCES

T. L. Hill, *Statistical Mechanics* (McGraw-Hill, 1956).
A. R. Ubbelohde, *Melting and Crystal Structure* (Clarendon, 1965).

ON THE GIBBS PHASE RULE

D. RUELLE
Institut des Hautes Etudes Scientifiques,
Bures-sur-Yvette (S. et O.), France

1. INTRODUCTION

Consider a one-component thermodynamic system. For given values of two independent thermodynamic variables describing the system an equilibrium state should (in general) be defined. More precisely the *Gibbs phase rule* tells us that in the plane of two intensive thermodynamic parameters almost all points correspond to a pure thermodynamic phase, that there are *curves* ‡ of coexistence of two phases (and isolated *points* of coexistence of three phases). For instance, using the variables p, T one obtains the familiar phase diagram (fig. 1) †.

The aim of the present lectures will be to try to understand the Gibbs phase rule from the point of view of statistical mechanics. To prove a strong version of the Gibbs phase rule from statistical mechanics is tantamount of making a theory of phase transitions and appears to be a rather formidable problem. We shall here only

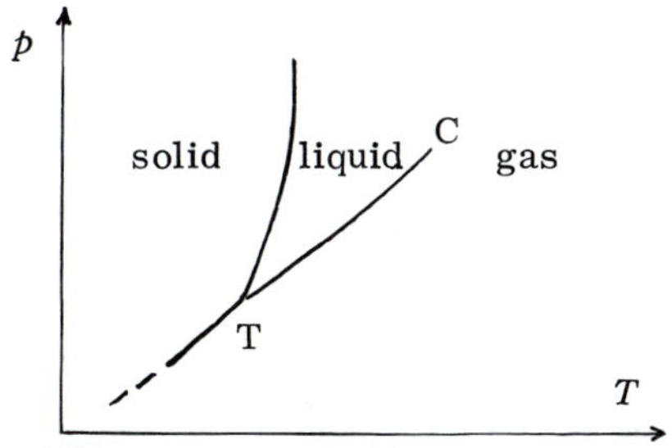

Fig. 1.

‡ To be precise one should assert certain smoothness (differentiability) properties, for these curves, we have indeed to reject things like the Peano curve.

† In the case of a system with several components, the Gibbs phase rule has been refined by L.S. Palatnik and A.I. Landau, a result useful in the study of alloys, see [1].

prove a weak phase rule, saying essentially that the set of points in the (p, T) diagram which do not correspond to a pure thermodynamic phase has (Lesbesgue-) measure zero. This statement has to be qualified by precising the systems and interactions for which it holds: we shall restrict our attention to lattice systems, and to a "large" class of interactions, to be described later. We shall also precise what we understand by a "pure thermodynamic phase".

Our approach to the problem of the Gibbs phase rule will, among other tools, use the ideas of the C^*-algebra approach to statistical mechanics and the information - theoretical notion of entropy. This will give us the opportunity to review a number of questions which are currently under intensive investigation.

2. DESCRIPTION OF LATTICE SYSTEMS

We denote by Z^ν the space of all ν-tuples $x = (x^1, \ldots, x^\nu)$ of integers $\gtrless 0$. We shall consider the points of Z^ν to be the sites of a ν-dimensional *lattice*. Every $a \in Z^\nu$ defines a *translation* of the lattice

$$x \to x + a \; . \tag{2.1}$$

We assume that every site of the lattice Z^ν may be occupied by either zero or one particle (not more). We are thus dealing with an infinite system with an invariance group $G = Z^\nu$ (the translations of the lattice). A *configuration* of the system is described by giving the set X of occupied lattice sites. Alternatively one may give the occupation number 0 or 1 of each point of Z^ν, the space of all configurations of the system is then identified with the product

$$K = \{0,1\}^{Z^\nu} \tag{2.2}$$

of one copy of the set $\{0,1\}$ for each lattice site. An element X of K is thus a function on Z^ν with values 0 or 1, it is the characteristic function of the corresponding subset of Z^ν, also noted X.

If, under certain conditions to be precised, our system is in thermodynamic equilibrium, its *state* will be described by a probability distribution on the space K of all configurations. Mathematically, this probability distribution is a positive measure on K, normalized to have total mass 1. We shall study the measures on K in the next section. Notice now that an equilibrium state of our system will naturally be invariant under translations of the lattice. This invariance, which is not possessed by finite systems enclosed in a box, will play an important role in our later considerations.

It is interesting to realize that for $\nu = 1$, our system (i. e. a one-dimensional lattice) may be identified with the usual set-up of information theory. In this case, a configuration is interpreted as a *message* written with a binary alphabet (one symbol 0 or 1 for each point of Z^1). A *source* of information is described by the sort of messages it sends out, and more precisely again by a probability distribution on the space K of all messages. Usually the source is taken to be *stationary* i.e. invariance under the translations of Z^1 is assumed [2].

3. STATES OF LATTICE SYSTEMS

We have defined the set K of all configurations of our lattice system as the product (2.2) of sets $\{0,1\}$. We want now to introduce a *topology* ‡ on K, i.e., roughly speaking, a notion of convergence. There is a natural topology on $\{0,1\}$, the *discrete topology*, for which every subset is open. We put on K the *product topology* corresponding to the factorization (2.2), the associated notion of convergence is the following: elements X of K, considered as functions on Z^ν with values 0, 1 converge in the sense of the product topology if they converge pointwise †. By the Cantor diagonal process, it is easy to check that from every sequence (X_n) a convergent subsequence (X_{i_n}) can be extracted: this shows that K is *sequentially compact*. Actually K is also *compact* (i.e. the Heine-Borel-Lebesgue property is satisfied) by Tychonov's theorem. It is readily seen that the translations (2.1) of the lattice define a group of *homeomorphisms* of the compact K. (A homeomorphism is a continuous mapping with an inverse which is also continuous.)

We call $\mathcal{C}(K)$ the space of continuous functions on K; $\mathcal{C}(K)$ is a commutative algebra with respect to the usual operations on functions, it has an involution $A \to A^*$ (complex conjugation) and a norm:

$$\|A\| = \sup_{X \in K} |A(X)| . \tag{3.1}$$

There is a topology on $\mathcal{C}(K)$ associated with the norm and the cor-

‡ A detailed understanding of the first part of this section is not required for the rest of the lectures. On the other hand a mathematical physicist cannot survive today without some knowledge of general topology [3].

† It can be shown that K is homeomorphic to a Cantor set.

responding notion of convergence is the uniform convergence.

A (Radon) *measure* μ on a compact set K is by definition a continuous linear form $A \to \mu(A)$ on $\mathcal{C}(K)$. One may use the notation

$$\mu(A) = \int A(X)\, d\mu(X) , \tag{3.2}$$

which shows that a measure permits to integrate continuous functions. A measure permits also to measure sets (and this point of view if often used to *define* measures). A measure μ is *positive* if $A \geq 0$ implies $\mu(A) \geq 0$. Let 1 denote the constant function with value 1 on K (the unit element of the algebra $\mathcal{C}(K)$), we say that $\mu \geq 0$ is normalized if $\mu(1) = 1$. A positive normalized measure μ is a *probability measure*, it may be considered as defining an *expectation value* on the algebra $\mathcal{C}(K)$. According to section 2 we want to represent a physical state of a lattice system by a probability measure on K. It is interesting to have a physical interpretation for the corresponding expectation value on $\mathcal{C}(K)$. An element A of $\mathcal{C}(K)$ is a function of $X \in K$ where X is itself a function of Z^ν. We examine first the situation where A depends only on the values of X at a finite number of lattice sites $a_1, \ldots, a_n$:

$$A(X) = \varphi(X(a_1), \ldots, X(a_n)) . \tag{3.3}$$

Since the n arguments of φ take each only two values, φ is an array of 2^n complex numbers. The functions of the form (3.3) form a $*$-algebra on K, include the constants and separate the points of K, therefore by the theorem of Stone-Weierstrass they are dense in $\mathcal{C}(K)$, i.e. every element of $\mathcal{C}(K)$ can be approximated uniformly by elements of the form (3.3). On the other hand every self-adjoint element of the form (3.3) may be considered as an "observable" taking a finite number of values depending on the occupation status of the sites $a_1, \ldots, a_n$ of the lattice. For instance A may take the value one if $a_1, \ldots, a_n$ are all occupied, zero otherwise. In this last case the expectation value $\mu(A)$ will just be the probability that the sites $a_1, \ldots, a_n$ be simultaneously occupied in the state described by μ.

As shown above, a physical state of a lattice system may be described by an expectation value functional on the algebra $\mathcal{C}(K)$, which is itself generated by observables. This viewpoint may be generalized: one assumes that the observables of a system generate a C^*-algebra $\mathfrak{A}$ and a physical state is described by a normalized positive linear functional on $\mathfrak{A}$. (A C^*-algebra is an algebra over the complex numbers, with a norm and a $*$-operation, which can be realized as a norm-closed self-adjoint algebra of operators in a Hilbert space [4, 5].) The lectures by Hugenholtz develop this point of view, which is specially interesting for quantum systems where $\mathfrak{A}$ is not abelian and the states of the system are not naturally described by the measures on some compact space.

4. ERGODIC STATES AND PURE PHASES

Let $A \in \mathcal{C}(K)$ and $a \in Z^\nu$, then a translated function $\tau_a A \in \mathcal{C}(K)$ is defined by

$$\tau_a A(X) = A(X - a) \, , \tag{4.1}$$

where $X - a$ is obtained by translating the configuration X by $-a$. Let J be the set of probability measures μ on K which are invariant under translations of the lattice Z^ν, i.e., such that

$$\mu(\tau_a A) = \mu(A) \text{ for all } a \in Z^\nu \, . \tag{4.2}$$

The elements of J are candidates for describing thermal equilibrium states of an infinite lattice system (in later sections, it will be indicated how these are determined). An extremal point of J, i.e., an invariant probability measure μ on K such that

$$\mu = \tfrac{1}{2}\mu_1 + \tfrac{1}{2}\mu_2 \tag{4.3}$$

(where μ_1, μ_2 are invariant probability measures) implies $\mu_1 = \mu_2$, is called an *ergodic measure* (with respect to the group of homeomorphisms of K corresponding to the translations of the lattice). It is often assumed in Information Theory that the measure describing the source is ergodic. Let us try to find the physical significance of ergodicity of an invariant measure μ representing a lattice system.

We define the average $\langle A \rangle_n \in \mathcal{C}(K)$ of $\mathfrak{A}$ over translations in a cube with side n:

$$\langle A \rangle_n = n^{-\nu} \sum_{x^1=1}^{n} \dots \sum_{x^\nu=1}^{n} \tau_x A \, . \tag{4.4}$$

We shall use the following form of the mean ergodic theorem: *let $\mu \in J$, μ is ergodic if and only if, for all $A \in \mathcal{C}(K)$,*

$$\lim_{n\to\infty} \mu([\langle A \rangle_n]^2) = [\mu(A)]^2 \, , \tag{4.5}$$

or (using $\mu(\langle A\rangle_n) = \mu(A)$):

$$\lim_{n\to\infty} \mu([\langle A\rangle_n - \mu(A)]^2) = 0 , \tag{4.6}$$

The observable A will in general fluctuate, i.e. the measure ν_A defined on the spectrum of A by $\nu_A(\varphi) = \mu(\varphi(A))$ is not carried by a point. Equation (4.6) means however that $\langle A\rangle_n$ fluctuates very little if n is large, i.e., if μ is ergodic, the space average of an observable over a large region is almost constant. We can interpret the fact that the space average of each observable tends to a constant to mean that the state we consider is a pure thermodynamic phase. A phase is indeed characterized by the fact that coarse-grained (i.e. space averaged) quantities do not fluctuate. On the contrary for a mixture (say of water and ice) some coarse-grained quantity (say the density) fluctuates. In fact if ρ is a mixture, the kth moment of the distribution of the coarse-grained average of the observable A is

$$\lim_{n\to\infty} \rho([\langle A\rangle_n]^k) . \tag{4.7}$$

Given a mixture μ it should, on physical grounds, have a unique decomposition into pure phases, i.e. a unique integral representation on the set $\mathcal{E}(J)$ of extremal points of J (i.e. ergodic measures). Actually a decomposition of μ can be obtained from the moments (4.7): if $\hat{A}$ is the function on J defined by

$$\hat{A}(\nu) = \nu(A) \qquad \text{for all } \nu \in J , \tag{4.8}$$

there exists a unique measure m_μ on J such that

$$m_\mu(\hat{A}_1 \cdot \ldots \cdot \hat{A}_k) = \lim_{n\to\infty} \mu(\langle A_1\rangle_n \cdot \ldots \cdot \langle A_k\rangle_n) . \tag{4.9}$$

(We consider here Radon measures on J where J is given the w*-topology (see section 8) which makes this set compact.) The measure m_μ is a probability measure, carried by $\mathcal{E}(J)$, with resultant μ, it is unique with these properties. We have thus obtained a unique integral representation of the invariant measures on K in terms of ergodic measures [6] and interpreted this decomposition as a decomposition of a mixture into pure thermodynamic phases. The corresponding problem for quantum systems has recently been intensively studied [7-10].

5. INTERACTIONS

To determine the equilibrium state of a lattice system we first have to precise the interaction. An interaction is a sequence of k-body potentials: $\varphi = (\varphi^{(k)})_{k \geq 1}$, which are assumed to be real, symmetric in their arguments and invariant under translations of the lattice, $\varphi^{(k)}$ depends thus only on the set $S = \{s_1, \ldots, s_k\}$ of its arguments and we shall write

$$\varphi(S) = \varphi^{(k)}(s_1, \ldots, s_k) . \tag{5.1}$$

Given a finite set $X = \{x_1, \ldots, x_n\}$ of occupied sites, the potential energy is

$$U_\varphi(X) = \sum_{S \subset X,\, S \neq \phi} \varphi(S) . \tag{5.2}$$

We assume that the interaction satisfies

$$\|\varphi\| = \sum_{S \ni 0} \frac{|\varphi(S)|}{N(S)} < +\infty , \tag{5.3}$$

where the sum extends over the sets S which contain the origin 0 of Z^ν and $N(S)$ is the number of points in S. The interactions φ form a real Banach space $\mathcal{B}$ with respect to the norm (5.3) (a real vector space with a norm is a Banach space if it is complete, i.e. if every Cauchy sequence is convergent) and the following "stability" condition is satisfied

$$|U_\varphi(X)| \leq N(X)\|\varphi\| , \tag{5.4}$$

as one checks immediately.

We shall define equilibrium by use of the grand canonical ensemble, the thermodynamic parameters will thus - for fixed interaction - be the chemical potential μ and the inverse temperature β. The chemical potential is readily absorbed in the interaction as "1-body potential" by putting $\varphi^{(1)} = -\mu$:

$$\varphi = (-\mu, \varphi'), \quad \text{where } \varphi' = (\varphi^{(2)}, \ldots, \varphi^{(k)}, \ldots) .$$

The inverse temperature on the other hand occurs just as a multiplicative factor for φ and may thus also be absorbed in the interaction. For each region Λ (= finite subset of Z^ν) we define a partition function by

$$Z_\Lambda(\varphi) = \sum_{X \subset \Lambda} e^{-U_\varphi(X)} , \tag{5.5}$$

and we put

$$P_\Lambda(\varphi) = N(\Lambda)^{-1} \log Z_\Lambda(\varphi) . \tag{5.6}$$

The grand partition function is thus

$$\Xi(\beta, \mu, \varphi') = Z_\Lambda\ (\beta(-\mu, \varphi')) , \tag{5.7}$$

and the pressure is $\beta^{-1}P$.

We shall in the next section show that $P_\Lambda(\varphi)$ tends to a limit $P(\varphi)$ when $\Lambda \to \infty$ and study the properties of this limit. Physically, the problem is to study the existence and properties of the pressure in the limit of an infinite volume (thermodynamic limit). Before we do this we note some preliminary results.

Proposition 5.1. *If* $\varphi \in \mathcal{B}$, *then*

$$0 \leqslant P_\Lambda(\varphi) \leqslant \log 2 + \|\varphi\| . \tag{5.8}$$

By (5.5) and (5.4) we have

$$Z_\Lambda(\varphi) = \sum_{X\subset\Lambda} e^{-U_\varphi(X)} \geqslant e^{-U_\varphi(\phi)} = 1 , \tag{5.9}$$

$$Z_\Lambda(\varphi) = \sum_{X\subset\Lambda} e^{-U_\varphi(X)} \leqslant \sum_{X\subset\Lambda} e^{N(X)\|\varphi\|} \leqslant 2^{N(\Lambda)}\, e^{N(\Lambda)\|\varphi\|} . \tag{5.10}$$

Therefore (5.6) yields (5.8).

Proposition 5.2. *If* $\varphi, \psi \in \mathcal{B}$, *then*

$$|P_\Lambda(\varphi) - P_\Lambda(\psi)| \leqslant \|\varphi - \psi\| . \tag{5.11}$$

By (5.4), we have

$$|U_\varphi(X) - U_\psi(X)| = |U_{\varphi-\psi}(X)| \leqslant N(X)\, \|\varphi - \psi\| , \tag{5.12}$$

and therefore if $X \subset \Lambda$

$$U_\psi(X) - N(\Lambda)\|\varphi - \psi\| \leqslant U_\varphi(X) \leqslant U_\psi(X) + N(\Lambda)\, \|\phi - \psi\| , \tag{5.13}$$

hence by the definition (5.5) of Z_Λ:

$$e^{-N(\Lambda)\|\varphi-\psi\|}\, Z_\Lambda(\psi) \leqslant Z_\Lambda(\psi) \leqslant e^{N(\Lambda)\|\varphi-\psi\|}\, Z_\Lambda(\psi) , \tag{5.14}$$

and (5.11) follows by the definition (5.6) of P_Λ.

Proposition 5.3. *The function* $\varphi \to P_\Lambda(\varphi)$ *is convex on* $\mathcal{B}$.
This is readily seen by computing a second derivative, from (5.6) and (5.5) we get indeed:

$$\frac{d^2}{d\lambda^2} P_\Lambda(\varphi + \lambda\psi)\Big|_{\lambda=0}$$

$$= N(\Lambda)^{-1} \Big[Z_\Lambda(\varphi)^{-1} \sum_{X\subset\Lambda} U_\psi(X)^2 \, e^{-U_\varphi(X)}$$

$$- \Big(Z_\Lambda(\varphi)^{-1} \sum_{X\subset\Lambda} U_\psi(X) \, e^{-U_\varphi(X)}\Big)^2\Big]$$

$$= N(\Lambda)^{-1} Z_\Lambda(\varphi)^{-2} \sum_{X\subset\Lambda} \sum_{Y\subset\Lambda} e^{-U_\varphi(X)-U_\varphi(Y)}$$

$$\times \tfrac{1}{2}[U_\psi(X)^2 + U_\psi(Y)^2 - 2U_\psi(X)\, U_\psi(Y)] \geqslant 0 . \qquad (5.15)$$

6. THERMODYNAMIC LIMIT FOR THE PRESSURE

We want to see that $P_\Lambda(\varphi)$ has a finite limit when the region Λ "tends to infinity". To give a precise meaning to this notion, we shall here restrict ourselves to regions Λ of the form

$$\Lambda(a) = \{x \in Z^\nu : 0 \leqslant x^i < a^i \qquad \text{for } i = 1, \ldots, \nu\} , \qquad (6.1)$$

where $a \in Z^\nu$ has components $a^i > 0$. We put

$$V(a) = N(\Lambda(a)) = \prod_{i=1}^{\nu} a^i$$

and we write $a \to \infty$ if $a^1, \ldots, a^\nu \to +\infty$ so that in particular $V(a) \to +\infty$.

Proposition 6.1. *Let* $a' = (a'^1, \ldots, a'^\nu)$, $a'' = (a''^1, \ldots, a''^\nu)$, $a = (a^1, \ldots, a^\nu)$, *where* $a^i = a'^i = a''^i$ *for* $i \neq k$ *and* $a^k = a'^k + {}''^k$, *then*

$$\left|V(a)\; P_{\Lambda(a)}(\varphi) - V(a')\; P_{\Lambda(a')}(\varphi) - V(a'')\; P_{\Lambda(a'')}(\varphi)\right| \leqslant \epsilon(a)\; V(a)\;, \tag{6.2}$$

where

$$\lim_{a\to\infty} \epsilon(a) = 0\;.$$

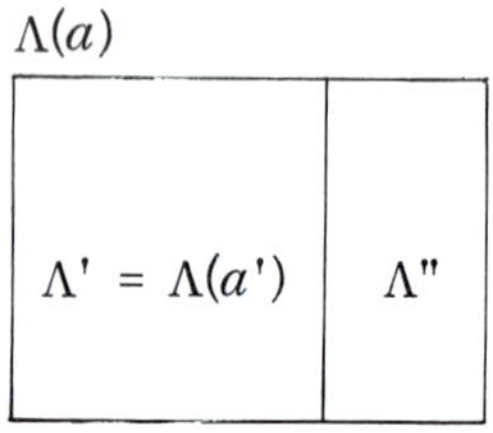

Fig. 2.

Let $\Lambda' = \Lambda(a')$ and Λ'' be the translate of $\Lambda(a'')$ by the vector $(0,\ldots,a'^k,\ldots,0)$, so that $\Lambda(a) = \Lambda' \cup \Lambda''$, $\Lambda' \cap \Lambda'' = \phi$ (see fig. 2). If $X' \subset \Lambda'$, $X'' \subset \Lambda''$ we have:

$$U_\varphi(X' \cup X'') = U_\varphi(X') + U_\varphi(X'') + \sum_{x\in X'\cup X''} \sum_{S\ni x}{}^{*} \frac{\varphi(S)}{N(S)}\;, \tag{6.3}$$

where $\sum^*$ extends over $S \ni x$, $S \subset X \cup X'$, $S \cap X' \neq \phi$, $S \cap X'' \neq \phi$. We put

$$\epsilon(a) = \sup_{k=1,\ldots,\nu}\; \sup_{a'^k=0,\ldots,a^k-1} V(a)^{-1} \sum_{x\in\Lambda(a)} \sum_{S\ni x}{}^{**} \frac{|\varphi(S)|}{N(S)} \tag{6.4}$$

where $\sum^{**}$ extends over $S \ni x$, $S \subset \Lambda(a)$, $S \cap \Lambda' \neq \phi$, $S \cap \Lambda'' \neq \phi$. Since the norm (5.3) of φ is finite, the sum $\sum^{**}$ is small when x is far from the "boundary" between Λ' and Λ'' (i.e. $|x^k - a^k| \to \infty$). If $a \to \infty$ this is the case for most x hence

$$\lim_{a\to\infty} \epsilon(a) = 0\;. \tag{6.5}$$

On the other hand (6.3), (6.4) give

$$\left|U_\varphi(X' \cup X'') - U_\varphi(X') - U_\varphi(X'')\right| < \epsilon(a)\; V(a)\;, \tag{6.6}$$

hence by (5.5)

$$e^{\epsilon(a)V(a)} \geq \frac{Z_{\Lambda(a)}(\varphi)}{Z_{\Lambda'}(\varphi)\, Z_{\Lambda''}(\varphi)} \geq e^{-\epsilon(a)V(a)} , \tag{6.7}$$

and by (5.6)

$$\epsilon(a)V(a) \geq V(a)\, P_{\Lambda(a)}(\varphi) - V(a')\, P_{\Lambda(a')}(\varphi) - V(a'')\, P_{\Lambda(a'')}(\varphi) \geq -\epsilon(a)V(a) , \tag{6.8}$$

which yields (6.2).

Proposition 6.2. *Let* $b = (b^1, \ldots, b^\nu)$, *where* $b^i = n^i a^i$, *the* n^i *and* a^i *being positive integers, then*

$$\left| P_{\Lambda(b)}(\varphi) - P_{\Lambda(a)}(\varphi) \right| \leq \epsilon_1(a) , \tag{6.9}$$

where

$$\lim_{a \to \infty} \epsilon_1(a) = 0 .$$

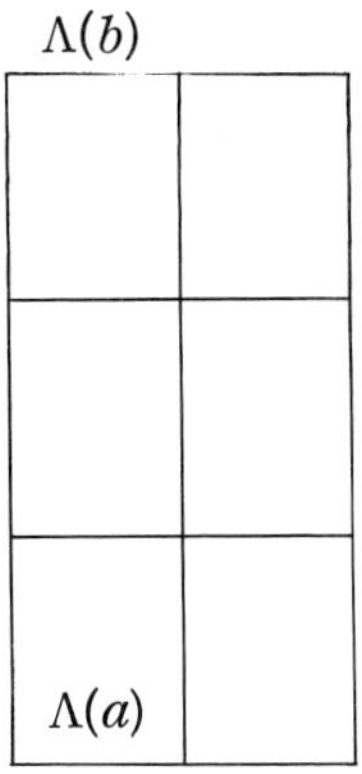

Fig. 3.

We only sketch the proof which is analogous to that of proposition 6.1. $\Lambda(b)$ is decomposed into a union of

$$\prod_{k=1}^{\nu} n^k$$

disjoint translates of $\Lambda(a)$ (see fig. 3). For each $X \subset \Lambda(b)$, $U_\varphi(X)$ is written as a sum of contributions $U_\varphi(X_i)$ of these translates plus a correction

$$\sum_{x \in X} \sum_{S \ni x}^{*} \frac{\varphi(S)}{N(S)} ,$$

where S must contain a point in a translate different from that containing x. If $a \to \infty$, most points x are far from the boundaries between different translates and therefore the correction is majorized by $V(b).\epsilon_1(a)$, where

$$\lim_{a \to \infty} \epsilon_1(a) = 0 .$$

The argument used in the proof of proposition 6.1 between (6.6) and (6.8) now yields

$$\epsilon_1(a)V(b) \geqslant V(b)\, P_{\Lambda(b)}(\varphi) - \sum_{\text{translates}} V(a)\, P_{\Lambda(a)}(\varphi) \geqslant -\epsilon_1(a)\, V(b) , \tag{6.10}$$

and since the number of translates times $V(a)$ is $V(b)$, (6.9) follows.

Theorem 6.3. *The limit*

$$P(\varphi) = \lim_{a \to \infty} P_{\Lambda(a)}(\varphi) \tag{6.11}$$

exists and satisfies the following properties

(i)
$$0 \leqslant P(\varphi) \leqslant \log 2 + \|\varphi\| , \tag{6.12}$$

(ii)
$$|P(\varphi) - P(\psi)| \leqslant \|\varphi - \psi\| , \tag{6.13}$$

in particular the functional $P(\ )$ *is continuous on the Banach space* $\mathcal{B}$.

(iii) *The functional* $P(\ )$ *is convex on* $\mathcal{B}$.

Given $\epsilon > 0$ we choose a_0 so that $\epsilon_1(a_0) < \epsilon/2$. For any $a = (a^1, \ldots, a^\nu)$, let $a^i = n^i a_0^i + r^i$ with integers $n^i > 0$, $r^i > 0$ such that $r^i < a_0^i$. Let also $b = (b^1, \ldots, b^\nu)$, with $b^i = n^i a_0^i$. From proposition 6.2 we have

$$|P_{\Lambda(b)}(\varphi) - P_{\Lambda(a_0)}(\varphi)| < \epsilon/2 . \tag{6.14}$$

On the other hand proposition 6.1 may be used to compare $P_{\Lambda(a)}(\varphi)$ and $P_{\Lambda(b)}$ in ν successive steps (see fig. 4). Using the bound

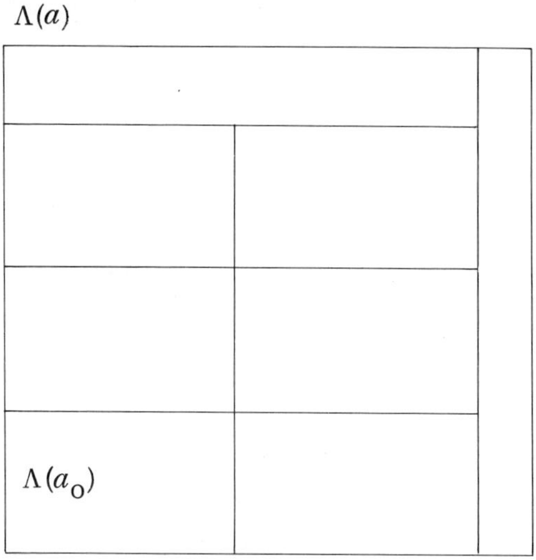

Fig. 4.

(5.8) one sees readily that if a is sufficiently large,

$$\left|P_{\Lambda(a)}(\varphi) - P_{\Lambda(b)}(\varphi)\right| < \epsilon/2 \ , \tag{6.15}$$

and (6.11) follows.

The properties (i), (ii) and (iii) follow from propositions 5.1, 5.2 and 5.3 respectively.

Remark. Instead of enclosing the system in a rectangular box $\Lambda(a)$ one often uses a "periodic box" or torus $\widetilde{\Lambda}(a)$, obtained by identifying opposite faces of $\Lambda(a)$. One has to be somewhat careful in the definition of the potential energy of a configuration because an occupied site in $\Lambda(a)$ interacts not only with the other occupied sites in $\Lambda(a)$ but also with their periodic images and with its own periodic images. Using a variation of proposition 6.1 one checks that the pressure $P_{\widetilde{\Lambda}(a)}(\varphi)$ defined with the periodic box $\widetilde{\Lambda}(a)$ is close to $P_{\Lambda(a)}(\varphi)$, when a is large, therefore (6.11) gives

$$P(\varphi) = \lim_{a \to \infty} P_{\widetilde{\Lambda}(a)}(\varphi) \ . \tag{6.16}$$

Naturally one is interested in proving the existence of the thermodynamic limit not only for lattice systems but also in the more difficult case of continuous systems. It is then again essential to make proper assumptions about the interaction (otherwise a large system may simply not behave thermodynamically). If we assume that only a pair potential Φ is present one has to assume that it vanishes at infinity in a suitable manner (in the lattice case the corresponding assumption was (5.3)). One has also to assume that the potential is not too attractive, to prevent collapse of the system, this is obtained by the following *stability condition:*

$$\sum_{1 \leq i < j \leq n} \Phi(x_j - x_i) \geq -nB\ . \tag{6.17}$$

It is thus assumed that there exists a constant $B \geq 0$ such that the potential energy of n particles is always larger than $-nB$ (in the lattice case (5.4) shows that this is automatically satisfied because there can be only one particle per lattice site).

For a proof of the existence of the thermodynamic limit for continuous systems see [11-13] and for an extension of these proofs to other ensembles and quantum systems see [14]. A particularly interesting and difficult problem is that of proving a "stability condition" for quantum systems with Coulomb interactions, for this see [15,16].

7. CORRELATION FUNCTIONS

We consider again a system enclosed in a region Λ (finite subset of Z^ν) and described by the grand canonical ensemble. As explained in section 5 we absorb the parameters μ and β in the interaction φ. The probability that the points $x_1, \ldots, x_m$ of Λ are simultaneously occupied is then given by

$$\rho_\varphi^\Lambda(X) = Z_\Lambda(\varphi)^{-1} \sum_{Y \subset \Lambda / X} e^{-U_\varphi(X \cup Y)}\ , \qquad X = \{x_1, \ldots, x_m\}\ , \tag{7.1}$$

where Λ/X is the complement of X in Λ (the difference between Λ and X). We shall describe ρ_φ^Λ, considered as a function of the subset X of Λ, as *the correlation function* of the system. (Usually one decomposes ρ_φ^Λ into several correlation functions, one for each value of m, this decomposition is especially natural for continuous systems but would be a nuisance for our present purposes.) In the case of dilute systems one can show that ρ_φ^Λ has a limit ρ_φ when $\Lambda \to \infty$ in a suitable manner (see [17] for continuous systems and [18] for lattice systems, for quantum systems see [19]), however it is expected that ρ_φ^Λ does not have a limit under conditions corresponding to the coexistence of several phases and we shall, in what follows, adopt an approach which proves the existence of a thermodynamic limit for ρ_φ^Λ in general, but not always.

First we extend the definition (7.1) to all finite $X \subset Z^\nu$ by putting

$\rho_\varphi^\Lambda(X) = 0$ if $X \not\subset \Lambda$. By averaging over translations we get

$$\bar{\rho}_\varphi^\Lambda(\{x_1,\ldots,x_m\}) = N(\Lambda)^{-1} \sum_{x\in Z^\nu} \rho_\varphi^\Lambda(\{x_1+x,\ldots,x_n+x\}) \ , \qquad (7.2)$$

so that for each $\psi \in \mathcal{B}$:

$$\sum_{X\ni 0} \bar{\rho}_\varphi^\Lambda(X) \frac{\psi(X)}{N(X)} = N(\Lambda)^{-1} \sum_{x\in Z^\nu} \sum_{X\ni x} \rho_\varphi^\Lambda(X) \frac{\psi(X)}{N(X)}$$

$$= N(\Lambda)^{-1} \sum_{X\neq\phi} \rho_\varphi^\Lambda(X)\psi(X) = Z_\Lambda(\varphi)^{-1} N(\Lambda)^{-1} \sum_{X\neq\phi} \sum_{Y\subset\Lambda, Y\supset X} e^{-U_\varphi(Y)}\psi(X)$$

$$= Z_\Lambda(\varphi)^{-1} N(\Lambda)^{-1} \sum_{Y\subset\Lambda} e^{-U_\varphi(Y)} \sum_{X\subset Y, X\neq\phi} \psi(X)$$

$$= Z_\Lambda(\varphi)^{-1} N(\Lambda)^{-1} \sum_{X\subset\Lambda} e^{-U_\varphi(X)} U_\psi(X) \ , \qquad (7.3)$$

where we have used (5.2). Clearly, $\bar{\rho}_\varphi^\Lambda$ is known if the left hand side of (7.3) is known for all $\psi \in \mathcal{B}$. The right hand side of (7.3) is, according to (5.5), (5.6), related to the functional derivative of $P_\Lambda(\varphi)$ with respect to φ. It is therefore natural to try to define the infinite volume correlation function as functional derivative of P. This point of view has in particular been used by M. E. Fisher [20].

Here we shall follow the recent work of G. Gallavotti and S. Miracle [21]. We define $D \subset \mathcal{B}$ to be the set of all interactions φ such that the graph of $P(\ )$, i.e. the set of points $(\varphi, P(\varphi))$ in $\mathcal{B}\times R$, has a unique tangent plane corresponding to φ. We assume thus that if $\varphi \in D$, there exists a unique α_φ in the dual $\mathcal{B}^*$ of $\mathcal{B}$ - the dual of a Banach space $\mathcal{B}$ is the space of continuous linear functionals on such that for all $\psi \in \mathcal{B}$

$$P(\varphi+\psi) \geq P(\varphi) - \alpha_\varphi(\psi) \ . \qquad (7.4)$$

Following Gallavotti-Miracle we prove now:

Theorem 7.1. *If* $\varphi \in D$ *then, with the notations of* (7.2), (7.4), *we have for all* $\psi \in \mathcal{B}$

$$\lim_{a\to\infty} \sum_{X\ni 0} \bar{\rho}_\varphi^{\Lambda(a)}(X) \frac{\psi(X)}{N(X)} = \alpha_\varphi(\psi) \ . \qquad (7.5)$$

Let us immediately notice that this implies the existence of the following limit:

$$\lim_{a\to\infty} \bar{\rho}_\varphi^{\Lambda(a)}(X) = \bar{\rho}_\varphi(X) , \tag{7.6}$$

defining the infinite volume correlation function $\bar{\rho}_\varphi$.

For finite Λ, the graph of $P_\Lambda(\ )$ has a unique tangent plane corresponding to φ for each $\varphi \in \mathcal{B}$:

$$P_\Lambda(\varphi+\psi) \geqslant P_\Lambda(\varphi) - \alpha_\varphi^\Lambda(\psi) , \tag{7.7}$$

where (see (7.3)):

$$\alpha_\varphi^\Lambda(\psi) = Z_\Lambda(\varphi)^{-1} N(\Lambda)^{-1} \sum_{X\subset\Lambda} e^{-U_\varphi(X)} U_\psi(X) = \sum_{X\ni 0} \bar{\rho}_\varphi^\Lambda(X) \frac{\psi(X)}{N(X)} . \tag{7.8}$$

Let now $\varphi \in D$, $\psi \in \mathcal{B}$ and $(\varphi+\psi, \xi)$ be a point strictly above the graph of $P(\ )$. For sufficiently large a, $(\varphi+\psi, \xi)$ will also be above the graph of $P_{\Lambda(a)}(\ )$ and therefore above the tangent plane to this graph:

$$\xi \geqslant P_\Lambda(\varphi+\psi) \geqslant P_\Lambda(\varphi) - \alpha_\varphi^\Lambda(\psi) . \tag{7.9}$$

Therefore the tangent plane to $P_{\Lambda(a)}(\ )$ tends to the tangent plane to $P(\ )$ because of the uniqueness of the latter:

$$\forall\, \psi \in \mathcal{B}: \lim_{a\to\infty} \alpha_\varphi^\Lambda(\psi) = \alpha_\varphi(\psi) , \tag{7.10}$$

which is just (7.5) in view of (7.8).

Theorem 7.1 shows that the infinite volume correlation function is well defined provided the interaction φ belongs to the set D. We would like to argue now that D is a "large" set, that it contains "almost all" interactions. Theorem 7.2 below is in this direction. There is at least one natural notion of "large" set in a Banach space, namely the notion of a Baire set. We say that B is a Baire set if it contains a countable intersection of dense open sets. The usefulness of this notion comes from the theorem (of Baire) stating that in a complete metric space a Baire set is dense. Since any finite or countable intersection of Baire sets is a Baire set and therefore dense, it is natural to consider a Baire set as "large", one should however remark that there exist Baire sets in the interval $[0,1]$ with Lebesgue-measure zero. One may also hope that D is a "large" set in the sense that its complement has measure zero with respect to a suitable measure on $\mathcal{B}$, this yields results like part (ii) of theorem 7.2, unfortunately this approach suffers from the defect that there is no natural notion of "measure zero" on Banach spaces with infinite dimension.

Let $\mathcal{B}'$ be the subspace of $\mathcal{B}$ formed by those $\varphi = (\varphi^{(k)})_{k\geq 1}$ such that $\varphi^{(1)} = 0$. We may identify $\mathcal{B}'$ with a space of sequences $\varphi = (\varphi^{(k)})_{k\geq 2}$.

Theorem 7.2. (i) *D is a Baire set,* (ii) *There exists a dense subset D' of $\mathcal{B}'$ such that for $\varphi' \in D'$ and for* (*Lebesgue*-) *almost every $(\beta, \mu) \in R_+ \times R$, the point $\beta(-\mu, \varphi')$ belongs to D.*

(i) follows by inspection of the proof of V. 9.8 in [22]. For the proof of (ii) we refer the reader to [21].

Remark 1. In the situation of a periodic box discussed in section 6, one can also (with some care) define a correlation function $X \to \rho_\varphi^{\tilde{\Lambda}(a)}(X)$ where $X \subset Z^\nu$ and prove by analogy with (7.5) that if $\varphi \in D$ and $\psi \in \mathcal{B}$

$$\lim_{a\to\infty} \sum_{X\ni 0} \rho_\varphi^{\tilde{\Lambda}(a)}(X) \frac{\psi(X)}{N(X)} = \alpha_\varphi(\psi) \,. \tag{7.11}$$

Here $\rho_\varphi^{\tilde{\Lambda}(a)}$ is automatically invariant under translations and does not have to be averaged.

Remark 2. The functional α_φ^Λ defined by (7.8) belongs to the dual $\mathcal{B}^*$ of $\mathcal{B}$, in fact $\|\alpha_\varphi^\Lambda\| \leq 1$. For general $\psi \in \mathcal{B}$ (not necessarily in D), one can still choose a sequence (a_n), $a_n \to \infty$ such that $\lim \alpha_\varphi^{\Lambda(a_n)}(\psi)$ exists for all ψ ($\alpha_\varphi^{\Lambda(a_n)}$ converges weakly). The argument in the proof of theorem 7.1 shows that this limit defines a tangent plane to the graph of $P(\ )$.

8. CONNEXION BETWEEN CORRELATION FUNCTION AND STATE OF A LATTICE SYSTEM

We revert now to the study of states of lattice systems started in sections 3 and 4. Let again $\mathcal{J}$ be the set of all probability measures on K which are invariant under translations of the lattice Z^ν, $\mathcal{J}$ is a convex set in the space $\mathcal{M}(K)$ of continuous linear functionals on $\mathcal{C}(K)$ ($\mathcal{M}(K)$ is the space of measures on K, the dual of $\mathcal{C}(K)$.) The notion of pointwise convergence of elements of $\mathcal{M}(K)$, considered as functions on $\mathcal{C}(K)$, corresponds to the w*-topology (or weak topology) on $\mathcal{M}(K)$. With respect to this topology, $\mathcal{J}$ is compact (this follows from the theorem of Alaoglu-Bourbaki [3].)

Let Λ be a finite subset of Z^ν, and Z^ν/Λ its complement, we may write

$$K = \{0,1\}^{Z^v} = \{0,1\}^{\Lambda} \times \{0,1\}^{Z^v/\Lambda} . \tag{8.1}$$

If μ is any probability measure on K, it may be projected on the first factor of (8.1) as follows: if X is a subset of Λ, identified to an element of $\{0,1\}^{\Lambda}$, we write

$$\mu_{\Lambda}(X) = \mu(X \times \{0,1\}^{Z^\nu/\Lambda}) . \tag{8.2}$$

We may also write (8.2) as $\mu_{\Lambda}(X) = \mu(A_{\Lambda}^{X})$ where A_{Λ}^{X} is the characteristic function of $X \times \{0,1\}^{Z^\nu/\Lambda}$ and $A_{\Lambda}^{X} \in \mathcal{C}(K)$. In particular for fixed X and Λ, $\mu_{\Lambda}(X)$ is a continuous function of μ for the w*-topology on $\mathcal{M}(K)$.

The number $\mu_{\Lambda}(X) \geq 0$ represents the probability that in the region Λ the points of X are occupied and the points of Λ/X empty, irrespective of what happens outside of Λ. Suppose that $\Lambda = \{a_1, \ldots, a_n\}$ and that $A \in \mathcal{C}(K)$ is given by (3.3), then

$$\mu(A) = \sum_{X \subset \Lambda} \mu_{\Lambda}(X)\, A(X) . \tag{8.3}$$

This shows that the $\mu_{\Lambda}(X)$ determine μ completely, because the A of the form (3.3) are dense in $\mathcal{C}(K)$. We list now the properties of the $\mu_{\Lambda}(X)$ when μ is invariant under lattice translations, i.e. $\mu \in J$.

i)
$$\sum_{X \subset \Lambda} \mu_{\Lambda}(X) = 1 \tag{8.4}$$

ii) *compatibility*: if

$$\Lambda' \supset \Lambda, \quad \mu_{\Lambda}(X) = \sum_{Y \subset \Lambda'/\Lambda} \mu_{\Lambda'}(X \cup Y) \tag{8.5}$$

iii) *invariance*: if

$$a \in Z^\nu : \mu_{\Lambda}(X) = \mu_{\Lambda+a}(X+a) \tag{8.6}$$

If μ represents an equilibrium state of a lattice system, the numbers $\mu_{\Lambda}(X)$ have a significance close to that of the quantities $Z_{\Lambda}(\varphi)^{-1}\, e^{-U_\varphi(X)}$, the latter describing a finite system and the former a finite subsystem of an infinite system. By analogy with (7.1), we can thus associate to each $\mu \in Z$ a correlation function ρ such that

$$\rho(X) = \sum_{Y \subset \Lambda/X} \mu_{\Lambda}(X \cup Y) \text{ if } X \subset \Lambda . \tag{8.7}$$

Taking $\Lambda = \phi$, this becomes

$$\rho(X) = \mu_X(X) \;, \tag{8.8}$$

and the independence on Λ of the right-hand side of (8.7) follows from (8.5). Clearly, ρ is invariant under translations. The relations (8.7) may be inverted, yielding

$$\mu_\Lambda(X) = \sum_{Y \subset \Lambda/X} (-1)^{N(Y)} \rho(X \cup Y) \quad \text{if} \quad X \subset \Lambda \;. \tag{8.9}$$

This shows that the mapping $\mu \to \rho$ is one-to-one.

We present now the relation between a state $\mu \in J$ and its correlation function ρ in a different form. Given $\psi \in \mathcal{B}$, we define $A_\psi \in \mathcal{C}(K)$ by

$$A_\psi(X) = \sum_{Y \ni 0,\, Y \subset X} \frac{\psi(Y)}{N(Y)} \;. \tag{8.10}$$

In particular $A_\psi(X) = 0$ if $0 \notin X$. From (5.3), we see that $\psi \to A_\psi$ is norm decreasing from $\mathcal{B}$ to $\mathcal{C}(K)$. From (8.10) it follows that $A_\psi(X)$ may be considered as the contribution to the energy of the configuration X of the site 0 (origin). In particular, using (4.1) and (5.2), we obtain if X is finite

$$\sum_{x \in Z^\nu} \tau_x A_\psi(X) = \sum_{x \in X} \tau_x A_\psi(Y)$$

$$= \sum_{x \in X} \sum_{Y \ni x,\, Y \subset X} \frac{\psi(Y)}{N(Y)} = \sum_{Y \subset X} \psi(Y) = U_\psi(X) \;. \tag{8.11}$$

If ψ is such that $\psi(X) = 0$ if $X \not\subset \Lambda = \{a_1, \ldots, a_n\}$, then A_ψ is of the form (3.3).

Proposition 8.1. *If* $\mu \in J$ *and* ρ *is the correlation function associated to* μ *by* (8.7), *then for all* $\psi \in \mathcal{B}$

$$\mu(A_\psi) = \sum_{X \ni 0} \rho(X) \frac{\psi(X)}{N(X)} \;. \tag{8.12}$$

This relation determines ρ *from* μ *and conversely.*

We may restrict ourselves to the case where $\psi(X) = 0$ if $X \not\subset \Lambda = \{a_1, \ldots, a_n\}$, (8.3) gives then

$$\mu(A_\psi) = \sum_{X\subset\Lambda} \mu_\Lambda(X) \sum_{Y\ni 0,\, Y\subset X} \frac{\psi(Y)}{N(Y)}$$

$$= \sum_{Y\ni 0,\, Y\subset\Lambda} \frac{\psi(Y)}{N(Y)} \sum_{X\supset Y,\, X\subset\Lambda} \mu_\Lambda(X) = \sum_{Y\ni 0} \rho(Y) \frac{\psi(Y)}{N(Y)} . \tag{8.13}$$

It is also clear that (8.12) determines ρ from μ and we have already seen that the mapping $\rho \to \mu$ is one-to-one.

9. THE INFORMATION - THEORETICAL ENTROPY NOTION

If a stationnary source of information is given by a measure on the space of all messages, then for every finite interval Λ of Z^1 an entropy $S_\Lambda(\mu)$ is defined and the following limit exists

$$S(\mu) = \lim_{\Lambda\to\infty} \frac{S_\Lambda(\mu)}{N(\Lambda)} . \tag{9.1}$$

defining the *entropy* (or *information*) per symbol produced by the source [2]. We generalize now these notions and results to the ν-dimensional situation of statistical mechanics.

Let $\mu \in J$, if Λ is a finite subset of Z^ν we define

$$S_\Lambda(\mu) = - \sum_{X\subset\Lambda} \mu_\Lambda(X) \log \mu_\Lambda(X) . \tag{9.2}$$

Clearly, the entropy S_Λ is invariant under translations of Λ.

Proposition 9.1. *If* $\Lambda_1 \cap \Lambda_2 = \phi$, *then*

$$S_{\Lambda_1\cup\Lambda_2}(\mu) \leqslant S_{\Lambda_1}(\mu) + S_{\Lambda_2}(\mu) . \tag{9.3}$$

We have indeed from (8.5) and the inequality $\log t \geqslant 1 - 1/t$ for $t > 0$:

$$S_{\Lambda_1\cup\Lambda_2}(\mu) - S_{\Lambda_1}(\mu) - S_{\Lambda_2}(\mu) =$$

$$= - \sum_{X_1\subset\Lambda_1} \sum_{X_2\subset\Lambda_2} \mu_{\Lambda_1\cup\Lambda_2}(X_1\cup X_2) \log \frac{\mu_{\Lambda_1\cup\Lambda_2}(X_1\cup X_2)}{\mu_{\Lambda_1}(X_1)\,\mu_{\Lambda_2}(X_2)}$$

$$\leqslant \sum_{X_1\subset\Lambda_1} \sum_{X_2\subset\Lambda_2} \mu_{\Lambda_1\cup\Lambda_2}(X_1\cup X_2) \left(\frac{\mu_{\Lambda_1}(X_1)\,\mu_{\Lambda_2}(X_2)}{\mu_{\Lambda_1\cup\Lambda_2}(X_1\cup X_2)} - 1 \right) = 0 . \tag{9.4}$$

Theorem 9.2. *If* $\mu \in J$, *the following limit exists*

$$s(\mu) = \lim_{a\to\infty} V(a)^{-1} S_{\Lambda(a)}(\mu) = \inf_{a} V(a)^{-1} S_{\Lambda(a)}(\mu) . \tag{9.5}$$

The functional $s(\ )$ *is affine and upper semi-continuous on* J. (A real or complex function f defined on a convex subset C of a real vector space is affine if $f(\alpha x + [1-\alpha)y] = \alpha f(x) + (1-\alpha) f(y)$ whenever x, y C and $0 \leq \alpha \leq 1$.) (f is upper semi-continuous (u.s.c.) if for every real number a, the set $\{\mu : s(\mu) \geq a\}$ is closed. A function is upper semi-continuous if it is the lower bound of a family of continuous functions. If f is upper semi-continuous on a compact C there exists $x \in C$ such that $f(x) = \sup_{y\in C} f(y)$; an u.s.c. function reaches its sup.)

Let us write

$$F(a^1, \ldots, a^\nu) = S_{\Lambda(a)}(\mu) . \tag{9.6}$$

From (9.3) and the invariance of the entropy under translations we obtain that F is a subadditive function of each of its ν arguments (a function f of one argument is subadditive if $f(a+b) \leq f(a) + f(b)$). This is sufficient to prove (9.5) as we now show. Let

$$\inf_{a} V(a)^{-1} S_{\Lambda(a)}(\mu) = s(\mu) . \tag{9.7}$$

Given $\epsilon > 0$ we can find b such that

$$V(b)^{-1} S_{\Lambda(b)}(\mu) < s(\mu) + \epsilon . \tag{9.8}$$

For any $a = (a^1, \ldots, a^\nu)$ we decompose $\Lambda(a)$ into two sorts of parallelopipeds: n_1 translates of $\Lambda(b)$ and n_2 translates of the unit parallelopiped $\Lambda(1) = \{0\}$.

From the subadditivity of the entropy and (9.8) we obtain then

$$\begin{aligned} S_{\Lambda(a)}(\mu) &\leq n_1 S_{\Lambda(b)}(\mu) + n_2 S_{\Lambda(1)}(\mu) \\ &\leq n_1 V(b) (s(\mu) + \epsilon) + n_2 S_{\Lambda(1)}(\mu) , \end{aligned} \tag{9.9}$$

but if $a \to \infty$ one can assume that

$$\frac{n_1 V(b)}{V(a)} \to 1 , \qquad \frac{n_2}{V(a)} \to 0 \tag{9.10}$$

and it follows that

$$\limsup_{a\to\infty} V(a)^{-1} S_{\Lambda(a)}(\mu) \leqslant s(\mu) , \tag{9.11}$$

which, together with (9.7), proves (9.5).

Consider now a convex linear combination

$$\mu = \alpha\mu_1 + (1-\alpha)\mu_2 \text{ of } \mu_1, \mu_2 \in J \text{ with } 0 < \alpha < 1 .$$

For any finite region Λ and $X \subset \Lambda$ we have correspondingly

$$\mu_\Lambda(X) = \alpha\mu_{1\Lambda}(X) + (1-\alpha)\,\mu_{2\Lambda}(X)$$

and therefore the inequalities

$$\begin{aligned}
&\alpha S_\Lambda(\mu_1) + (1-\alpha)S_\Lambda(\mu_2)\\
&= -\sum_{X\subset\Lambda} [\alpha\mu_{1\Lambda}(X)\log\mu_{1\Lambda}(X) + (1-\alpha)\mu_{2\Lambda}(X)\log\mu_{2\Lambda}(X)]\\
&\leqslant \sum_{X\subset\Lambda} [(\alpha\mu_{1\Lambda}(X) + (1-\alpha)\mu_{2\Lambda}(X))\log(\alpha\mu_{1\Lambda}(X) + (1-\alpha)\mu_{2\Lambda}(X))]\\
&= S_\Lambda(\mu)\\
&\leqslant -\sum_{X\subset\Lambda} [\alpha\mu_{1\Lambda}(X)\log(\alpha\mu_{1\Lambda}(X)) + (1-\alpha)\mu_{2\Lambda}(X)\log(1-\alpha)\mu_{2\Lambda}(X))]\\
&= \alpha S_\Lambda(\mu_1) + (1-\alpha)\,S_\Lambda(\mu_2) - \alpha\log\alpha - (1-\alpha)\log(1-\alpha)\\
&\leqslant \alpha S_\Lambda(\mu_1) + (1-\alpha)\,S_\Lambda(\mu_2) + \log 2 .
\end{aligned} \tag{9.12}$$

In conclusion

$$0 \leqslant S_\Lambda(\mu) - \alpha S_\Lambda(\mu_1) - (1-\alpha)S_\Lambda(\mu_2) \leqslant \log 2 . \tag{9.13}$$

Dividing by $N(\Lambda)$ and using (9.5) we get

$$s(\alpha\mu_1 + (1-\alpha)\mu_2) = \alpha s(\mu_1) + (1-\alpha)s(\mu_2) , \tag{9.14}$$

which shows that $s(\ )$ is affine.

We have seen (after equation (8.2)) that $\mu_\Lambda(X)$ is a continuous function of μ for the w*-topology on $\mathcal{M}(K)$, by (9.2) the same is therefore true of $S_\Lambda(\mu)$ and it follows from (9.5) that $s(\ )$ is upper semi-continuous on J.

Remark. A stronger version of theorem 9.2 and an analogous result for continuous systems (rather than lattice systems) have been proved [23].

10. A VARIATIONAL PROPERTY

In this and the next section we follow closely D. Ruelle [24]. We recall that to each interaction $\varphi \in \mathcal{B}$, an element A_ψ of $\mathcal{C}(K)$ has been associated, with the interpretation of the contribution of the origin to the potential energy.

Theorem 10.1. *Let* $\varphi \in \mathcal{B}$, *then*

$$P(\varphi) = \sup_{\mu \in J} \left[s(\mu) - \mu(A_\varphi)\right] . \tag{10.1}$$

We first prove that for each $\mu \in J$ we have

$$P(\varphi) \geqslant s(\mu) - \mu(A_\varphi) . \tag{10.2}$$

We have seen in section 3 that any element of $\mathcal{C}(K)$, in particular A_φ may be approximated by A of the form (3.3). Therefore according to (8.3) if a is sufficiently large, *for most* $x \in \Lambda(a)$ and for all $\mu \in J$, we have

$$\mu(A) = \sum_{X \subset \Lambda(a)} \mu_{\Lambda(a)}(X)\ \tau_x A(X) . \tag{10.3}$$

From this we obtain, uniformly in $\mu \in J$:

$$\mu(A_\varphi) = \lim_{a\to\infty} V(a)^{-1} \sum_{x \in \Lambda(a)} \sum_{X \subset \Lambda(a)} \mu_{\Lambda(a)}(X)\ \tau_x A_\varphi(X) , \tag{10.4}$$

so that, using (8.11),

$$\mu(A_\varphi) = \lim_{a\to\infty} V(a)^{-1} \sum_{X \subset \Lambda(a)} \mu_{\Lambda(a)}(X)\ U_\varphi(X) . \tag{10.5}$$

On the other hand by (9.5)

$$s(\mu) = \lim_{a\to\infty} V(a)^{-1} \left[- \sum_{X \subset \Lambda(a)} \mu_{\Lambda(a)}(X) \log \mu_{\Lambda(a)}(X)\right] , \tag{10.6}$$

therefore using the concavity of the logarithm and (6.11)

$$s(\mu) - \mu(A_\varphi) = \lim_{a\to\infty} V(a)^{-1}[-\sum_{X\subset\Lambda(a)} \mu_{\Lambda(a)}(X)(U_\varphi(X) + \log \mu_{\Lambda(a)}(X))]$$

$$= \lim_{a\to\infty} V(a)^{-1} \sum_{X\subset\Lambda(a)} \mu_{\Lambda(a)}(X) \log [\exp(-U_\varphi(X))/\mu_{\Lambda(a)}(X)]$$

$$\leqslant \lim_{a\to\infty} V(a)^{-1} \log \sum_{X\subset\Lambda(a)} e^{-U_\varphi(X)} = P(\varphi) , \tag{10.7}$$

which proves (10.2).

We show now that, given $\epsilon > 0$, $\mu \in J$ can be found such that

$$P(\varphi) < s(\mu) - \mu(A_\varphi) + \epsilon . \tag{10.8}$$

For this purpose μ is chosen to look as much as possible like an equilibrium state for the interaction φ. First we define a probability measure $\tilde{\mu}$ on K which is not invariant but only periodic under lattice translations with unit cell

$$\Lambda_n = \{x \in Z^\nu : 0 \leqslant x^i < n \quad \text{for } i = 1, \ldots, \nu\}. \tag{10.9}$$

For n large enough, (6.11) yields

$$|P(\varphi) - N(\Lambda_n)^{-1} \log Z_{\Lambda_n}(\varphi)| < \epsilon/2 . \tag{10.10}$$

The translates

$$\Lambda_n + nk , \tag{10.11}$$

of Λ_n, where $k \in Z^\nu$ form a partition of Z^ν. Let Λ be the union of a finite number of such translates

$$\Lambda = \bigcup_{i=1}^{N} [\Lambda_n + nk_i] . \tag{10.12}$$

If $X \subset \Lambda$ and $X_i = X \cap [\Lambda_n + nk_i]$ we put

$$\tilde{\mu}_\Lambda(X) = \prod_{i=1}^{N} [Z_{\Lambda_n}(\varphi)^{-1} \exp(-U_\varphi(X_i))] . \tag{10.13}$$

According to section 8, $\tilde{\mu}$ is completely determined by the $\tilde{\mu}_\Lambda(X)$. We obtain now $\mu \in J$ by averaging $\tilde{\mu}$ over translations:

$$\mu(A) = N(\Lambda_n)^{-1} \sum_{x \in \Lambda_n} \rho(\tau_x A) . \tag{10.14}$$

By easy estimates we get

$$\begin{aligned} s(\mu) &= \lim_{\Lambda \to \infty} N(\Lambda)^{-1} [- \sum_{X \subset \Lambda} \mu_\Lambda(X) \log \mu_\Lambda(X)] \\ &= \lim_{\Lambda \to \infty} N(\Lambda)^{-1} [- \sum_{X \subset \Lambda} \tilde{\mu}_\Lambda(X) \log \tilde{\mu}_\Lambda(X)] \\ &= N(\Lambda_n)^{-1} [- \sum_{X \subset \Lambda_n} \tilde{\mu}_{\Lambda_n}(X) \log \tilde{\mu}_{\Lambda_n}(X)] . \end{aligned} \tag{10.15}$$

Using (10.5) one checks also that for large n

$$\left| \mu(A_\varphi) - V(\Lambda_n)^{-1} \sum_{X \subset \Lambda_n} \tilde{\mu}_{\Lambda_n}(X) \, U_\varphi(X) \right| < \epsilon/2 . \tag{10.16}$$

From (10.10), (10.15), (10.16) we get

$$\begin{aligned} s(\mu) - \mu(A_\varphi) + \epsilon/2 &> -N(\Lambda_n)^{-1} \sum_{X \subset \Lambda_n} \tilde{\mu}_{\Lambda_n}(X) [U_\varphi(X) + \log \tilde{\mu}_{\Lambda_n}(X)] \\ &= V(\Lambda_n)^{-1} \log Z_{\Lambda_n}(\varphi) > P(\varphi) - \epsilon/2 , \end{aligned}$$

which proves (10.8) and therefore theorem 10.1.

11. THE GIBBS PHASE RULE

We shall prove the following weak version of the Gibbs phase rule

Theorem 11.1. (i) *If φ belongs to the "large" set D of section* 7, *the function $\mu \to s(\mu) - \mu(A_\varphi)$ reaches its maximum $P(\varphi)$ at exactly one point $\mu^\varphi \in J$.*

(ii) *If $\varphi \in D$ and α_φ is the functional defined by* (7.4) *then for all $\psi \in \mathcal{B}$*

$$\mu^\varphi(A_\psi) = \alpha_\varphi(\psi) . \tag{11.1}$$

Therefore, in view of (7.5), (8.12), *μ^φ describes the "equilibrium state" corresponding to the interaction φ.*

(iii) *If $\varphi \in D$, the equilibrium state μ^φ is an extremal point of J (= ergodic measure on K = pure thermodynamic phase by section* 4).

Given $\varphi \in \mathcal{B}$, the functional $\mu \rightarrow s(\mu) - \mu(A_\varphi)$ is affine upper semi-continuous on K and reaches thus its maximum $P(\varphi)$ on a non-empty set Δ_φ which is convex and compact. The set Δ_φ has an extremal point (by the theorem of Krein-Milman [3]) and since Δ_φ is the set where an affine function assumes its maximum value, this extremal point is also an extremal point of J.

Given $\mu \in \Delta_\varphi$, we have for all $\psi \in \mathcal{B}$:

$$P(\varphi + \psi) \geq s(\mu) - \mu(A_{\varphi+\psi}) = s \qquad - \mu(A_\varphi) - \mu(A_\psi) = P(\varphi) - \mu(A_\psi) \,. \tag{11.2}$$

This shows that $\psi \rightarrow P(\varphi) - \mu(A_\psi)$ is a tangent plane to the graph of $P(\ )$ at φ. By proposition 8.1 two distinct elements of Δ_φ yield different tangent planes to the graph of $P(\ )$. If $\varphi \in D$, Δ_φ is thus reduced to one point μ^φ which is extremal in J and $\mu^\varphi(A_\psi) = \alpha_\varphi(\psi)$ for all $\psi \in \mathcal{B}$, proving (i), (ii) and (iii).

Let us mention that the temperature and chemical potential can be reintroduced explicitly in theorem 11.1. Using theorem 7.2 (ii) one sees in particular that *for a dense set of interactions in* $\mathcal{B}'$, *the set of* (μ, β) *points for which there is more than one phase is of Lebesgue-measure zero.*

12. CONCLUDING REMARKS

The approach to the Gibbs phase rule which has been described in these lectures is essentially based on convexity properties and one cannot hope to get results much more precise than theorem 11.1 with such techniques. The interest of the results obtained so far is mainly to connect the traditional ideas of statistical mechanics with new view points, like the description of a physical state by a state on a C^*-algebra, and new concepts like that of the entropy of a state. It would therefore be interesting to obtain results similar to those discussed here for continuous systems (rather than lattice systems). An extension in another direction was obtained by D. Robinson who obtained a variational property and a Gibbs phase rule for *quantum lattice systems* (spin systems treated quantum-mechanically) these results are based on a study of the entropy of states of quantum lattice systems by O. Lanford and D. Robinson and on the study of the thermodynamic limit for these systems by R. Griffiths [25].

REFERENCES

[1] See A. Prince, *Alloy Phase Equilibria* (Elsevier, Amsterdam 1966).
[2] A. I. Khinchin, *Mathematical Foundations of Information Theory* (Dover, 1957).
[3] G. Köthe, *Topologische Lineare Räume* I (Springer, 1960).
[4] J. Dixmier, *Les C*-algèbres et leur représentations* (Hermann, 1964).
[5] R. Kadison, *Cargèse Lecture Notes* (1965).
[6] K. Jacobs, Lectures on *Ergodic Theory*, Aarhus University (1962-1963).
[7] S. Doplicher, D. Kastler and D. Robinson, Commun. Math. Phys. 3 (1966) 1.
[8] D. Ruelle, Commun. Math. Phys. 3 (1966) 133.
[9] D. Kastler and D. Robinson, Commun. Math. Phys. 3 (1966) 151.
[10] O. Lanford and D. Ruelle, J. Math. Phys., 8 (1967) 1460.
[11] C. N. Yang and T. D. Lee, Phys. Rev. 87 (1952) 404.
[12] D. Ruelle, Helv. Phys. Acta 36 (1963) 183.
[13] M. Fisher, Arch. Rat. Mech. Anal. 17 (1954) 377.
[14] R. Griffiths, J. Math. Phys. 6 (1965) 1447.
[15] F. Dyson and A. Lenard, J. Math. Phys. 8 (1967) 423.
[16] F. Dyson, preprint.
[17] D. Ruelle, Ann. Phys. (N.Y.) 25 (1963) 109.
[18] G. Gallavotti and S. Miracle, Commun. Math. Phys., to be published.
[19] J. Ginibre, J. Math. Phys. 6 (1965) 238, 251, 1432.
[20] M. E. Fisher, J. Math. Phys. 6 (1965) 1443.
[21] G. Gallavotti and S. Miracle, Commun. Math. Phys. 5 (1967) 317.
[22] N. Dunford and J. Schwartz, *Linear Operators*, Vol. 1 (Interscience, 1958).
[23] D. Robinson and D. Ruelle, Commun. Math. Phys. 5 (1967) 288.
[24] D. Ruelle, Commun. Math. Phys. 5 (1967) 324.
[25] R. Griffiths, J. Math. Phys. 5 (1964) 1215.

PHOTON STATISTICS

R. J. GLAUBER
Harvard University, Cambridge, Mass.

1. INTRODUCTION

Statistical experiments on photon fields have only recently become a practical reality. To give some examples of what we mean by photon statistics let us think of some simple experiments which can be performed with a light source and a photon counter. We can imagine a shutter of some sort, either mechanical or electrical, to be placed between the source and the counter. We may open the shutter for a specified time and then close it again, observing how many photons are recorded by the counter in the interval. That number is clearly a random variable. Repetitions of the experiment will lead to a statistical distribution for the number of photons counted. A satisfactory theory should furnish us with a prediction of that distribution, its moments and other properties.

There are other kinds of experiments which can be performed with essentially the same apparatus. If the shutter is opened during two different time intervals, for example, we may ask for the joint distribution, or the degree of correlation of the random numbers of photons counted during the pairs of intervals. If we have a second photon counter available we may begin to inquire about spatial as well as temporal correlations. If the two counters are arranged to register in delayed coincidence with one another then we may measure the combined space-time correlations of their counting rates. The prediction of all of these statistical quantities should be the task of a well-formulated theory. But more important perhaps, the theory should be capable of dealing with a much broader variety of light sources than have been familiar in the past.

The light source in the experiment need not be of the usual gas discharge or incandescent type. It might be any of numerous varieties of laser oscillators or it might be one of the new types of amplifiers operating at optical frequencies. We might alternatively have a compound source which emits scattered light, the light from a laser, say, scattered by the density fluctuations of a fluid. All of these sources can emit fields which are rather different in character from one another. The best way of making these differences evident is through experiments of the kind we have noted, that is to say experiments which reveal a good deal more about the statistical properties of a field than the usual static measurements of its intensity and its spectrum.

It is only within roughly the last ten years that measurements of the sort we are discussing have actually been carried out. The first such experiment, which was performed by Hanbury Brown and Twiss [1], used a gas discharge tube as a source and revealed a distinct tendency for two photon detectors, placed in equivalent positions in the field, to register photons simultaneously. Many of the early experiments on photon statistics were in effect repetitions of this experiment with one or another improvement of method. More recently, as the technique of making statistical measurements upon fields has become refined and as the new coherent light sources have been developed, we have seen an increase in the number of such measurements and in their variety as well.

Since the applications of such experiments are likely to increase further it is worth developing a systematic way of discussing them. Let us begin therefore by noting that there are two rather different kinds of statistical uncertainty which are probed by each of the types of experiments we have mentioned. One kind of uncertainty concerns the state of the radiation source. The source is a macroscopic system which can rarely if ever be prepared in a pure quantum state. The best we can hope to do in general is to describe its uncertain state or the uncertain state of the field to which it leads by the methods of statistical mechanics and in that way to predict the associated fluctuations.

The second source of statistical uncertainty is inherent in the photodetection process. The photon counter is an intrinsically quantum mechanical instrument. Even if an incident field is fully predetermined in its behavior (a condition which can only be approached in the classical or strong field limit), the response of the photon counter to the field is never fully predictable. The random integers it records are numbers which are meaningless in the context of classical electromagnetic theory. The only language which can be used with logical consistency in discussing counting experiments is the language of quantum mechanics [2-6]. It will be evident nonetheless that many features of the quantum mechanical discussion exhibit simple correspondences with classical theory. In a sense what we propose to do is in fact to extend the classical theory of noise in electromagnetic fields down into the quantum domain and in so doing to take account also of the noise which is an unavoidable part of the detection process. It may provide some useful background therefore if we begin by saying a few words about classical noise theory.

2. CLASSICAL THEORY

In order to be able to describe the electromagnetic field in terms of a discrete set of variables let us consider the field inside a volume of finite dimensions. A set of homogeneous boundary conditions which we need not specify for present purposes is assumed to constrain the fields at the surface; they could be periodic boundary conditions for example, or the conditions appropriate to reflecting (i.e. perfectly conducting) walls. For an appropriate set of frequencies ω_k we can find a sequence of vector mode functions $\boldsymbol{u}_k(\boldsymbol{r})$ which satisfy the Helmholtz equation

$$(\nabla^2 + \frac{\omega_k^2}{c^2})\boldsymbol{u}_k(\boldsymbol{r}) = 0\,, \tag{2.1}$$

the transversality condition,

$$\nabla \cdot \boldsymbol{u}_k(\boldsymbol{r}) = 0\,, \tag{2.2}$$

and the boundary conditions on the electric fields. We can furthermore choose these functions to form a set which is complete and orthonormal,

$$\int \boldsymbol{u}^*{}_k(\boldsymbol{r}) \cdot \boldsymbol{u}_l(\boldsymbol{r})\,\mathrm{d}\boldsymbol{r} = \delta_{kl}\,. \tag{2.3}$$

The corresponding solutions to the time dependent wave equation take the form $\boldsymbol{u}_k(\boldsymbol{r})\exp[\pm\mathrm{i}\omega_k t]$.

It is often convenient in classical theory to express the field variables as the sum of two complex conjugate terms. We may write the electric field vector, for example, as

$$E(\boldsymbol{r}t) = E^{(+)}(\boldsymbol{r}t) + E^{(-)}(\boldsymbol{r}t)\,, \tag{2.4}$$

where $E^{(+)}$ is the positive frequency part of the field, i.e. the sum of all terms varying with time as $\exp[-\mathrm{i}\omega_k t]$ for all $\omega_k > 0$, and $E^{(-)}$ is the negative frequency part of the field,

$$\boldsymbol{E}^{(-)} = (\boldsymbol{E}^{(+)})^*\,. \tag{2.5}$$

Let us now expand a component of the field $\boldsymbol{E}^{(+)}$ in terms of the set of mode functions $\boldsymbol{u}_k$. To avoid the use of vector indices we can write the appropriate vector components as $E^{(+)}$ and u_k respectively. If the Maxwell equations are homogeneous, i.e. all sources are turned off after they have radiated, then there must exist a set of complex constants $\mathcal{C}_k$ such that

$$E^{(+)}(\boldsymbol{r}t) = \sum_k \mathcal{C}_k u_k(\boldsymbol{r}) \exp[-\mathrm{i}\omega_k t]. \tag{2.6}$$

The set of all solutions to the Maxwell equations is described by giving all possible complex values to the $\mathcal{C}_k$.

In actual practice we rarely know the set of numbers $\mathcal{C}_k$ with more than limited certainty; the fields we examine are usually radiated by systems whose behavior can only be described in statistical terms. The best we can do then is to describe our knowledge of the coefficients $\mathcal{C}_k$ by means of a probability distribution $\mathcal{P}(\{\mathcal{C}_k\})$ defined over the space of the $\mathcal{C}_k$. The real and imaginary parts of each $\mathcal{C}_k$ can be varied independently, so that the differential element of the subspace corresponding to one coefficient is

$$\mathrm{d}^2\mathcal{C}_k = \mathrm{d}(\mathrm{Re}\ \mathcal{C}_k)\,\mathrm{d}(\mathrm{Im}\ \mathcal{C}_k). \tag{2.7}$$

and the normalization condition on the probability distribution is

$$\int \mathcal{P}(\{\mathcal{C}_k\}) \prod_k \mathrm{d}^2\mathcal{C}_k = 1. \tag{2.8}$$

When we measure the intensity of a field we are in effect measuring the ensemble average

$$\{|E^{(+)}(\boldsymbol{r}t)|^2\}_{\mathrm{av.}} = \{E^{(-)}(\boldsymbol{r}t)\ E^{(+)}(\boldsymbol{r}t)\}_{\mathrm{av.}}. \tag{2.9}$$

In interference experiments one or another means is often used to superpose fields which correspond to different positions or times before the total field intensity is measured. In such cases the intensity may be expressed directly, as a rule, in terms of a correlation function for the field which we may define as

$$G^{(1)}_{\mathrm{class.}}(\boldsymbol{r}t, \boldsymbol{r}'t') = \{E^{(-)}(\boldsymbol{r}t)\ E^{(+)}(\boldsymbol{r}'t')\}_{\mathrm{av.}}. \tag{2.10}$$

To illustrate the meaning of the ensemble average more explicitly let us note that it is given by the integral

$$G^{(1)}_{\mathrm{class.}}(\boldsymbol{r}t, \boldsymbol{r}'t') = \int \mathcal{P}(\{\mathcal{C}_k\})\ E^{(-)}(\boldsymbol{r}t)\ E^{(+)}(\boldsymbol{r}'t') \prod_k \mathrm{d}^2\mathcal{C}_k, \tag{2.11}$$

in which the field $E^{(+)}$ is a linear form in the $\{\mathcal{C}_k\}$ and $E^{(-)}$ a linear form in the $\{\mathcal{C}_k^*\}$.

When the coefficients $\mathcal{C}_k$ are regarded as random the function $E^{(+)}(\boldsymbol{r}t)$ becomes a species of time-dependent stochastic process. The detailed study of such processes is the province of classical noise theory and a vast literature exists on that subject. We shall see presently that a large part of quantum mechanical noise theory can be discussed in closely analogous terms.

3. QUANTUM THEORY: INTRODUCTION

In quantum mechanics the fields $\boldsymbol{E}$ and $\boldsymbol{B}$ must be regarded not as numbers but as operators in the space of states which describe the field. The Hamiltonian operator for the free field is

$$H = \tfrac{1}{2} \int (\boldsymbol{E}^2 + \boldsymbol{B}^2)\, \mathrm{d}\boldsymbol{r}\,, \tag{3.1}$$

(in rationalized units). When sources are coupled to the field the vector $\boldsymbol{E}$ in this expression must be interpreted as the transverse part of the electric field, and the source and interaction Hamiltonians must be added to H.

The operator $\boldsymbol{E}$, like its classical counterpart, can be separated into its positive and negative frequency parts

$$\boldsymbol{E} = \boldsymbol{E}^{(+)} + \boldsymbol{E}^{(-)}. \tag{3.2}$$

The requirement that $\boldsymbol{E}$ be Hermitian is secured by requiring $\boldsymbol{E}^{(\pm)}$ to be mutually adjoint,

$$\boldsymbol{E}^{(-)} = (\boldsymbol{E}^{(+)})^{\dagger}. \tag{3.3}$$

Since $E^{(\pm)}$ represent intrinsically complex fields neither is Hermitian. It is perhaps worth remarking at this early point that many of the most familiar features of quantum mechanical calculations depend on the use of Hermitian operators. In order to make the most direct use of the non-Hermitian fields $\boldsymbol{E}^{(\pm)}$ we shall see that it is necessary to introduce some rather interesting and unfamiliar mathematical methods.

If the operator $E^{(+)}$, which represents a positive frequency field component, is expanded in terms of the set of mode functions discussed earlier, then the role of the classical Fourier coefficients $\{\mathcal{C}_k\}$ must be taken by a sequence of quantum mechanical amplitude operators. It is conventional to write those amplitude operators as $\{a_k\}$ and to normalize them so that the expansion of $E^{(+)}$ takes the form

$$E^{(+)}(\boldsymbol{r}t) = \mathrm{i} \sum_k (\tfrac{1}{2}\hbar\omega_k)^{\frac{1}{2}}\, a_k u_k(\boldsymbol{r}) \exp[-\mathrm{i}\omega_k t]. \tag{3.4}$$

The canonical commutation relations which the fields must satisfy then reduce to the simple relations

$$\begin{aligned} [a_k, a_{k'}^{\dagger}] &= \delta_{kk'} \\ [a_k, a_{k'}] = [a_k^{\dagger}, a_{k'}^{\dagger}] &= 0 \end{aligned} \tag{3.5}$$

between the operators $\{a_k\}$ and their adjoints. The operators a_k and $a^\dagger_k$ which correspond to any mode have exactly the same algebraic properties as the complex amplitude operators which are used to describe the quantum mechanical form of the simple harmonic oscillator. The commutation relations (3.5) thus define the amplitude operators for an infinite set of oscillators, one for each of the field modes.

When the fields in the Hamiltonian (3.1) are expressed in terms of the amplitude operators we find that the Hamiltonian reduces to the form

$$H = \sum_k \hbar\omega_k a_k^\dagger a_k + \text{const.} \tag{3.6}$$

This operator, apart from the additive constant, is simply the Hamiltonian of a set of dynamically independent harmonic oscillators. What we have shown is just the observation with which Dirac began the development of quantum electrodynamics; the free electromagnetic field in a vacuum is equivalent in its dynamical properties to an infinite sequence of harmonic oscillators.

One of the familiar properties of the amplitude operators for harmonic oscillators is the fact that the products $a^\dagger_k a_k$ have as eigenvalues the integers $n_k = 0, 1, 2, \ldots$ It follows then from eq. (3.6) that each mode has equally spaced energy levels; the integer n_k is evidently just the number of photons occupying the kth mode.

The eigenstates of the Hamiltonian (3.6) may easily be constructed from oscillator eigenstates. The ground state of the field is defined as the state $|\text{vac}\rangle$ which satisfies the conditions

$$a_k |\text{vac}\rangle = 0 \tag{3.7}$$

for all modes k. It is clearly the state for which all of the n_k vanish; we can regard it simply as a product of ground state vectors of all of the individual mode oscillators.

Any set of integers $\{n_k\}$ which specifies the number of photons in each mode defines an eigenstate of the Hamiltonian. It is easy to show that this eigenstate, which we shall write as $|\{n_k\}\rangle$, can be generated from the vacuum state by writing

$$|\{n_k\}\rangle = \prod_k \frac{(a_k^\dagger)^{n_k}}{n_k!} |\text{vac}\rangle . \tag{3.8}$$

This relation illustrates one of the most important properties of the operators $a^\dagger_k$; they are photon creation operators. It follows from the commutation rules (3.5) that the adjoint operators a_k are photon annihilation operators. The definition of the vacuum state given by eq. (3.7), for example, is simply the condition that no photons are available for annihilation.

The state vectors $|\{n_k\}\rangle$ for all values of the integers $\{n_k\}$ form a complete orthogonal set; the space they span is the familiar Fock space of quantum field theory. Although other sets of basis vectors are available in the Fock space virtually all quantum electrodynamical calculations to date have been based on the states $|\{n_k\}\rangle$. Their use is quite natural, for example, in calculations based on time-dependent perturbation theory, since they are after all the stationary states of the free-field Hamiltonian. Stationary states, however, are characteristically ones in which we lack all knowledge of the phase associated with the motion of a system. To deal with any phenomena which involve the phase of oscillation of the electromagnetic field, for example, we must be prepared to deal with superpositions of states $|\{n_k\}\rangle$ having different sets of quantum numbers. In general the more accurate and specific our phase information is, the more different states must be superposed. Nearly all of the applications of quantum electrodynamics to date have been to elementary particle problems and these have rarely required that we deal with more than one or two photons at a time; phase information, for example, tends to play a minimal role in such problems.

The classical limit of quantum electrodynamics represents the opposite extreme. In strong fields the quantum numbers n_k are not only large, they are typically quite uncertain. To use the states $|\{n_k\}\rangle$ in discussing the classical limit would be to obscure precisely what is simple about that limit. What we shall do instead is to make use of a rather different set of basis states, ones which are equally applicable to the discussion of weak and strong fields but which make the classical limit of the theory a great deal clearer. Before introducing these states, however, it will be useful to survey some of the quantities which can be measured in the simplest photon counting experiments and to use them in defining what we mean by the coherence properties of fields.

4. INTENSITY AND COINCIDENCE MEASUREMENTS

Photon counters function by absorbing photons from the field. What they count are, strictly speaking, not photons but atomic photoabsorption processes rendered individually detectable by an amplification mechanism of some sort. In most of the devices used experimentally the fundamental absorption process is the photoemission of an electron. The amplifying mechanism for the photoelectrons is usually a cascade multiplier. The precise way in which amplification is achieved need not concern us here however, since it plays no direct role in determining what the counter detects.

An ideal photon counter would be one which is quite small in size and has a sensitivity independent of photon frequencies (at least over the spectral range of the incident field). Both requirements can be met rather well, in fact, by the simplest possible model of a counter, a single atom which we observe to see whether and when it emits a photoelectron. Since the atom is quite small compared with the wavelength of visible light most of the transitions it can undergo may be treated by means of the electric dipole approximation. In this approximation the atom is coupled to the field through the interaction Hamiltonian

$$H_\mathrm{I} = -\mathrm{e} \sum_{\gamma} \boldsymbol{q}_{\gamma} \cdot \boldsymbol{E}(\boldsymbol{r}, t) , \qquad (4.1)$$

in which $\boldsymbol{q}_{\gamma}$ is the spatial coordinate of the γth electron of the atom relative to its nucleus which is located at $\boldsymbol{r}$.

The intensity of a field is proportional to the rate at which a counter records photons, or in the case of our single-atom counter to the probability per unit time of our observing a photoabsorption process. Let us suppose that the initial state of the field is given by the state vector $|i\rangle$ and that the atom is initially in its ground state. Then it is quite a simple matter to use first order perturbation theory to find the photoabsorption probability. There is no need for us to go through more than one or two of the details of that calculation here since we are only concerned with the way in which the field operators enter it. (The remaining details can be found for example in ref. [5] p. 78-84.)

The transition amplitude, calculated in lowest order perturbation theory, is proportional simply to the matrix element of H_I between the initial and final states. If the final state of the field is $|f\rangle$ the contribution of the field to the amplitude is a factor $\langle f|E(\boldsymbol{r}t)|i\rangle$. Let us recall now that the field E is the sum of an annihilation operator $E^{(+)}$ (a linear combination of the $\{a_k\}$) and a creation operator $E^{(-)}$ which is its Hermitian adjoint. The fact that the atom began in its ground state means that it can only undergo transitions to states of higher energy. Since energy is conserved in these transitions (to an excellent approximation) photon emission processes are excluded. The effect of energy conservation, in other words, is to insure that only the annihilation operator $E^{(+)}$ contributes to the desired transition amplitude which is thus proportional to $\langle f|E^{(+)}(\boldsymbol{r}t)|i\rangle$.

The final state of the field usually remains unobserved. The transition probability we seek is therefore proportional to a sum of the squared moduli of the matrix elements taken over a complete set of final states,

$$\sum_{f} |\langle f|E^{(+)}(\boldsymbol{r}, t)|i\rangle|^2 = \langle i|E^{(-)}(\boldsymbol{r}t)\, E^{(+)}(\boldsymbol{r}t)|i\rangle . \qquad (4.2)$$

The transition rate which we measure is proportional to the expectation value in the initial state of the field of the product $E^{(-)}E^{(+)}$, which we may note is a Hermitian operator with eigenvalue zero in the vacuum state.

The initial state of the field is rarely if ever a pure quantum state. Our knowledge of the many parameters upon which the field depends is usually quite incomplete and some or all of them must be regarded as random. All field measurements must be carried out therefore as averages over the ensemble of ways in which the field can be prepared. (The need to interpret measurements as ensemble averages is of course already intrinsic in our use of a quantum mechanical detector; there are simply more things we must average over because the source behaves randomly as well as the detector.)

The ensemble-averaged intensity is proportional to the expression (4.2) averaged with the appropriate weight over all possible initial states $|i\rangle$. If we introduce the density operator

$$\rho = \{|i\rangle\langle i|\}_{\text{av.}}, \tag{4.3}$$

to express our knowledge of the state of the field, we may write the average intensity as

$$\{\langle i|E^{(-)}(\boldsymbol{r}t)\,E^{(+)}(\boldsymbol{r}t)|i\rangle\}_{\text{av.}} = \mathrm{Tr}\{\rho E^{(-)}(\boldsymbol{r}t)\,E^{(+)}(\boldsymbol{r}t)\}, \tag{4.4}$$

where Tr stands for trace.

The average intensity is the particular value for $\boldsymbol{r} = \boldsymbol{r}'$ and $t = t'$ of a more general function

$$G^{(1)}(\boldsymbol{r}t, \boldsymbol{r}'t') = \mathrm{Tr}\{\rho E^{(-)}(\boldsymbol{r}t)\,E^{(+)}(\boldsymbol{r}'t')\}, \tag{4.5}$$

which we shall call the first order correlation function for the field. It is a complex-valued function in general and satisfies wave equations in both the primed and unprimed variables. As we shall indicate in the next section it is in fact this function which we determine when we superpose the fields at two different points before measuring the field intensity.

There are two other details of the calculation of the photoabsorption probabilities which are worth at least mentioning. First, the electric field is of course a vector quantity and so the correlation function $G^{(1)}$ should strictly speaking be defined as a second rank tensor. Adding all of the tensor indices to the functions we shall discuss however would be a real encumbrance of the notation. It will be simpler instead to omit the indices and to understand that each specification of a space-time coordinate $\boldsymbol{r}, t$ carries with it implicitly the specification of a vector index. Alternatively the need for vector indices can be avoided altogether by assuming that our detectors are fitted with polarizing filters so that they are sensitive to only one component of the field.

A second point worth noting is that our rather schematic argument may appear not to have used the assumption that the sensitivity of the counter is independent of frequency. Implicit use of that assumption has in fact been made, however. It is only when the frequency sensitivity is flat that we can say in general that the transition probability at time t depends only on the field operators at time t and no other times. To give the argument more fully would require our going through the time-dependent perturbation calculation in some detail. It is sufficient for present purposes to note that when we do so [5] we find that the probability that an atom has undergone a detected photoabsorption process between times 0 and t can be written as

$$p^{(1)}(t) = s \int_0^t G^{(1)}(\boldsymbol{r}t', \boldsymbol{r}t')\,\mathrm{d}t', \tag{4.6}$$

where s, the sensitivity factor, has been assumed to be frequency-independent. (If, by contrast, the sensitivity varies appreciably with frequency then the transition probability is given by a double time integral which involves the correlation function $G^{(1)}$ at pairs of unequal time arguments.)

Let us suppose now that we use not just one atom as a detector but n atoms whose nuclei are at the positions $\boldsymbol{r}_1 \ldots \boldsymbol{r}_n$. We want to calculate the probability that all n of the atoms undergo detected photoabsorption process, more specifically that atom 1 does so between times 0 and t_1, atom 2 between times 0 and t_2 etc. The calculation must be carried out in nth order perturbation theory but it differs from carrying out the first order perturbation calculation n times in only one essential respect. The absorption of n photons is regarded as a single compound process. No observations are made of the state of the field in between the individual absorption processes. Hence there is no way of telling in general which atom has absorbed which photon. The amplitudes corresponding to processes in which the photons are permuted contribute coherently in general and interfere with one another.

On the basis of the same assumptions that we made in discussing the single atom it is easy to show (ref. [5], p. 84-88) that the compound transition probability can be written as

$$p^{(n)}(t_1 \ldots t_n) =$$

$$= s^n \int_0^{t_1} \ldots \int_0^{t_n} G^{(n)}(\boldsymbol{r}_1 t_1' \ldots \boldsymbol{r}_n t_n', \boldsymbol{r}_n t_n' \ldots \boldsymbol{r}_1 t_1') \prod_j \mathrm{d}t_j', \tag{4.7}$$

where the function $G^{(n)}$ is the nth order correlation function defined by

$$G^{(n)}(\boldsymbol{r}_1 t_1 \ldots \boldsymbol{r}_n t_n, \boldsymbol{r}_{n+1} t_{n+1} \ldots \boldsymbol{r}_{2n} t_{2n}) \tag{4.8}$$

$$= \mathrm{Tr}\{\rho E^{(-)}(\boldsymbol{r}_1 t_1) \ldots E^{(-)}(\boldsymbol{r}_n t_n)\, E^{(+)}(\boldsymbol{r}_{n+1} t_{n+1}) \ldots E^{(+)}(\boldsymbol{r}_{2n} t_{2n})\}\,.$$

The n atoms may be regarded as n independent counters consisting of a single atom each. In that case their joint counting rate corresponding to the set of times $t_1 \ldots t_n$ is given by

$$w(t_1 \ldots t_n) \equiv \frac{\partial^n}{\partial t_1 \ldots \partial t_n} p^{(n)}(t_1 \ldots t_n)$$

$$= s^n G^{(n)}(\boldsymbol{r}_1 t_1 \ldots \boldsymbol{r}_n t_n, \boldsymbol{r}_n t_n \ldots \boldsymbol{r}_1 t_1)\,. \tag{4.9}$$

For $n = 1$ the function is simply the intensity noted earlier. Evidently for $n > 1$ the function $G^{(n)}$ with its $2n$ space-time arguments set equal in pairs tells us a time-delayed joint counting rate, one which we could measure in principle by using n-fold coincidence counting techniques. Two-fold coincidence experiments of the type which measure the function $G^{(2)}$ are not especially difficult to perform and a considerable number of them have been carried out in the last few years. Triple and higher order coincidence experiments are considerably more difficult, however, and the functions $G^{(n)}$ for $n \geqslant 3$ are best evaluated in practice by other means. They have an important bearing as we shall see on the statistical distributions of counted photons.

One of the properties that the correlation functions $G^{(n)}$ all share is that they are the averages of operator products which are ordered in a similar way. This scheme, usually referred to as normal order, is one in which annihilation operators (factors of $E^{(+)}$) precede creation operators (factors of $E^{(-)}$) when we read from right to left. The occurrence of this particular ordering is a direct consequence of the fact that we are discussing experiments in which photons are absorbed. There are other ways in which photons could be detected in principle, even if they would be rather hard to apply in practice. The presence of photons of the appropriate frequency, for example, will tend to stimulate the radiative decay of an initially excited atom. It is easy to see that the correlation functions which would be measured by detectors based in this way on stimulated emission [7] are ones in which the roles of the operators $E^{(+)}$ and $E^{(-)}$ are interchanged, that is to say they would be the expectation values of antinormally ordered products. Still other experiments could be devised to measure the averages of symmetrically ordered products, etc. The expectation values which correspond to other orderings can all be found, of course, from the set of normally ordered ones by applying the commutation rules for the creation and annihilation operators.

5. FIRST AND HIGHER ORDER COHERENCE

To illustrate the meaning of the correlation function $G^{(1)}$ let us imagine an interference experiment similar to the famous one carried out by Young. We let a beam of light fall on an opaque screen with two tiny identical pinholes one at position $\boldsymbol{r}_1$ the other at $\boldsymbol{r}_2$. Some distance behind the screen at a position $\boldsymbol{r}$ we place a photon counter. If the distances from the counter to the two pinholes are s_1 and s_2 respectively then the field which arrives at the counter at time t is a superposition of two fields, a field proportional to the one at pinhole 1 at time $t_1 = t - (s_1/c)$ and another proportional to the field at pinhole 2 at time $t_2 = t - (s_2/c)$. The intensity detected by the counter,

$$I(\boldsymbol{r}t) = G^{(1)}(\boldsymbol{r}t, \boldsymbol{r}t)$$
$$= \mathrm{Tr}\{\rho E^{(-)}(\boldsymbol{r}t)\, E^{(+)}(\boldsymbol{r}t)\},$$

is therefore proportional to

$$\mathrm{Tr}\{\rho[E^{(-)}(x_1) + E^{(-)}(x_2)][E^{(+)}(x_1) + E^{(+)}(x_2)]\}$$
$$= G^{(1)}(x_1, x_1) + G^{(1)}(x_2, x_2) + 2\,\mathrm{Re}\, G^{(1)}(x_1, x_2), \qquad (5.1)$$

where we have written x_1 for the coordinates $\boldsymbol{r}_1, t_1$ and x_2 for $\boldsymbol{r}_2, t_2$.

It is clear that in the expression (5.1) the terms $G^{(1)}(x_1, x_1)$ and $G^{(1)}(x_2, x_2)$ represent the intensities which would be contributed by each pinhole in the absence of the other. The remaining cross-correlation term is the contribution of interference to the total intensity. If we write the complex function $G^{(1)}$ as $|G^{(1)}| \exp(i\varphi)$ where φ is an appropriate phase function then the interference term is

$$2\,\mathrm{Re}\, G^{(1)}(x_1, x_2) = 2|G^{(1)}(x_1, x_2)| \cos\varphi. \qquad (5.2)$$

The oscillations of $\cos\varphi$ correspond to the intensity fringes which are observed as the counter is moved about behind the screen. The degree of contrast of these fringes is clearly governed by the function $|G^{(1)}(x_1, x_2)|$. If the fields incident on the two pinholes were uncorrelated in any way, for example, then we should have $G^{(1)}(x_1, x_2) = 0$ and no fringes present. The two incident fields are usually referred to in that case as being incoherent.

We can define the opposite condition, that is to say coherence, by asking how great the contrast of the fringe pattern can be made. Since $|\cos \varphi|$ never exceeds one the contrast of the fringes is maximized by making the function $|G^{(1)}(x_1, x_2)|$ as large as possible. It is easy to see that this function possesses an upper bound however, since the total intensity given by eq. (5.1) can never become negative, nor can the intensity in any sort of superposition experiment with arbitrary superposition coefficients. This positive definiteness condition on the intensity shows us directly (see ref. [3], for example) that the function $G^{(1)}$ obeys the Schwarz inequality

$$G^{(1)}(x_1, x_1)\, G^{(1)}(x_2, x_2) \geqslant |G^{(1)}(x_1, x_2)|^2 . \tag{5.3}$$

The condition for maximum fringe contrast therefore corresponds to the bound

$$G^{(1)}(x_1, x_1)\, G^{(1)}(x_2, x_2) = |G^{(1)}(x_1, x_2)|^2 . \tag{5.4}$$

Let us now consider the set of fields for which the function $G^{(1)}(x_1, x_2)$ factorizes into the product of a function of x_1 and a function of x_2. We can easily use the general properties of $G^{(1)}$ to show that any such factorized form must be expressible as

$$G^{(1)}(x_1, x_2) = \mathcal{E}^*(x_1)\, \mathcal{E}(x_2) , \tag{5.5}$$

where the function $\mathcal{E}(x)$ is a solution of the Maxwell equations containing positive frequency components only. If the correlation function factorizes in this way it is obvious that it satisfies the condition (5.4) for maximum fringe contrast. Factorization of the function $G^{(1)}$, in other words, implies precise coherence in the sense which has become familiar in optical interference experiments.

Not only does the factorization condition eq. (5.5) imply maximum fringe contrast, but the converse statement holds as well. By making use of no relations more complicated than the Schwarz inequality it is possible to show [8] that if the maximum fringe contrast condition holds for all x_1 and x_2 then the first order correlation function must fall into the factorized form of eq. (5.5). The factorization condition on $G^{(1)}$ and the maximum fringe contrast conditions in other words furnish equivalent definitions of fields which are optically coherent. Since this form of coherence constrains only the function $G^{(1)}$ we shall refer to it as first order coherence.

By analogy with the factorization condition for first order coherence we may construct a succession of further factorization conditions which require that the higher order correlation functions take the form

$$G^{(n)}(x_1 \ldots x_{2n}) = \mathcal{E}^*(x_1) \ldots \mathcal{E}^*(x_n)\, \mathcal{E}(x_{n+1}) \ldots \mathcal{E}(x_{2n}), \qquad (5.6)$$

in which the function $\mathcal{E}(x)$ is the same for all values of n. If the functions $G^{(j)}$ satisfy these conditions for $j \leqslant n$ we shall speak of the field as having nth order coherence. For $n > 1$ this is of course a much more restrictive definition than first order coherence, which it implies. If the condition (5.6) holds for all values of n we shall speak of the field as being fully coherent.

That fully coherent fields do exist becomes immediately evident if we consider the classical limit. For any classical field which has completely predetermined behavior, i.e. no statistical uncertainty, the probability density $\mathcal{P}(\{\mathcal{C}_k\})$ is a delta-function and the correlation functions $G^{(n)}{}_{\text{class.}}$ all factorize into the form of eq. (5.6). The function $\mathcal{E}(x)$ is then simply the positive frequency part of the field itself. This simple example also illustrates a close relation between coherence and noiselessness; the sense in which we are extending the optical definition of the term "coherence" corresponds closely to the sense in which pure signals are referred to as "coherent" in communication theory. In the next section we shall show that there also exists a considerable variety of quantum mechanical fields which satisfy the conditions for full coherence.

The higher order coherence properties of a field have a direct bearing on the results of coincidence counting experiments performed in it. The n-fold joint counting rate, as we have noted in eq. (4.9), is given by an nth order correlation function with its space-time arguments occurring in repeated pairs. If the field possesses coherence of order $m \geqslant n$ then the joint counting rate obeys the identity

$$G^{(n)}(x_1 \ldots x_n, x_n \ldots x_1) = \prod_{j=1}^{n} G^{(1)}(x_j, x_j)\,; \qquad (5.7)$$

that is to say the joint counting rate is equal to the product of the counting rates which would be measured by each of the n counters in the absence of all the others. The responses of the counters in other words, are statistically independent of one another; if the average intensity of the field is independent of time for example, the counters detect no tendency toward any sort of correlation in the arrival times of photons.

Fields which obey the higher order coherence conditions can only be provided by rather special sources in general. We have already noted that when a two-fold coincidence experiment is performed with an ordinary source such as a gas discharge tube the counters detect a distinct tendency for the photons to arrive in correlated pairs. That result means that the field of an ordinary source never has second or higher order coherence. When two-fold coincidence experiments are performed in the output beam of a stably operating laser however, the results are rather different; they do indeed show statistical independence in the responses of the counters. It is now a good deal more than a surmise [2] that laser beams can possess coherence of quite a high order; all present theories of the laser indicate that and several experimental results as well.

High order coherence implies the existence of detailed space-time correlations in the behavior of the field amplitude, so it may perhaps be a little surprising that it also implies the absence of photon counting correlations. One of the things the higher order coherence conditions do is to restrict greatly the amount of randomness or noisiness which can be present in the field. As we shall see presently the photon correlations which lead to breakdown of relations such as eq. (5.7) are an expression of noisiness of the field, precisely the sort of noisiness which higher order coherence eliminates.

A number of simple theorems which characterize fields possessing first and higher order coherence have been proved by Titulaer and Glauber [8, 11]. A few of them are perhaps worth mentioning here:

1. The condition for first order coherence, if it holds at all space-time points, goes a long way toward implying the higher order coherence conditions (5.6). It implies that the higher order correlation functions factorize into the forms

$$G^{(n)}(x_1 \dots x_{2n}) = g_n\, \mathcal{E}^*(x_1) \dots \mathcal{E}^*(x_n)\, \mathcal{E}(x_{n+1}) \dots \mathcal{E}(x_{2n}), \tag{5.8}$$

which differ in general form those of eq. (5.6) by the presence of the constant factor g_n. The numbers of g_n are real and non-negative ($g_1 \equiv 1$). The condition for full coherence is that they all be equal to one.

2. By using linear combinations of the solutions $u_k(\boldsymbol{r}) \exp[-i\omega_k t]$ of the wave equation we can form other sets of mode functions which need not in general be monochromatic. By taking suitable linear combinations of the operators, a_k, we can form annihilation operators appropriate to these generalized field modes. The first order coherence condition is equivalent to the requirement that there be a single generalized mode of the field which contains all of the photons present; all orthogonal modes are empty. Precise first or-

der coherence, in other words, means that all of the photons in the field have identical wave packets.

3. The higher order coherence conditions put constraints only on the statistical distribution of the number of identical photons present; i.e. on the diagonal elements of the density operator in the representation based on the appropriate choice of a generalized mode. In particular the conditions for full coherence of the field simply require that there be a Poisson distribution of the identical photons present. We shall shortly see an illustration of this theorem.

6. THE COHERENT STATES

Let us try to construct some of the states of the field which satisfy the conditions for full coherence. We shall begin by seeking pure states since they are likely to be the simplest. For a pure state represented by the state vector $|\ \rangle$ the density operator is

$$\rho = |\ \rangle\langle\ | \tag{6.1}$$

and the nth order correlation function is

$$G^{(n)}(x_1 \ldots x_{2n})$$

$$= \langle|E^{(-)}(x_1)\ldots E^{(-)}(x_n)\, E^{(+)}(x_{n+1})\ldots E^{(+)}(x_{2n})|\rangle . \tag{6.2}$$

We must now ask how we can choose the state $|\ \rangle$ so that $G^{(n)}$ separates into the factorized form

$$G^{(n)}(x_1 \ldots x_{2n}) = \mathcal{E}^*(x_1)\ldots \mathcal{E}^*(x_n)\, \mathcal{E}(x_{n+1})\ldots \mathcal{E}(x_{2n}) . \tag{6.3}$$

The fact that the fields $E^{(+)}$ and $E^{(-)}$ do not commute means that we cannot solve the problem by finding states $|\ \rangle$ which are simultaneously eigenstates of $E^{(+)}$ and $E^{(-)}$; no such states exist.

What we can do instead is to exploit the fact that the operator products occurring in the correlation functions are all in normally ordered form. It is quite sufficient, for that reason, to require only that the unknown state satisfy the condition

$$E^{(+)}(\boldsymbol{r}t)|\ \rangle = \mathcal{E}(\boldsymbol{r}t)|\ \rangle , \tag{6.4}$$

since that condition implies the adjoint relation

$$\langle\ |E^{(-)}(\boldsymbol{r}t) = \mathcal{E}^*(\boldsymbol{r}t)\langle\ | \tag{6.5}$$

and the two relations together imply that the functions $G^{(n)}$ factorize to precisely the form of eq. (6.3).

The function $\mathcal{E}(\boldsymbol{r}, t)$ must be a positive frequency solution of the Maxwell equations which satisfies the same boundary conditions as the field $E^{(+)}(\boldsymbol{r}t)$. We can thus expand it in terms of the mode functions, just as we expanded $E^{(+)}$ in eq. (3.4) by writing

$$\mathcal{E}(\boldsymbol{r}t) = \mathrm{i} \sum_k (\tfrac{1}{2}\hbar\omega_k)^{\frac{1}{2}} \alpha_k u_k(\boldsymbol{r}) \exp[-\mathrm{i}\omega_k t] , \tag{6.6}$$

where the α_k are a suitable set of complex numbers. Since the mode functions $u_k(r)$ form an orthogonal set the eigenstate condition (6.4) implies that we must have

$$a_k | \ \rangle = \alpha_k | \ \rangle , \tag{6.7}$$

for all values of the mode index k.

Let us define for the kth mode alone a state $|\alpha_k\rangle_k$ with the property

$$a_k | \alpha_k\rangle_k = \alpha_k | \alpha_k\rangle_k . \tag{6.8}$$

Once we have found such states for the individual modes we can use their products to form the states

$$|\{\alpha_k\}\rangle = \prod_k |\alpha_k\rangle_k , \tag{6.9}$$

which satisfy eq. (6.7) and therefore obey the full set of coherence conditions. We shall refer to these as coherent states of the field.

One problem is now to find the state of each mode which satisfies the condition (6.8). Since the problem is essentially the same for all modes we can consider a single arbitrary mode and drop the use of its index k. We want then to find the harmonic oscillator state $|\alpha\rangle$ which satisfies the condition

$$a\,|\alpha\rangle = \alpha\,|\alpha\rangle . \tag{6.10}$$

One of the simplest ways of solving for the state is to expand it in terms of the n-quantum states. It is clear that the state $|\alpha\rangle$ can not have any precisely defined value of n, in general, since eq. (6.10) requires that the state remain unchanged when a quantum is removed. To find the appropriate expansion we note the relations

$$a^\dagger |n\rangle = \sqrt{n+1}\ |n+1\rangle ,$$

$$\langle n | a = \sqrt{n+1}\ \langle n+1 | ,$$

which follow from the commutation rules (3.5) and the definition of the states $|n\rangle$. By using the latter of these relations we may write eq. (6.10) in the form

$$\langle n|\,a\,|\alpha\rangle = \sqrt{n+1}\;\langle n+1\,|\,\alpha\rangle = \alpha\,\langle n\,|\alpha\rangle\,,$$

which implies

$$\langle n|\,\alpha\rangle = \frac{\alpha}{\sqrt{n}}\langle n-1\,|\,\alpha\rangle$$

and therefore

$$\langle n|\,\alpha\rangle = \frac{\alpha^n}{\sqrt{n!}}\;\langle 0\,|\alpha\rangle\,. \tag{6.11}$$

It follows that the expansion of the state $|\,\alpha\rangle$ takes the form

$$|\,\alpha\rangle = \sum_n |n\rangle\langle n\,|\,\alpha\rangle$$

$$= \langle 0|\,\alpha\rangle \sum_n \frac{\alpha^n}{\sqrt{n!}}\;|\,n\rangle\,. \tag{6.12}$$

Hence by fixing the arbitrary phase factor so that

$$\langle 0|\,\alpha\rangle = \exp[-\tfrac{1}{2}|\,\alpha\,|^2]\,, \tag{6.13}$$

we find that a coherent state of a single mode is given by the expansion

$$|\,\alpha\rangle = \exp[-\tfrac{1}{2}|\,\alpha\,|^2]\sum_n \frac{\alpha^n}{\sqrt{n!}}\;|n\rangle\,. \tag{6.14}$$

It is worth noting in particular that the state $|n=0\rangle$ is identical to the state $|\,\alpha=0\rangle$; the ground state of a mode is one of the coherent states.

Clearly a coherent state of a mode can be constructed for all values of the complex parameter α. Hence a coherent state of the field can be constructed, at least formally, for any sequence of complex parameters $\{\alpha_k\}$ or equivalently any eigenvalue function $\mathcal{E}(\boldsymbol{r},t)$. It follows then that there is a one-to-one correspondence between the positive frequency solutions to the Maxwell equations $\mathcal{E}(\boldsymbol{r},t)$ and the coherent states of the field $|\{\alpha_k\}\rangle$. There are just as many coherent states as there are classical (i.e. non-operator) solutions to the Maxwell equations.

When a mode is in a coherent state the probability that it contains n quanta is

$$|\langle n\,|\,\alpha\rangle|^2 = \frac{|\,\alpha\,|^{2n}}{n!}\exp[-|\,\alpha\,|^2]\,, \tag{6.15}$$

which is a Poisson distribution, in agreement with the theorem stated at the end of the last section. Indeed when each mode has a Poisson distribution it is easy to see that the total photon population of the field also has one. The mean number of quanta in the distribution (6.15) is

$$\langle n\rangle = |\alpha|^2 . \tag{6.16}$$

The mode excitations which are strong enough to correspond to the limit of classical theory must therefore have $|\alpha|^2 \gg 1$. The classical fields to which these states correspond can be found by comparing the expansion (2.6) for the classical field $E^{(+)}$ with the expansion (6.6) for the eigenvalue field $\mathcal{E}(\boldsymbol{r},t)$. We see then the correspondence

$$\mathcal{C}_k \leftrightarrow \mathrm{i}(\tfrac{1}{2}\hbar\omega_k)^{\frac{1}{2}}\,\alpha_k \tag{6.17}$$

between the classical expansion coefficients $\mathcal{C}_k$ and the coherent state amplitudes α_k. Since $\hbar$ is in effect negligibly small for classical fields we see once again that the classical limit requires $|\alpha_k|^2 \gg 1$.

While the expansion (6.14) tells us most of what we need to know about the coherent states, there is a much more compact way of generating them which is often useful. Let us define a unitary operator $D(\beta)$ which carries out the translation

$$D^{-1}(\beta)\,a\;D(\beta) = a + \beta\,, \tag{6.18}$$

$$D^{-1}(\beta)\,a^\dagger D(\beta) = a^\dagger + \beta^* \tag{6.19}$$

on the operators a and $a^\dagger$. It is easy to verify by using the commutation relations that a suitable operator is

$$D(\beta) = \exp[\beta a^\dagger - \beta^* a] \tag{6.20}$$

and its unitarity is evident from the relations

$$D(-\beta) = D^{-1}(\beta) = D^\dagger(\beta)\,. \tag{6.21}$$

Now if we begin with the definition

$$a|\alpha\rangle = \alpha|\alpha\rangle$$

of a coherent state, multiply both sides of the equation by $D^{-1}(\alpha)$ and then use the relation (6.18), we see that

$$a\,D^{-1}(\alpha)|\alpha\rangle = 0\,. \tag{6.22}$$

Since the ground state of the oscillator is the unique solution of eq. (6.22) we must have

$$D^{-1}(\alpha)|\alpha\rangle = |0\rangle\,,$$

or

$$|\alpha\rangle = D(\alpha)|0\rangle\,. \tag{6.23}$$

The coherent states, in other words, are just displaced versions of the ground state of the harmonic oscillator. When we introduce position and momentum operators for the oscillator we will see that the ground state wave packet is displaced in general both in position and momentum space. It was because of this simple property that the states were first discussed by Schrödinger [9] in 1926.

The coherent states have an extremely simple form of time dependence. If we subtract the additive constant from the Hamiltonian (3.6) (a change which only alters a phase factor in the state vector) we find that the time dependent form of a coherent state is

$$\begin{aligned} |\alpha, t\rangle &= \exp[-\mathrm{i}\frac{Ht}{\hbar}]\,|\alpha\rangle \\ &= \exp[-\mathrm{i}a^\dagger a\omega t]\,\exp[-\tfrac{1}{2}|\alpha|^2]\sum_n \frac{\alpha^n}{\sqrt{n!}}\,|n\rangle \\ &= |\alpha\,\exp[-\mathrm{i}\omega t]\rangle\,. \end{aligned} \tag{6.24}$$

The state remains a coherent one at all times. Its complex amplitude, $\alpha(t) = \alpha\,\exp[-\mathrm{i}\omega t]$, simply describes circles in the complex plane; the projections of this motion upon the real or imaginary axes form the familiar picture of simple harmonic motion.

We may digress for a moment by asking a somewhat broader question about time-dependent states. If the field Hamiltonian were different from that of a set of free harmonic oscillators we could not expect in general that a state which is initially coherent would remain so at later times. Are there any other Hamiltonians that do have the property of preserving the coherent character of states which are initially coherent? The answer is that there are a good many, including ones for which the oscillators are coupled. It is easy to show [10] that all Hamiltonians taking the form

$$H = \sum_{jk} a_j^\dagger f_{jk}(t) a_k + \sum_k \{g_k(t) a_k^\dagger + g_k^*(t) a_k\}\,, \tag{6.25}$$

where $f_{jk} = f_{kj}^*$ and the f_{jk} and g_k are arbitrary functions of time, have the desired property. These Hamiltonians in fact form the most general class (apart from the trivial possibility of adding functions of the time to H) which preserves the coherent character of state. Whenever the Hamiltonian takes the form of eq. (6.25) the amplitudes $\{\alpha_k(t)\}$ of the time-dependent coherent states carry out the motion one would expect in a classical theory [10]. The wave packet of the system in such states is always similar to that of the ground state of the free oscillators and simply moves along the classical trajectories without spreading.

It may help in forming a picture of the coherent states to construct their wave functions in position and momentum space. The position q and momentum p which we introduce for this purpose can be defined by

$$q = (\hbar/2\omega)^{\frac{1}{2}}(a^\dagger + a)\,, \tag{6.26}$$

$$p = \mathrm{i}(\tfrac{1}{2}\hbar\omega)^{\frac{1}{2}}(a^\dagger - a)\,. \tag{6.27}$$

The condition which defines a coherent state can then be written as

$$a|\alpha\rangle = \frac{1}{(2\hbar\omega)^{\frac{1}{2}}}(\omega q + \mathrm{i}p)|\alpha\rangle = \alpha|\alpha\rangle\,. \tag{6.28}$$

In the representation in which q is diagonal this relation is the first order differential equation

$$\frac{1}{(2\hbar\omega)^{\frac{1}{2}}}\left(\omega q' + \hbar\frac{\mathrm{d}}{\mathrm{d}q'}\right)\langle q'|\alpha\rangle = \alpha\langle q'|\alpha\rangle\,. \tag{6.29}$$

It is convenient to write the wave function which solves the equation in the form

$$\langle q'|\alpha\rangle = \left(\frac{\omega}{\pi\hbar}\right)^{\frac{1}{4}}\exp\left[-\left(\left(\frac{\omega}{2\hbar}\right)^{\frac{1}{2}}q' - \alpha\right)^2 + \tfrac{1}{2}\alpha(\alpha - \alpha^*)\right]. \tag{6.30}$$

In so doing we have normalized the function so that

$$\int_{-\infty}^{\infty}|\langle q'|\alpha\rangle|^2\,\mathrm{d}q' = 1 \tag{6.31}$$

and chosen its phase in a special way which we shall explain presently. (What we have done is to require that $\exp[\frac{1}{2}|\alpha|^2]\langle q'|\alpha\rangle$ be a function of α and not of α^*.)

By an analogous calculation in the momentum representation we can show that the momentum space wave function for a coherent state is

$$\langle p'|\alpha\rangle = \frac{1}{(\pi\hbar\omega)^{\frac{1}{4}}}\exp\left[-\left(\frac{p}{(2\hbar\omega)^{\frac{1}{2}}} + \mathrm{i}\alpha\right)^2 - \tfrac{1}{2}\alpha(\alpha + \alpha^*)\right]. \tag{6.32}$$

(The phase of this wave function has been chosen so that $\exp[\frac{1}{2}|\alpha|^2]\langle p'|\alpha\rangle$ is independent of α^*).

If $\langle q\rangle$ and $\langle p\rangle$ are the mean values of the coordinate and momentum of a wave packet then the variances of these quantities are defined as

$$(\Delta q)^2 = \langle q^2\rangle - \langle q\rangle^2\,, \tag{6.33}$$

$$(\Delta p)^2 = \langle p^2\rangle - \langle p\rangle^2\,. \tag{6.34}$$

The uncertainty principle requires that for all quantum states we have

$$(\Delta q)^2(\Delta p)^2 \geq \tfrac{1}{4}\hbar^2. \tag{6.35}$$

It is easy to see from the wave functions (6.30) and (6.32) that for a coherent state we have

$$(\Delta p)^2 = \omega^2(\Delta q)^2 = \tfrac{1}{2}\hbar\omega \tag{6.36}$$

and hence

$$(\Delta q)^2(\Delta p)^2 = \tfrac{1}{4}\hbar^2. \tag{6.37}$$

The coherent states, in other words, reduce the uncertainty product to its minimum value and in that sense are as nearly classical in character as the quantum theory allows any state to be.

There exist other wave packets which satisfy the minimum uncertainty condition (6.37), in fact a whole family of them often referred to as Kennard [12] packets. The familiar derivation of the uncertainty principle [13] shows that the condition for the state $|\,\rangle$ to have the minimal value for $(\Delta q)^2(\Delta p)^2$ is that it satisfy the eigenstate relation

$$(p-\langle p\rangle)|\,\rangle = \mathrm{i}\,\mu(q-\langle q\rangle)|\,\rangle, \tag{6.38}$$

where μ is an arbitrary real positive constant. The ratio

$$\frac{(\Delta p)^2}{(\Delta q)^2} = \mu^2, \tag{6.39}$$

therefore may assume any positive value for these states. Only when $\mu = \omega$ is the state a coherent one in the sense that it is an eigenstate of the annihilation operator $a = (2\hbar\omega)^{-\frac{1}{2}}(\omega q + \mathrm{i}p)$. It is clear from eq. (6.38), on the other hand, that each member of the family of states is the eigenstate of an operator which we can define as

$$A(\mu) = (2\hbar\mu)^{-\frac{1}{2}}(\mu q + \mathrm{i}p) \tag{6.40}$$

$$= \tfrac{1}{2}(\mu\omega)^{-\frac{1}{2}}\{(\mu+\omega)a + (\mu-\omega)a^\dagger\}. \tag{6.41}$$

The operator $A(\mu)$ can be regarded as a species of annihilation operator, but for an oscillator with frequency μ rather than ω. We see thus that the minimum uncertainty states, as a broad class, are quite similar to the coherent states; their wave packets differ from those of the coherent states in general only by an obvious scale transformation, but their physical interpretation as harmonic oscillator states is rather different for $\mu \neq \omega$.

A somewhat different sense in which the coherent states are minimum uncertainty states can be constructed as a corollary of the uncertainty principle. The arithmetic mean-geometric mean inequality tells us that

$$\tfrac{1}{2}\{(\Delta p)^2 + \omega^2(\Delta q)^2\} \geqslant \omega\{(\Delta q)^2(\Delta p)^2\}^{\frac{1}{2}}$$

and hence

$$\tfrac{1}{2}\{(\Delta p)^2 + \omega^2(\Delta q)^2\} \geqslant \tfrac{1}{2}\hbar\omega. \tag{6.42}$$

The lower bound for the uncertainty in this relation is only attained for the coherent states of the oscillator. That is clear, for example, if $\langle q\rangle = 0$ and $\langle p\rangle = 0$ since the lower bound condition is then the definition of the ground state of the oscillator. If $\langle q\rangle$ and $\langle p\rangle$ are different from zero we can always find a unitary operator $D(\alpha)$ which displaces the state without changing its variances $(\Delta q)^2$ and $(\Delta p)^2$ so that $\langle q\rangle$ and $\langle p\rangle$ are reduced to zero. It follows then that the only states which satisfy the minimum condition in eq. (6.43) are displaced forms of the ground state, or coherent states as we see from eq. (6.23).

7. EXPANSIONS IN TERMS OF THE COHERENT STATES

The coherent states do not form an orthogonal set as the eigenstates of Hermitian operators do. We easily see from the expansion (6.14) for a coherent state of a single mode that the scalar product of two such state vectors is

$$\langle\alpha|\beta\rangle = \exp[\alpha^*\beta - \tfrac{1}{2}|\alpha|^2 - \tfrac{1}{2}|\beta|^2].$$

The squared modulus of this expression,

$$|\langle\alpha|\beta\rangle|^2 = \exp[-|\alpha-\beta|^2],$$

shows us, in fact, that no two coherent states are completely orthogonal, even though their scalar product is extremely small for $|\alpha-\beta| \gg 1$.

The coherent states do, however, form a complete set. There exists a very simple way of resolving the unit operator into a sum of projection operators of the form $|\alpha\rangle\langle\alpha|$, and that is all we need to demonstrate completeness [14, 4]. To show this we begin by writing the (real) element of area of the complex plane as

$$d^2\alpha = d(\mathrm{Re}\,\alpha)\, d(\mathrm{Im}\,\alpha) \tag{7.1}$$

and noting the integral identity

$$\int(\alpha^*)^n\alpha^m \exp[-|\alpha|^2]\,d^2\alpha = 2\pi\delta_{nm}\int_0^\infty |\alpha|^{2n+1} \exp[-|\alpha|^2]\,d|\alpha|$$

$$= \pi n!\;\delta_{nm}. \tag{7.2}$$

It is clear therefore, from the expansion (6.14), that if we integrate the dyadic $|\alpha\rangle\langle\alpha|$ over the complex plane we have

$$\frac{1}{\pi}\int |\alpha\rangle\langle\alpha|\, d^2\alpha = \frac{1}{\pi}\sum_{nm} (n!\, m!)^{-\frac{1}{2}} \int(\alpha^*)^n\alpha^m \exp[-|\alpha|^2]\, d^2\alpha\, |m\rangle\langle n|$$

$$= \sum_n |n\rangle\langle n|$$

$$= 1\,. \tag{7.3}$$

The completeness of the coherent st ıtes follows in the last step from the known completeness of the n-quantum states.

The completeness relation assures us that the set of coherent states is sufficiently large so that any state vector can be expanded in terms of them. The fact of completeness does not tell us, however, that the expansion can be carried out in only one way; the set of states may be larger than is necessary for completeness and some of its states may be expressible as linear combinations of others. The two-dimensional continuum of coherent states $|\alpha\rangle$ is in fact vastly overcomplete in this sense. Discrete sequences of the coherent states can be shown to be sufficient to form complete sets [15], so there must be a great deal of linear dependence present in the full two-dimensional set. We shall exhibit some of this linear dependence shortly.

An arbitrary state $|\ \rangle$ of a single mode can be expanded in terms of the n-quantum states by writing

$$|\ \rangle = \sum_n f_n |n\rangle$$

$$= \sum_n \frac{f_n (a^\dagger)^n}{(n!)^{\frac{1}{2}}} |0\rangle\,, \tag{7.4}$$

where the coefficients f_n obey the normalization condition $\Sigma |f_n|^2 = 1$. We can use the series which occurs in eq. (7.4) to define a function of a complex variable z as follows:

$$f(z) = \sum_{n=0}^{\infty} \frac{f_n z^n}{(n!)^{\frac{1}{2}}}. \tag{7.5}$$

It is clear from the normalization condition on the f_n that this series converges for all values of z in the finite plane, and therefore represents an entire function. Bargmann [16] has discussed the representation of the quantum states of an oscillator which is furnished by these functions. He has shown that the functions form the elements of a Hilbert space and has developed rules for calculating with them. We can make use of the function f to write the arbitrary state in eq. (7.4) as

$$|\,\rangle = f(a^\dagger)\,|0\rangle\,. \tag{7.6}$$

To expand the arbitrary state in terms of the coherent states we can multiply it from the left by the resolution of the unit operator given by eq. (7.3). We then have

$$|\,\rangle = \frac{1}{\pi}\int |\beta\rangle\langle\beta|\,\rangle\, d^2\beta\,, \tag{7.7}$$

and the expansion coefficient in the integrand is given by

$$\begin{aligned}\langle\beta|\,\rangle &= \langle\beta|f(a^\dagger)\,|0\rangle \\ &= f(\beta^*)\langle\beta|0\rangle = f(\beta^*)\,\exp[-\tfrac{1}{2}|\beta|^2]\,.\end{aligned} \tag{7.8}$$

The expansion coefficient, apart from the factor $\exp[-\frac{1}{2}|\beta|^2]$ which serves as a weight function, is the entire function which represents the state in the Bargmann space. (We have already made use of the condition represented by eq. (7.8) in determining the wave functions $\langle q'|\alpha\rangle$ and $\langle p'|\alpha\rangle$ in the preceding section. The phase factors left undetermined by the differential equations and normalization conditions for these functions were chosen so that the complex conjugate form of eq. (7.8) is satisfied.)

Since there is clearly a one-to-one correspondence between the functions f and the states of the mode, we have exhibited a construction for the coherent state expansion which is unique. The essential step in our construction is the use of the resolution of unity given by eq. (7.3) and it is this which is responsible for the uniqueness of the resulting expansion. Because the coherent states form an overcomplete set other dyadic resolutions of unity exist, and their use could lead to different kinds of expansions of the same state. In practice, one universally applicable way of expanding the state vectors seems to be sufficient, at least for present purposes.

To give a simple illustration of the expansion (7.7) we may take the arbitrary state to be a coherent state. We then have the relation

$$\begin{aligned}|\alpha\rangle &= \frac{1}{\pi}\int |\beta\rangle\langle\beta|\alpha\rangle\, d^2\beta \\ &= \frac{1}{\pi}\int |\beta\rangle\,\exp[\beta^*\alpha - \tfrac{1}{2}|\alpha|^2 - \tfrac{1}{2}|\beta|^2]\, d^2\beta\,,\end{aligned} \tag{7.9}$$

which shows explicitly that each coherent state may be expressed as a linear combination of all of the others.

As we have mentioned earlier, our knowledge of the state of a field usually requires that we describe it as a mixed state rather than a pure one. It is quite important therefore to be able to expand the density operator in terms of the coherent states. The most direct procedure for expanding the operator ρ is the immediate generalization of the one we have used for pure states. By multiplying ρ from both sides by the resolution of the unit operator in eq. (7.3) we may write

$$\rho = \frac{1}{\pi^2} \int |\alpha\rangle\langle\alpha|\rho|\beta\rangle\langle\beta| \, d^2\alpha \, d^2\beta . \tag{7.10}$$

which is an expansion in terms of the dyadic products $|\alpha\rangle\langle\beta|$. To find the properties of the weight function in this expansion we write the density operator in terms of its expansion in the n-quantum states

$$\begin{aligned} \rho &= \sum_{n,m} |n\rangle \, \rho_{nm} \langle m| \\ &= \sum_{n,m} \rho_{nm} (n!m!)^{-\frac{1}{2}} (a^\dagger)^n \, |0\rangle\langle 0| a^m . \end{aligned} \tag{7.11}$$

The form taken by this sum suggests that we define the complex-valued function

$$R(\alpha^*, \beta) = \sum_{n,m} \rho_{nm} (n!m!)^{-\frac{1}{2}} (\alpha^*)^n \beta^m , \tag{7.12}$$

which permits us to write the weight function in eq. (7.10) as

$$\begin{aligned} \langle\alpha|\rho|\beta\rangle &= R(\alpha^*, \beta)\langle\alpha|0\rangle\langle 0|\beta\rangle \\ &= R(\alpha^*, \beta) \exp[-\tfrac{1}{2}|\alpha|^2 - \tfrac{1}{2}|\beta|^2] . \end{aligned} \tag{7.13}$$

Since the density operator is a bounded operator in the sense that

$$\mathrm{Tr}\,\{\rho^2\} = \sum_{n,m} |\rho_{nm}|^2 \leqslant 1 , \tag{7.14}$$

the series (7.12) for the function R converges for all finite values of α^* and β and R must be an entire function of both its arguments. Dyadic expansions of the type we are considering may be made for other types of operators than density operators. The weight function can be shown to retain the same analyticity property for a broad class of unbounded operators as well.

The advantage of expanding the density operator in terms of the coherent states becomes evident when we calculate the correlation functions $G^{(n)}$. The matrix element between two coherent states of any normally ordered product of field operators reduces to a product of the appropriate field eigenvalues, multiplied by the scalar product of the states. The non-commuting operators, in other words, become replaced by ordinary complex numbers and evaluating the average value of their product boils down to a problem of integration. There are evidently two complex amplitudes to be integrated over for each mode of the field.

The unit operator also possesses a simple dyadic expansion analogous to the one we have found for ρ. We may write it as

$$1 = \frac{1}{\pi^2}\int |\alpha\rangle\langle\alpha|\beta\rangle\langle\beta|\, d^2\alpha\, d^2\beta$$

$$= \frac{1}{\pi^2}\int |\alpha\rangle \exp[\alpha^*\beta - \tfrac{1}{2}|\alpha|^2 - \tfrac{1}{2}|\beta|^2]\langle\beta|\, d^2\alpha\, d^2\beta . \qquad (7.15)$$

Because this is an expansion in terms of products of the form $|\alpha\rangle\langle\beta|$, it has a structure rather different from the resolution of unity which we have used earlier. We may make use of eq. (7.9), however, to carry out the integration over the variable α and in this way arrive once more at the reduced or "diagonal" form of the expansion

$$1 = \frac{1}{\pi}\int |\beta\rangle\langle\beta|\, d^2\beta ,$$

which we derived in eq. (7.3). In effect the linear dependence of the coherent states supplies us with an alternative expansion for the unit operator and a much simpler one which happens already to be familiar to us.

The existence of a reduced or "diagonal" form for the coherent state expansion of the unit operator suggests that an analogous representation might exist for density operators. We may, accordingly, look for a representation of the density operator which takes the simple form

$$\rho = \int P(\alpha)\, |\alpha\rangle\langle\alpha|\, d^2\alpha . \qquad (7.16)$$

The Hermitian character of the expansion is secured by requiring that the weight function $P(\alpha)$ be real-valued, and the normalization of the density operator, $\mathrm{Tr}\,\rho = 1$, is secured by the condition

$$\int P(\alpha)\, d^2\alpha = 1 . \qquad (7.17)$$

An example of a density operator which clearly possesses a representation of this type is the one for a pure coherent state $|\beta\rangle\langle\beta|$. For this case it suffices to let the function P be a two-dimensional delta function

$$P(\alpha) = \delta^{(2)}(\alpha - \beta) , \tag{7.18}$$

which we may define via

$$\delta^{(2)}(\alpha) = \delta(\mathrm{Re}\,\alpha)\;\delta(\mathrm{Im}\,\alpha) . \tag{7.19}$$

Other examples of the representation (7.16), which we have called the P-representation, are easy to construct and we shall see some useful ones presently. The existence of these examples does not, however, answer the question of whether it it possible to construct a P-representation for all density operators. We shall postpone discussing that question until we can consider it in a somewhat broader context in the next section. For the present let us simply assume that a P-representation does exist and note some of its properties.

The role of the function $P(\alpha)$ as a weight function in a sum over the projection operators $|\alpha\rangle\langle\alpha|$ gives it some of the appearance of a probability density, as does the normalization condition (7.17). A literal interpretation of $P(\alpha)$ as a probability density however, would require that the projection operators $|\alpha\rangle\langle\alpha|$ form an orthogonal set, and we know that no such thing is true. There are many more or less related ways of seeing that $P(\alpha)$ is not in general a probability density. Perhaps the simplest is through the positive-definiteness property of the density operator, or the condition that its eigenvalues can never be negative. It is easy to see that functions $P(\alpha)$ which are nowhere negative lead to positive-definite density operators, but the converse statement is, in fact, false. The positive definiteness conditions on ρ freely admit the possibility that the function $P(\alpha)$ takes on negative values (though not over very broad areas of the α-plane). It is worth noting too that there is no physical sense in which we could truly measure a probability distribution for the complex eigenvalues of the operator a. The real and imaginary parts of a correspond to the position and momentum of an oscillator and since these operators do not commute they can not simultaneously be measured precisely.

In spite of all the reasons why $P(\alpha)$ is not a probability density there are a number of calculations which can be carried out very much as if it were one. Let us ask for the average value of a normally ordered operator such as $(a^\dagger)^n a^m$. Then we can use the P-representation to write

$$\mathrm{Tr}\{\rho(a^\dagger)^n a^m\} = \int P(\alpha)(\alpha^*)^n \alpha^m \, \mathrm{d}^2\alpha , \tag{7.20}$$

and the latter expression looks a great deal like a classical average calculated with a probability density $P(\alpha)$.

A state of affairs not unlike the one we are describing has arisen at least once before in quantum mechanics. In 1932 Wigner [17] introduced a function which serves as a quantum mechanical analogue of the classical phase space distribution in evaluating the averages of a particular class of ordered operators. He also showed that the function may take on negative values. The function $P(\alpha)$ is different from Wigner's function, though related to it, and evidently also deserves to be called a quasi-probability density, as the Wigner function is. We shall discuss the interrelationship of these functions and some others as well in the next section.

When the P-representation is used for the density operator of a multimode field it takes the form

$$\rho = \int P(\{\alpha_k\}) \, |\{\alpha_k\}\rangle\langle\{\alpha_k\}| \prod_k \mathrm{d}^2\alpha_k . \tag{7.21}$$

If we now use this expression for the density operator in evaluating the correlation functions $G^{(n)}$, we find that they reduce to integrals of precisely the same form as we encountered in our discussion of classical correlation functions. If the state $|\{\alpha_k\}\rangle$ satisfies the eigenstate condition

$$E^{(+)}(x) \, |\{\alpha_k\}\rangle = \mathcal{E}(x\{\alpha_k\}) \, |\{\alpha_k\}\rangle , \tag{7.22}$$

then we see that the function $G^{(1)}$, for example, is given by

$$G^{(1)}(x_1, x_2) = \int P(\{\alpha_k\}) \, \mathcal{E}^*(x_1\{\alpha_k\}) \, \mathcal{E}(x_2\{\alpha_k\}) \prod_k \mathrm{d}^2\alpha_k , \tag{7.23}$$

which is an integral having exactly the same structure as the classical one in eq. (2.11). Part of this correspondence has already been indicated in eq. (6.17); the classical amplitudes $\mathcal{C}_k$ correspond to $\mathrm{i}(\frac{1}{2}\hbar\omega_k)^{\frac{1}{2}}\alpha_k$. The remainder of it lies in the correspondence of the classical probability density $\mathcal{P}(\{\mathcal{C}_k\})$ with the quasi-probability density $P(\{\alpha_k\})$. Indeed, in the classical limit of strongly excited fields we may expect the function $P(\{\alpha_k\})$ to become more and more accurately interpretable as a probability density and to approach the function $\mathcal{P}(\{\mathcal{C}_k\})$ asymptotically.

What is remarkable about eq. (7.23) and the corresponding expression for all of the functions $G^{(n)}$ is that anything like the classical prescription for averaging field quantities should hold, no matter how weak the fields may be, e.g. even when the fields contributed by the light sources are much weaker than the vacuum fluctuations that are always present. The fields dealt with in optical experiments are typically just that weak, but their intensities have traditionally been evaluated as averages of the products of classical fields. The existence of the P-representation for such fields evidently gives that familiar procedure a precise and non-trivial justification. It works for intensity measurements based on the absorption of photons since those are the ones described by normally ordered operator products (and therefore ones to which vacuum fluctuations do not contribute).

8. CHARACTERISTIC FUNCTIONS AND QUASI-PROBABILITY DENSITIES

We have emphasized the special role played by the weight function P in the averaging of normally ordered products. It is interesting to discuss two other forms of operator ordering and the corresponding weight functions which are associated with them. These are antinormal ordering, which is simply the reverse of normal ordering and symmetrical ordering, which we may define as the arithmetic mean of all the different orderings of a given product.

Before discussing questions of operator ordering, let us recall that one of the familiar ways of calculating classical probability densities is through the evaluation of their Fourier transforms or characteristic functions. The corresponding function, for quantum mechanical purposes, must be a kind of Fourier transform of the density operator ρ. We can define such a function as the average value of the unitary exponential operator D given by eq. (6.20). The characteristic function then is

$$\begin{aligned}\chi(\lambda) &= \mathrm{Tr}\{\rho\, D(\lambda)\}\\ &= \mathrm{Tr}\{\rho \exp[\lambda a^\dagger - \lambda^* a]\}. \end{aligned} \tag{8.1}$$

An upper bound for this function can be found by evaluating the trace in the representation which diagonalizes the density operator. It is clear then that the trace is the average value of the diagonal elements of $D(\lambda)$ and each of these has modulus less than or equal to one since $D(\lambda)$ is unitary. We thus have

$$|\chi(\lambda)| \leqslant 1\,, \tag{8.2}$$

for all values of the complex variable λ.

In close correspondence with $\chi(\lambda)$ we can define another species of characteristic function which is the average of the normally ordered form of the exponential function,

$$\chi_N(\lambda) = \mathrm{Tr}\{\rho \exp[\lambda a^\dagger] \exp[-\lambda^* a]\}\,, \tag{8.3}$$

and still another in which the exponential function is antinormally ordered,

$$\chi_A(\lambda) = \mathrm{Tr}\{\rho \exp[-\lambda^* a] \exp[\lambda a^\dagger]\}. \tag{8.4}$$

We shall refer to χ_N and χ_A as the normally ordered and antinormally ordered characteristic functions respectively.

The three characteristic functions we have defined may be regarded as generating functions for the averages of the appropriately ordered products. By expanding $\chi_N(\lambda)$ in powers of λ and λ^*, for example, we find

$$\frac{\partial^n}{\partial\lambda^n}\frac{\partial^m}{\partial(-\lambda^*)^m}\chi_N(\lambda)\Big|_{\lambda=0} = \mathrm{Tr}\{\rho(a^\dagger)^n a^m\}, \tag{8.5}$$

i.e. χ_N generates the averages of normally ordered products. Likewise by differentiating $\chi_A(\lambda)$ and $\chi(\lambda)$ we see that χ_A generates the averages of antinormally ordered products and χ generates the averages of products that are symmetrically ordered.

The identity

$$e^A e^B = e^{A+B+\frac{1}{2}[A,B]}, \tag{8.6}$$

which holds for all operators A and B which commute with their commutator $[A,B]$, can be used to express each characteristic function in terms of the others. The relationship we find is

$$\chi_N(\lambda) = \exp[\tfrac{1}{2}|\lambda|^2]\chi(\lambda) = \exp[|\lambda|^2]\chi_A(\lambda). \tag{8.7}$$

The inequality (8.2) then shows that $|\chi_A(\lambda)|$ always decreases at least as fast as $\exp[-\frac{1}{2}|\lambda|^2]$ for $|\lambda| \to \infty$, while $|\chi_N(\lambda)|$, on the other hand, may diverge as rapidly as $\exp[\frac{1}{2}|\lambda|^2]$ in that limit.

Let us turn away from the characteristic functions for a moment to examine the properties of some diagonal matrix elements of the density operator. The matrix elements $\langle\alpha|\rho|\alpha\rangle$, which are expectation values of ρ in the coherent states, have a number of interesting properties [5,18,19]. Like all diagonal elements of ρ they are real and constrained by the inequalities

$$1 \geqslant \langle\alpha|\rho|\alpha\rangle \geqslant 0. \tag{8.8}$$

They may be expressed in terms of the analytic function R defined in eq. (7.12) as

$$\langle\alpha|\rho|\alpha\rangle = R(\alpha^*, \alpha) \exp[-|\alpha|^2]. \tag{8.9}$$

The most useful property of the function $\langle\alpha|\rho|\alpha\rangle$ becomes evident when we evaluate the average of an antinormally ordered product. Let us consider an average of the form

$$\mathrm{Tr}\{\rho\, U(a)\, V(a^\dagger)\},$$

where U and V are arbitrary functions of the annihilation and creation operators, respectively. By using our resolution of unity and the cyclical property of the trace we may write the average as

$$\begin{aligned}\mathrm{Tr}\{V(a^\dagger)\,\rho U(a)\} &= \frac{1}{\pi}\int \mathrm{Tr}\{|\alpha\rangle\langle\alpha|\,V(a^\dagger)\,\rho U(a)\}\,\mathrm{d}^2\alpha \\ &= \frac{1}{\pi}\int \langle\alpha|\,V(a^\dagger)\,\rho U(a)\,|\alpha\rangle\,\mathrm{d}^2\alpha \\ &= \frac{1}{\pi}\int \langle\alpha|\rho|\alpha\rangle\, U(\alpha)\;V(\alpha^*)\,\mathrm{d}^2\alpha\,. \end{aligned} \tag{8.10}$$

Thus we see that the function $\pi^{-1}\langle\alpha|\rho|\alpha\rangle$ plays the same role in averaging antinormally ordered products as $P(\alpha)$ plays in averaging normally ordered ones.

By setting $U = V = 1$ in eq. (8.10) we see that the normalization condition on the density operator may be written as

$$\begin{aligned} 1 &= \mathrm{Tr}\{\rho\} \\ &= \frac{1}{\pi}\int \langle\alpha|\rho|\alpha\rangle\,\mathrm{d}^2\alpha\,. \end{aligned} \tag{8.11}$$

The function $\pi^{-1}\langle\alpha|\rho|\alpha\rangle$ evidently has many properties in common with a probability density, including the fact that it can never be negative. Still, the fact that it refers to a continuum of non-orthogonal states means that there is no well defined sense in which it can be measured physically as a probability density. We must regard it as still another form of quasi-probability function to add to our growing list.

We can now use the rule given by eq. (8.10) as a means of evaluating the antinormally ordered characteristic function χ_A. What we find is

$$\chi_A(\lambda) = \frac{1}{\pi}\int \exp[\lambda\alpha^* - \lambda^*\alpha]\langle\alpha|\rho|\alpha\rangle\,\mathrm{d}^2\alpha\,. \tag{8.12}$$

We note that the argument $\lambda\alpha^* - \lambda^*\alpha$ of the exponential function which occurs in the integrand is purely imaginary. In fact, if we use the real and imaginary parts of α and λ as coordinates in writing the integral we see that it is just a two-dimensional Fourier integral; $\chi_A(\lambda)$ is the Fourier transform of $\langle\alpha|\rho|\alpha\rangle$.

The inversion of the transform may be carried out by using the Fourier integral representation of the delta function,

$$\frac{1}{\pi^2}\int \exp[\lambda(\alpha^* - \beta^*) - \lambda^*(\alpha - \beta)]\,\mathrm{d}^2\lambda = \delta^{(2)}(\alpha - \beta)\,, \tag{8.13}$$

which is easily proved by introducing real and imaginary parts of the variables. The solution for $\langle\alpha|\rho|\alpha\rangle$ is simply

$$\langle\alpha|\rho|\alpha\rangle = \frac{1}{\pi}\int \exp[\alpha\lambda^* - \alpha^*\lambda]\chi_A(\lambda)\, d^2\lambda\,, \tag{8.14}$$

and we see that it is a two-dimensional Fourier transform of the antinormally ordered characteristic function.

Let us turn now to the characteristic function $\chi(\lambda)$ and make use of its Fourier transform to define the function

$$W(\alpha) = \frac{1}{\pi^2}\int \exp[\alpha\lambda^* - \alpha^*\lambda]\chi(\lambda)\, d^2\lambda\,. \tag{8.15}$$

This is in fact the quasi-probability function defined by Wigner [17, 20]. As usually encountered it is expressed somewhat less compactly in terms of coordinate and momentum variables. By integrating by parts and using the identity (8.13) we see that the complex moments of the function W may be written as

$$\begin{aligned}\int (\alpha^*)^n \alpha^m W(\alpha)\, d^2\alpha &= \int \delta^{(2)}(\lambda)\frac{\partial^n}{\partial\lambda^n}\frac{\partial^m}{\partial(-\lambda^*)^m}\chi(\lambda)\, d^2\lambda \\ &= \frac{\partial^n}{\partial\lambda^n}\frac{\partial^m}{\partial(-\lambda^*)^m}\chi(\lambda)\Big|_{\lambda=0}\,,\end{aligned} \tag{8.16}$$

and by evaluating these derivatives as we did in the case of eq. (8.5) we see that they are the averages of the symmetrically ordered products of n factors of $a^\dagger$ and m factors of a. The Wigner function is thus the quasi-probability density appropriate to symmetric ordering. In particular for $n = m = 0$ we see that it obeys the normalization condition

$$\int W(\alpha)\, d^2\alpha = \chi(0) = 1\,. \tag{8.17}$$

After these brief surveys of the treatment of antinormal and symmetric ordering let us return once more to the discussion of normal ordering. If we assume that the density operator ρ possesses a P-representation then we can write its normally ordered characteristic function as

$$\begin{aligned}\chi_N(\lambda) &= \int P(\alpha)\langle\alpha|\exp[\lambda a^\dagger]\exp[-\lambda^* a]|\alpha\rangle\, d^2\alpha \\ &= \int \exp[\lambda\alpha^* - \lambda^*\alpha]\, P(\alpha)\, d^2\alpha\,,\end{aligned} \tag{8.18}$$

which is the two-dimensional Fourier transform of $P(\alpha)$. The inverse of this transform, when it possesses one, is given according to eq. (8.13), by the Fourier integral

$$P(\alpha) = \frac{1}{\pi^2}\int \exp[\alpha\lambda^* - \alpha^*\lambda]\chi_N(\lambda)\, d^2\lambda\,, \tag{8.19}$$

which is an explicit construction of the function $P(\alpha)$. If the function χ_N is quadratically integrable, as it is in many examples, then we see that the weight function $P(\alpha)$ must be quadratically integrable as well. The cases in which χ_N is not quadratically integrable, however, take us out of the classic domain of Fourier integral theory. For the ground state of the oscillator, for example, it is easy to see that $\chi_N(\lambda) = 1$ for all λ. The integral formula (8.13) shows that in this case $P(\alpha)$ is the delta function $\delta^{(2)}(\alpha)$. The delta function and its derivatives are examples of a class of generalized functions known as tempered distributions; they are more or less familiar to physicists to the extent that they can be dealt with as ordinary functions. It can be shown that $P(\alpha)$ is a tempered distribution whenever the moduli of $\chi_N(\lambda)$ and of its derivatives are dominated for $\lambda \to \infty$ by real-valued polynomials of finite degree in λ and λ^*. It is altogether possible for $|\chi_N(\lambda)|$ to increase a good deal more rapidly, however. The relations (8.2) and (8.7) allow $|\chi_N(\lambda)|$ to increase as $\exp[\frac{1}{2}|\lambda|^2]$ for larger $|\lambda|$ and states for which this actually happens are not too hard to find [21]. For such states the integral (8.19) is so highly divergent that no simple summation technique can handle it; $P(\alpha)$ becomes, in effect, more singular than any tempered distribution.

It is not difficult to see, at least, in rather crude physical terms, why the P-representation exhibits this singular behavior. The characteristic functions are Fourier transforms of the quasi-probability densities and so we may surmise that their remaining large for $|\lambda| \to \infty$ is connected with sharply peaked behavior of the quasi-probability densities over small intervals in the α-plane. Let us therefore consider as an example one of the minimum uncertainty states we discussed earlier. It is sufficient to consider the eigenstate of the operator $A(\mu)$ defined by eq. (6.40) which corresponds to the eigenvalue zero. The states with non-zero eigenvalues are simply displaced forms of the same state. If the state $|\,\rangle$ satisfies

$$A(\mu)|\,\rangle = 0\,, \tag{8.20}$$

then a very simple calculation shows that,

$$\begin{aligned}\chi_N(\lambda) &= \langle\,|\exp[\lambda a^\dagger]\exp[-\lambda^* a]|\,\rangle \\ &= \exp\left\{\frac{\omega-\mu}{2}\left[\frac{(\mathrm{Re}\,\lambda)^2}{\omega} - \frac{(\mathrm{Im}\,\lambda)^2}{\mu}\right]\right\}.\end{aligned} \tag{8.21}$$

For $\omega > \mu$ this expression diverges as $|\mathrm{Re}\,\lambda| \to \infty$; for $\mu > \omega$ it diverges as $|\mathrm{Im}\,\lambda| \to \infty$. In neither case does it possess a Fourier transform in any of the familiar senses.

The state we are trying to represent differs from the coherent states in that it has the variances

$$(\Delta p)^2 = \tfrac{1}{2}\hbar\mu, \qquad (\Delta q)^2 = \tfrac{1}{2}\frac{\hbar}{\mu}. \tag{8.22}$$

According to whether μ is greater or less than ω either one of these variances or the other must be smaller than the values given by eq. (6.36), which are the variances common to all of the coherent states. The physical reason for the breakdown of the P-representation in these cases is now evident. What we are trying to do is to find something akin to a probabilistic mixture of states which has altogether $(\Delta p)^2 = \frac{1}{2}\hbar\mu$ when each of the states that we are mixing has $(\Delta p)^2 = \frac{1}{2}\hbar\omega$. For $\mu < \omega$ no such mixture exists; the divergence of χ_N shows up, as we expect, in its dependence on Re λ, which is the variable conjugate to the momentum. Similarly we see that for $\mu > \omega$ we are trying to represent a state with a coordinate variance smaller than that of the coherent states.

While the example of the minimum uncertainty states is a rather special one there is nothing unphysical about it. The eigenstate defined by eq. (8.20) for example is the ground state the mode oscillator would have if its frequency were changed from ω to μ. It could, in other words, represent the ground state of the same field mode in a dielectric medium rather than the vacuum for which our coherent states have been defined.

In view of the difficulty of extending the P-representation to all states of the field two approaches seem open. One is to further broaden the class of mathematical structures we are willing to accept in place of a function $P(\alpha)$. Sudarshan [22, 19, 24], Klauder [21, 23, 24] and their coworkers have pointed out that a class of generalized distributions has been defined which is extensive enough to admit all Fourier transforms such as that of eq. (8.19). The implicit drawback of this approach is that it sacrifices the last resemblances between $P(\alpha)$ and an ordinary function. Statistical averages taken over $P(\alpha)$ can only be defined rather abstractly as linear functionals, and no means is known of calculating them in general apart from carrying out limiting processes in each case.

The alternative approach is to acknowledge that the quasi-probability distribution $P(\alpha)$ is mainly of use as a computational device and as a way of seeing correspondences with the classical limit. When the function ceases to serve either of these ends and becomes too difficult to define we may as well say that the P-representation ceases to exist. The general coherent state representation given by eq. (7.10) of course, still offers a convenient way of averaging normally ordered operators for these cases.

Fortunately the P-representation exists and is quite well behaved, at least for many of the fields which interest us. When the Fourier transform which defines $P(\alpha)$ does exist we can exhibit some very simple relations between $P(\alpha)$ and the other two quasi-probability densities we have discussed. Eq. (8.7) shows that the three different characteristic functions are related to one another through Gaussian factors. By taking Fourier transforms of the characteristic functions and using the convolution theorem we find the three relations

$$\frac{1}{\pi}\langle\alpha|\rho|\alpha\rangle = \frac{2}{\pi}\int \exp[-2|\alpha - \beta|^2]\, W(\beta)\, d^2\beta, \tag{8.23}$$

$$W(\alpha) = \frac{2}{\pi}\int \exp[-2|\alpha - \beta|^2]\, P(\beta)\, d^2\beta \tag{8.24}$$

and

$$\frac{1}{\pi}\langle\alpha|\rho|\alpha\rangle = \frac{1}{\pi}\int \exp[-|\alpha - \beta|^2] P(\beta)\, d^2\beta. \tag{8.25}$$

The Wigner function W is a Gaussian average of the function P taken over a mean radius $|\alpha - \beta| \approx 2^{-\frac{1}{2}}$ and $\pi^{-1}\langle\alpha|\rho|\alpha\rangle$ is a similar average of W. It follows then, as we see in the third relation, that $\pi^{-1}\langle\alpha|\rho|\alpha\rangle$ is an average of P taken over a mean radius $|\alpha - \beta| \approx 1$. The averaging processes which take us from P to W to $\langle\alpha|\rho|\alpha\rangle$ tend to wash out any unruly behavior of P and transform it into the smooth and bounded behavior of $\langle\alpha|\rho|\alpha\rangle$. They also tend to wash away the areas of the α-plane in which P and W are negative; the function $\langle\alpha|\rho|\alpha\rangle$ which results from the averaging is nowhere negative. The functions $W(\alpha)$ and $\langle\alpha|\rho|\alpha\rangle$ exist and remain well defined for all quantum states. The cases in which $P(\alpha)$ becomes undefinable are precisely the ones in which the convolution integrals (8.24) and (8.25) can not be inverted.

9. SOME EXAMPLES

It may be of help in visualizing the quasi-probability functions to cite a few examples. The simplest mode excitation to discuss is a pure coherent state $|\beta\rangle$. For this state we clearly have

$$\begin{aligned}\chi_N(\lambda) &= \langle\beta|\exp[\lambda a^\dagger]\exp[-\lambda^* a]|\,\rangle \\ &= \exp[\lambda\beta^* - \lambda^*\beta], \end{aligned}\tag{9.1}$$

and the two other characteristic functions are then given by eq. (8.7). The quasi-probability functions which are their Fourier transforms are found to be

$$P(\alpha) = \delta^{(2)}(\alpha - \beta)\,, \tag{9.2}$$

$$W(\alpha) = \frac{2}{\pi} \exp[-2|\alpha - \beta|^2]\,, \tag{9.3}$$

$$\frac{1}{\pi} \langle\alpha|\rho|\beta\rangle = \frac{1}{\pi} \exp[-|\alpha - \beta|^2]\,. \tag{9.4}$$

The latter two relations, of course, follow immediately from the first if we use eqs. (8.24) and (8.25).

The state which is by far the most common of all states occurring in nature is best described as the state of chaos. For a single mode of the field, it is the state with the maximum value of the entropy,

$$S = -\operatorname{Tr}\{\rho \log \rho\}\,, \tag{9.5}$$

for a fixed mean occupation number

$$\langle n\rangle = \operatorname{Tr}\{\rho\, a^\dagger a\}\,. \tag{9.6}$$

The problem of maximizing the entropy subject to the condition of fixed $\langle n\rangle$ (and $\operatorname{Tr} \rho = 1$, of course) is a familiar one. Its solution can be written as

$$\rho = \exp[-\zeta_1 - \zeta_2\, a^\dagger a]\,, \tag{9.7}$$

where ζ_1 and ζ_2 are Lagrange multipliers to be determined from the normalization condition and eq. (9.6). When these constants are evaluated we find

$$\rho = \frac{1}{1+\langle n\rangle} \exp\{a^\dagger a \log (\frac{\langle n\rangle}{1+\langle n\rangle})\} \tag{9.8}$$

$$= \frac{1}{1+\langle n\rangle} \{\frac{\langle n\rangle}{1+\langle n\rangle}\}^{a^\dagger a} \tag{9.9}$$

$$= \frac{1}{1+\langle n\rangle} \sum_j \{\frac{\langle n\rangle}{1+\langle n\rangle}\}^j\, |j\rangle\langle j|\,, \tag{9.10}$$

where in the last relation we have expanded ρ in terms of the j-quantum states.

It is not at all difficult to construct the three differently ordered forms for the density operator ρ in this case, but instead of pausing to do so let us proceed directly to the evaluation of the quasi-probability functions. The function $R(\alpha^*, \beta)$ defined by eq. (7.12) is clearly given by

$$R(\alpha^*, \beta) = \frac{1}{1+\langle n\rangle} \exp\{\frac{\langle n\rangle}{1+\langle n\rangle}\, \alpha^* \beta\}\,, \tag{9.11}$$

hence the function $\langle\alpha|\rho|\alpha\rangle$, as we see from eq. (8.9), is given by

$$\langle\alpha|\rho|\alpha\rangle = \frac{1}{1+\langle n\rangle} \exp\{-\frac{|\alpha|^2}{1+\langle n\rangle}\}. \qquad (9.12)$$

The function $\chi_A(\lambda)$, which is the Fourier transform of this function, is

$$\chi_A(\lambda) = \exp[-(1+\langle n\rangle)|\lambda|^2] \qquad (9.13)$$

and that in turn determines $\chi(\lambda)$ and $\chi_N(\lambda)$ through eq. (8.7). When the functions $W(\alpha)$ and $P(\alpha)$ are determined by Fourier inversion we then find

$$W(\alpha) = \frac{1}{\pi(\frac{1}{2}+\langle n\rangle)} \exp\{-\frac{|\alpha|^2}{\frac{1}{2}+\langle n\rangle}\} \qquad (9.14)$$

and

$$P(\alpha) = \frac{1}{\pi\langle n\rangle} \exp\{-\frac{|\alpha|^2}{\langle n\rangle}\}. \qquad (9.15)$$

The definition of chaotic states which we have given is a great deal more general than the notion of thermal equilibrium; there are all sorts of ways of generating chaos which are not associated with any temperature at all. Thermal equilibrium states are chaotic states, however, with

$$\langle n\rangle = \frac{1}{\exp[\frac{\hbar\omega}{\kappa T}] - 1} \qquad (9.16)$$

and substituting this expression into eq. (9.8) or (9.10) leads to familiar forms for the thermal density operator.

If the field has a multimode excitation the chaotic state with mean occupation numbers $\langle n_k\rangle$ is represented by the P-function

$$P(\{\alpha_k\}) = \prod_k \frac{1}{\pi\langle n_k\rangle} \cdot \exp\{-\sum_k \frac{|\alpha_k|^2}{\langle n_k\rangle}\}. \qquad (9.17)$$

Chaos in electromagnetic fields is synonymous with noise; eq. (9.17) describes a field of pure noise with a spectrum specified by the $\{\langle n_k\rangle\}$. Any light source which is divisible in principle into many parts, each of which radiates independently of the others, will almost inevitably radiate a chaotic field. The range of applicability of eq. (9.17) is therefore presumably quite broad.

The higher order correlation functions have a particularly simple structure for chaotic fields. It is not difficult to show [5] that they can all be expressed in terms of the first order correlation function via the relations

$$G^{(n)}(x_1 \ldots x_n, y_1 \ldots y_n) = \sum_P \prod_{j=1}^{n} G^{(1)}(x_j, y_{Pj}), \tag{9.18}$$

in which the sum is carried out over all permutations Pj of the integers $j = 1 \ldots n$. The particular value of $G^{(2)}$ which corresponds to a two-fold joint counting rate, for example, can be written as

$$G^{(2)}(x_1 x_2, x_2 x_1) = G^{(1)}(x_1, x_1)\, G^{(1)}(x_2, x_2) + |G^{(1)}(x_1, x_2)|^2. \tag{9.19}$$

If the field possessed second order coherence the second of the terms on the right would be absent, and two counters would record photons at the same rates they would measure independently. The second term is in fact equal to the first for $x_1 = x_2$ and decreases to zero in general as x_1 recedes from x_2. It represents the positive correlation in the delayed coincidence rate first detected by Hanbury Brown and Twiss [1]. Its origin as we see, lies entirely in the noisiness of the field, i.e. the fluctuations of its amplitudes.

A simple way of generating other examples of fields is through superposition. The fields generated by two sources may be superposed linearly as long as neither source can be affected by the field produced by the other. To describe the superposed field we must be able to construct its density operator from a knowledge of the density operators of the two component fields. The rule which does this can be shown [4] to correspond to convolution of the functions $P(\alpha)$. If the functions $P_1(\alpha)$ and $P_2(\alpha)$ describe single mode excitations produced by the first and second sources respectively, then the total excitation is described by the function

$$P(\alpha) = \int P_1(\alpha')\, P_2(\alpha - \alpha')\, d^2\alpha'. \tag{9.20}$$

Other ways of stating the rule, in terms of the characteristic functions and quasi-probability functions, are readily found.

An example which frequently occurs nowadays is the superposition of a coherent signal and noise. For single mode excitations a coherent signal corresponds to a pure coherent state $|\beta\rangle$, and this is represented by the delta function given in eq. (9.2). The noise excitation is represented by the Gaussian function given in eq. (9.15). The superposed fields are therefore represented by the convolution integral

$$P(\alpha) = \int \delta^{(2)}(\alpha' - \beta)\, \frac{1}{\pi\langle n\rangle} \exp\left[-\frac{|\alpha - \alpha'|^2}{\langle n\rangle}\right] d^2\alpha'$$

$$= \frac{1}{\pi\langle n\rangle} \exp\left[-\frac{|\alpha - \beta|^2}{\langle n\rangle}\right], \tag{9.21}$$

which is simply a displaced form of the Gaussian function.

Apart from rather idealized cases, such as those we have been discussing, there is no simple way in general of guessing what density operator characterizes a field. Our understanding of what makes a density operator is still so elementary that the only quantum mechanically reliable way we have of finding ρ is to construct a mathematical model for the source and its coupling to the field and to solve for the time dependent description of the whole system in statistical terms. Analyses of this type have been and are being made for a number of elementary models of laser oscillators, amplifiers and attenuators. The ways of representing density operators which we have been discussing here have turned out to be quite useful in this new area of statistical mechanics.

10. PHOTON COUNTING DISTRIBUTIONS

Let us return now to the problem we stated at the outset. A photon counter is exposed to a given field from time zero to time t. What is the statistical distribution of the number of photons it records when this experiment is repeated many times? It is the tendency of the photon counts to occur at correlated times which makes an interesting question of this, and one which is not always easy to answer.

To find the distribution, let us begin by defining an operator $C(t)$ which represents the number of photons counted in time t. Such an operator clearly has integer eigenvalues $n = 0, 1, 2, \ldots$ It will depend on the variables which describe the counter system, as opposed to the field, but we do not need for present purposes any explicit construction of the operator. The introduction of an appropriately defined generating function makes it possible to give a fairly concise derivation of the unknown distribution. We let λ be an arbitrary parameter and define the generating function

$$\begin{aligned} Q(\lambda, t) &= \mathrm{Tr}\{\rho(1-\lambda)^{C(t)}\} \\ &\equiv \langle(1-\lambda)^{C(t)}\rangle, \end{aligned} \tag{10.1}$$

where ρ is a density operator which describes the counter as part of the system.

As a complete set of states to use in describing the final state of the counter and field it is clearly convenient to take the eigenstates of $C(t)$. These are defined to satisfy

$$C(t)|n, l, t\rangle = n|n, l, t\rangle, \tag{10.2}$$

in which l represents all of the quantum numbers other than the eigenvalue of $C(t)$. The trace which defines the generating function may then be expressed in terms of the set of numbers

$$p(n,t) = \sum_l \langle n,l,t|\rho|n,l,t\rangle\,, \tag{10.3}$$

for $n = 0, 1, 2, \ldots$. These are simply the probabilities that the counter records n photons in time t. The generating function reduces therefore to the simple expression

$$Q(\lambda,t) = \sum_{n=0}^{\infty} (1-\lambda)^n\, p(n,t)\,. \tag{10.4}$$

If we are able to solve directly for the generating function we can use it to evaluate the probabilities since they are the coefficients

$$p(n,t) = \frac{(-1)^n}{n!}\frac{\mathrm{d}^n}{\mathrm{d}\lambda^n}\, Q(\lambda,t)\Big|_{\lambda=1}\,, \tag{10.5}$$

in its power series expansion about $\lambda = 1$. We can also find the factorial moments of the probability distribution easily since they are just the derivatives

$$\langle C(C-1)\ldots(C-n+1)\rangle = (-1)^n \frac{\mathrm{d}^n}{\mathrm{d}\lambda^n}\, Q(\lambda,t)\Big|_{\lambda=0}\,, \tag{10.6}$$

of the generating function evaluated at $\lambda = 0$.

To begin the process of finding the generating function, let us note that the sensitive element of a photon counter consists of many atoms ($N \gg 1$), any of which may undergo a photoabsorption process and give rise to a detected photon count. The operator C may be written therefore, as a sum

$$C = \sum_{j=1}^{N} c_j \tag{10.7}$$

of operators c_j which represent the contributions of the individual atoms. The operators c_j are time-dependent in general, as is C, and have eigenvalues one and zero; one for final states to which jth atom has contributed a photon count and zero for states to which it has contributed none.

Since the operators c_j refer to the variables of altogether different atoms they commute, and we can write

$$(1-\lambda)^C = (1-\lambda)^{\Sigma c_j}$$

$$= \prod_j (1-\lambda)^{c_j}\,.$$

The fact that the eigenvalues of the c_j are only zero and one means that

$$(1-\lambda)^{c_j} = 1 - \lambda c_j \tag{10.8}$$

and the generating function may be written therefore as

$$Q(\lambda, t) = \langle \prod_j (1-\lambda c_j) \rangle \tag{10.9}$$

$$= 1 - \lambda \sum_j \langle c_j \rangle + \lambda^2 \sum_{j \neq l} \langle c_j c_l \rangle - \ldots . \tag{10.10}$$

Because the operators c_j are, in effect, projection operators on the states for which the jth atom has contributed a count, an expectation value $\langle c_1 c_2 \ldots c_n \rangle$ is simply the probability that all of the atoms $1, 2 \ldots n$ have contributed counts in time t. That probability is one we have already calculated (in nth order perturbation theory) and stated in eq. (4.7). It is

$$\langle c_1 \ldots c_n \rangle = p^{(n)}(t \ldots t) \tag{10.11}$$

$$= s^n \int_0^t \ldots \int_0^t G^{(n)}(\boldsymbol{r}_1 t_1' \ldots \boldsymbol{r}_n t_n', \boldsymbol{r}_n t_n' \ldots \boldsymbol{r}_1 t_1') \prod_{j=1}^n \mathrm{d}t_j',$$

where $\boldsymbol{r}_1 \ldots \boldsymbol{r}_n$ are the positions of the n atoms.

We can simplify our calculation somewhat at this point by assuming that the geometry of our experiment is such that all the sensitive atoms of the detector experience the same field, e.g. that plane waves are incident normally on a plane photocathode. This assumption means simply that all atoms are equivalent in the summations in eq. (10.10). (The assumption can easily be removed by replacing the sums over atoms by integrations over the detector volume.) If we assume furthermore that $N \gg n$, then we find

$$Q(\lambda, t) = \sum_{n=0} \frac{(-\lambda N s)^n}{n!} \int_0^t \ldots \int_0^t G^{(n)}(t_1' \ldots t_n', t_n' \ldots t_1') \prod \mathrm{d}t_j' . \tag{10.12}$$

This expression for the generating function shows us immediately via eq. (10.6) that the factorial moments of the photon counting distribution are

$$\langle C(C-1) \ldots (C-n+1) \rangle =$$

$$= (Ns)^n \int_0^t \ldots \int_0^t G^{(n)}(t_1' \ldots t_n', t_n' \ldots t_1') \prod \mathrm{d}t_j' . \tag{10.13}$$

Since it is not too difficult, in practice, to measure the higher order moments of a distribution, we see that the integrals of the higher order correlation functions are more or less directly accessible to measurement.

We can now make use of our coherent state representation of the density operator to reduce the expression for the generating function to a closed form. Let us suppose, for example, that we can define a P-representation for the density operator. Then the function $G^{(1)}(t_1', t_1')$ may be expressed as an integral over the amplitude parameters $\{\alpha_k\}$, as we have shown in eq. (7.23), and analogous integral representations can be used for all of the higher order $G^{(n)}$. By substituting these representations in eq. (10.12), we find that the sum over n can be carried out trivially, and leaves

$$Q(\lambda, t) = \int P(\{\alpha_k\}) \exp[-\lambda\Omega(\{\alpha_k\}, t)] \prod_k \mathrm{d}^2\alpha_k \,, \qquad (10.14)$$

where Ω is a bilinear form in the amplitudes $\{\alpha_k\}$ given by

$$\Omega(\{\alpha_k\}, t) = Ns \int_0^t |\mathcal{E}(t', \{\alpha_k\})|^2 \,\mathrm{d}t' . \qquad (10.15)$$

The simplest illustration of this result is the case of a pure coherent state. In that case the function P is a product of delta functions which constrains the amplitudes $\{\alpha_k\}$ to have a particular set of values, say $\{\beta_k\}$. Then the generating function reduces to

$$Q(\lambda, t) = \exp[-\lambda\Omega(\{\beta_k\}, t)] \qquad (10.16)$$

and we see from eq. (10.5) that the $p(n, t)$ form a Poisson distribution,

$$p(n, t) = \frac{\langle C\rangle^n}{n!} \exp[-\langle C\rangle] \,, \qquad (10.17)$$

about the mean value $\langle C\rangle = \Omega(\{\beta_k\}, t)$. The mean counting rate is just proportional to the instantaneous field intensity in this case

$$\frac{\mathrm{d}}{\mathrm{d}t}\langle C(t)\rangle = Ns|\mathcal{E}(t, \{\beta_k\})|^2 \,,$$

and in coherent states that may vary arbitrarily with time.

To the extent that the function P resembles a probability distribution it is clear that the function $Q(\lambda)$ given by eq. (10.14) will generate a kind of averaged Poisson distribution. We have already noted, however, that the function P may take on negative values and singularities unlike those of any probability distribution. It is not too surprising to find, therefore, that the distribution $p(n, t)$

need not actually have much resemblance to a Poisson distribution.

There is another way of constructing a closed form for the generating function which does not rely on the use of the P-representation. The general form of the coherent state representation given by eq. (7.10) can be used in much the same way and leads to an integral representation for $Q(\lambda t)$ analogous to eq. (10.14), but one in which two complex amplitudes must be integrated for each mode. When the function P is highly singular, as it is for example when the field has a fixed number of photons in a mode, this general representation is a good deal simpler to use; it leads to a binomial or Bernoulli distribution for the number of photons counted.

Another particularly useful example is one in which a single mode of the field is excited to a chaotic, or noisy state. The function $P(\{\alpha_k\})$, then is given by eq. (9.17) with all of the $\{\langle n_k\rangle\}$ equal to zero save one. If we define an average counting rate w by writing

$$\langle C(t)\rangle = \langle \Omega(\{\alpha_k\}, t)\rangle = wt\,, \tag{10.18}$$

then we find in this case that the Gaussian integral for the generating function has the value

$$Q(\lambda, t) = \frac{1}{1+\lambda wt}. \tag{10.19}$$

It follows that the photon counting probabilities have the geometrical distribution

$$p(n, t) = \frac{(wt)^n}{(1+wt)^{n+1}}. \tag{10.20}$$

A somewhat less elementary case is the superposition of a coherent signal and a noise excitation in a single mode. For this case we use the P-function given by eq. (9.21). The average counting rate w defined by eq. (10.18) is seen to consist of two terms in this case, one contributed by the noise excitation and the other by the signal. We may write

$$w = w_{\mathcal{S}} + w_{\mathcal{N}}, \tag{10.21}$$

where $w_{\mathcal{S}}$ and $w_{\mathcal{N}}$ are proportional to the intensities of their respective fields. Their ratio, in other words, is

$$\frac{w_{\mathcal{S}}}{w_{\mathcal{N}}} = \frac{|\beta|^2}{\langle n\rangle}, \tag{10.22}$$

which is the signal-to-noise ratio. Now if we let

$$\begin{aligned} \mathcal{S} &= w_{\mathcal{S}}\, t, \\ \mathcal{N} &= w_{\mathcal{N}}\, t \end{aligned} \tag{10.23}$$

be the numbers of quanta which would be counted in the signal and noise fields measured individually, then we find on evaluating the elementary integral for the generating function that it can be written as

$$Q(\lambda, t) = \frac{1}{1+\lambda\mathcal{N}} \exp\left[-\frac{\lambda\mathcal{S}}{1+\lambda\mathcal{N}}\right]. \tag{10.24}$$

This expression may be recognized as the generating function for the Laguerre polynomials L_m. By expanding it about $\lambda = 1$ we find

$$p(m,t) = \frac{\mathcal{N}^m}{(1+\mathcal{N})^{m+1}} L_m\left[-\frac{\mathcal{S}}{\mathcal{N}(1+\mathcal{N})}\right] \exp\left[-\frac{\mathcal{S}}{1+\mathcal{N}}\right]. \tag{10.25}$$

When the signal strength $|\beta|^2$ vanishes this distribution assumes the geometrical form given in eq. (10.20). When, in the opposite extreme, the noise excitation $\langle n\rangle$ vanishes, $p(m,t)$ reduces to the Poisson distribution characteristic of a coherent state.

Measurements which have been made in the beam of a gas laser operating well above threshold yield photon counting distributions which can be compared directly with the distributions in eq. (10.25). The measured distributions resemble Poisson distributions a good deal more than the geometrical ones of eq. (10.20), but have appreciably greater variances than Poisson distributions. They are found to be in excellent agreement [25] with the distribution (10.25) when the laser beam is assumed to be the superposition of a strong coherent excitation and a relatively weak chaotic one. It is possible to increase the strength of the superposed noise by separating part of the beam with a beam splitter, passing it through a random scatterer such as a moving ground glass screen and then rejoining the transmitted beam with the original one. Photon counting distributions measured in such synthetically noisy beams are also in good agreement [26] with the distribution (10.25).

It is only by letting the number of excited modes become infinite that we can deal with fields of arbitrary spectral intensities. The formalism we have constructed deals with multimode fields of all sorts including the continuum limit, but to evaluate the expression (10.14) for the generating function requires, naturally enough, somewhat more mathematics for the multimode cases than for the single mode ones. When the function P behaves like a probability density, as it does for all chaotic fields, the problem is very closely related to ones which have already been discussed in detail in classical noise theory. For some of the simpler cases, it is possible to find a closed and exact expression for $Q(\lambda, t)$. This does not complete the solution of the problem however, since there remains the task of finding the probabilities $p(m, t)$. In practice it is only possible to carry out exactly the sequence of differentiations required by eq. (10.5) if the expression for the generating function is fairly simple in its analytic form. The alternative is to use approximation methods and several of these have been devised.

It is interesting to note that comparisons of the generating function and several of its properties may be made directly with experiment. Let us consider the value of the generating function

$$Q(1,t) = p(0,t). \tag{10.26}$$

That is the probability that no photons are recorded in the time t, a function which is quite measurable for sufficiently short time intervals, i.e. before it too closely approaches zero. Another type of experiment closely related to this one is the measurement of the distribution of the time intervals which elapse before the occurrence of the first photon count. If we call this interval distribution $w(t)$ we clearly have

$$w(t) = -\frac{\mathrm{d}}{\mathrm{d}t}\,p(0,t) = -\frac{\mathrm{d}}{\mathrm{d}t}\,Q(1,t). \tag{10.27}$$

If the field we are investigating has a mean intensity which varies with time, then we must be careful always to begin the measurements of our time intervals at the same zero of time relative to the development of the field. If the field is statistically stationary on the other hand, the zero points for the time intervals may be chosen arbitrarily.

It is interesting to contrast this experiment with another which measures simply the distribution of time intervals between successive photon counts. To find this distribution, let us assume that the field is stationary. Then the probability that an initial photon count occurs in the interval Δt_0 about time t_0 and that no further photons are counted prior to time t is

$$\Delta t_0 \frac{\mathrm{d}}{\mathrm{d}t_0}\, p(0, t-t_0).$$

The probability that an initial count occurs in Δt_0 and the first subsequent one in the interval Δt following time t is

$$-\Delta t_0 \frac{\mathrm{d}}{\mathrm{d}t_0}\frac{\mathrm{d}}{\mathrm{d}t}\,p(0,t-t_0)\,\Delta t = \Delta t_0 \frac{\mathrm{d}^2}{\mathrm{d}t^2}\,p(0,t-t_0)\,\Delta t. \tag{10.28}$$

The distribution function we seek is the conditioned probability density $w(t_0|t)$ that, given a count at time t_0, the next occurs at time t. If the average counting rate in the stationary field is w, then the probability for the compound event of a count in Δt_0, none in the interval from t_0 to t, and then a count in the interval Δt is evidently

$$w\,\Delta t_0\, w(t_0\,|\,t)\,\Delta t = \Delta t_0 \frac{d^2}{dt^2} p(0, t-t_0)\,\Delta t. \qquad (10.29)$$

It follows that

$$w(t_0\,|\,t) = \frac{1}{w}\frac{d^2}{dt^2} Q(1, t-t_0). \qquad (10.30)$$

The distributions $w(t)$ and $w(0\,|\,t)$ are the same for a stationary coherent field, but are quite different in general for other types of fields. For a single-mode chaotic excitation, for example, the generating function $Q(1, t)$ is given by eq. (10.19) and we find

$$w(t) = \frac{w}{(1+wt)^2}, \qquad (10.31)$$

$$w(0\,|\,t) = \frac{2w}{(1+wt)^3}. \qquad (10.32)$$

The conditioned distribution is twice as large as the unconditioned one for $t = 0$ and decreases more rapidly with increasing time. The physical reason for this behavior is easy to see. When we require that a photon be counted at $t = 0$ we are giving additional weight to those cases in which the field strength has fluctuated to abnormally large values at that time. When the field strength is large the second photoabsorption process tends naturally to be hastened.

ACKNOWLEDGEMENT

These notes are based on a set prepared at Noordwijk by H. Van Vianen and M. Renne and a set prepared from related lectures at Harvard by T. Von Foerster. The author is grateful to them for their generous help.

REFERENCES

[1] R. Hanbury Brown and R.Q. Twiss, Nature 177 (1956) 27; Proc. Roy. Soc. A242 (1957) 300; Proc. Roy. Soc. A243 (1957) 291.
[2] R.J. Glauber, Phys. Rev. Letters 10 (1963) 84.
[3] R.J. Glauber, Phys. Rev. 130 (1963) 2529.
[4] R.J. Glauber, Phys. Rev. 131 (1963) 2766.
[5] R.J. Glauber, in: Quantum Optics and Electronics, Les Houches, 1964, Edited by C. DeWitt et al., (Gordon and Breach, New York, 1965) p. 63.
[6] R.J. Glauber, in: Physics of Quantum Electronics, Conference Proceedings, San Juan, Puerto Rico, 1965, Edited by P. L. Kelley et al., (McGraw Hill, New York, 1966) p. 788.

[7] L. Mandel, Phys. Rev. 152 (1966) 438.
[8] U.M. Titulaer and R.J. Glauber, Phys. Rev. 140 (1965) B676.
[9] E. Schrödinger, Naturwissenschaften 14 (1926) 644.
[10] R.J. Glauber, Phys. Letters 21 (1966) 650.
[11] U.M. Titulaer and R.J. Glauber, Phys. Rev. 145 (1966) 1041.
[12] E.H. Kennard, Z. Physik 44 (1927) 326.
[13] E. Merzbacher, in: Quantum Mechanics (Wiley and Sons, New York, 1961) p. 154.
[14] J.R. Klauder, Ann. Phys. N.Y. 11 (1960) 123.
[15] K.E. Cahill, Phys. Rev. 138 (1965) B1566.
[16] V. Bargmann, Commun. Pure and Appl. Math. 14 (1961) 187; Proc. Natl. Acad. Sci. U.S. 48 (1962) 199.
[17] E. Wigner, Phys. Rev. 40 (1932) 749.
[18] Y. Kano, J. Math. Phys. 6 (1965) 1913.
[19] C.L. Mehta and E.C.G. Sudarshan, Phys. Rev. 138 (1965) B753.
[20] J.E. Moyal, Proc. Cambridge Phil. Soc. 45 (1948) 99.
[21] J.R. Klauder, J. McKenna and D.G. Currie, J. Math. Phys. 6 (1965) 734.
[22] E.C.G. Sudarshan, Phys. Rev. Letters 10 (1963) 277.
[23] J.R. Klauder, Phys. Rev. Letters 15 (1966) 534.
[24] J.R. Klauder and E.C.G. Sudarshan, in: Fundamentals of Quantum Optics, W.A. Benjamin, Inc., to be published.
[25] C. Freed and H.A. Haus, I.E.E.E. J. Quantum Electronics, QE-2 (1966) 190.
[26] F.T. Arecchi, A. Berné and P. Burlamacchi, Phys. Rev. Letters 16 (1966) 32.

ON BOSE-EINSTEIN CONDENSATION

H. B. G. CASIMIR

Philips Research Laboratories, N. V. Philips' Gloeilampenfabrieken, Eindhoven, The Netherlands

1. INTRODUCTION

Superfluidity and superconductivity are most remarkable properties of matter in bulk. For many years they defied theoretical explanation; to give such an explanation has been one of the main tasks of modern statistical mechanics and many particle theory.

It is often said, that the superfluid or superconductive state is characterized by a macroscopically coherent many particle wavefunction. Such a statement can be rendered more exact by means of the density matrix. In superfluidity the matrix elements of the one-particle density matrix, $\rho(x,x')$ remain of the order of magnitude of the total number of particles N even when the distance between x and x' is of the order of linear dimensions of the system:

$$\rho(x,x') \approx N , \qquad \text{even if } |x-x'| \approx L .$$

In superconductivity a similar statement obtains for the two-particle density. Now it stands to reason that there must be some limitations to this long range coherence. I have often emphasized the difficulty of believing in a coherent wavefunction extending over a mile of dirty lead wire. There must be some limitation: there should not be any coherence between the helium in two different Dewars.

One, admittedly rather unrealistic, model that shows long range order below a transition temperature is the ideal Bose gas; it is also the only model that can easily be treated with adequate rigour. Studying Bose-Einstein condensation under imperfect conditions would appear to be useful as an elementary contribution to the problem indicated above and the results, though not directly applicable to real systems, are at least suggestive.

2. BOSE-EINSTEIN CONDENSATION IN A CUBE

We shall in the following always work at constant density. We write a^{-3} for the number of particles per cc. Further we write $\eta = \hbar^2/m$. Since we are not interested in exact numerical results we use the following convention: a can mean any length differing from the one defined above by a numerical factor of order one. Similarly η can also mean $C\eta$, where the number C is of order one (say $0.1 < C < 10$). Of course some care must be exercised in using this convention, but it does save a lot of time. Let L be the edge of the cube. The energy levels are

$$\epsilon = \frac{\eta}{L^2}(\nu_1^2 + \nu_2^2 + \nu_3^2) + \epsilon_{000} \; .$$

The total number of levels with energies between ϵ and $\epsilon + \mathrm{d}\epsilon$ is

$$\left(\frac{\eta}{L^2}\right)^{-\frac{3}{2}} \epsilon^{\frac{1}{2}} \, \mathrm{d}\epsilon = \frac{L^3}{\eta^{\frac{3}{2}}} \epsilon^{\frac{1}{2}} \, \mathrm{d}\epsilon \; .$$

The number of particles in one quantum state with energy ϵ is

$$n(\epsilon) = \frac{1}{\exp[(\epsilon - \zeta)/kT] - 1} \; ,$$

where $\zeta \leqslant \epsilon_{000}$ and determined by the condition

$$\sum n(\epsilon) = \frac{L^3}{a^3} \; .$$

Replacing the sum by an integral we have

$$\frac{L^3}{\eta^{\frac{3}{2}}} (kT)^{\frac{3}{2}} \int_0^\infty \frac{\mathrm{d}x}{\exp[x + (|\zeta|/kT)] - 1} = \frac{L^3}{a^3} \; .$$

The maximum value of this integral is obtained when $\zeta = 0$. Therefore it is impossible to satisfy this equation when $T < T_\mathrm{B}$ where

$$kT_\mathrm{B} = \frac{\eta}{a^2} \; .$$

For $T = \theta T_\mathrm{B}$ with $0 < \theta < 1$ there are $(1 - \theta^{\frac{3}{2}})L^3/a^3$ particles that cannot be accounted for by the integral. They have to go into the lowest state. If we take $|\zeta|$ (or more rigorously $\alpha = |\zeta| - \epsilon_{000}$) slightly above zero the number of particles in the lowest state will be

$$n_{000} = \frac{1}{\exp(\alpha/kT) - 1} \approx \frac{\theta\eta/a^2}{\alpha} .$$

Since this has to be equal to $(1 - \theta^{\frac{3}{2}})L^3/a^3$ we find

$$\alpha = f(\theta)\left(\frac{\eta}{L^2}\right)\frac{a}{L} ,$$

with

$$f(\theta) = \frac{\theta}{(1 - \theta^{\frac{3}{2}})} .$$

It should be obvious that the fact that a finite fraction of all particles is in the lowest state leads to long range coherence.

Also, for large L, α is much smaller than the distance to the next excited level, which is η/L^2. Therefore the occupation number of excited states is, in the limit $L \to \infty$, the same as for $\alpha = 0$ and the integral is an adequate approximation for the total number of particles in excited states.

A word should be inserted about this limit $L \to \infty$. Strictly speaking, the limit for an infinite number of particles is entirely irrelevant: we are not interested in the behaviour of mathematical abstractions but in the behaviour of a concrete physical system, which does always contain a finite number of particles. However, 10^{18} is already a pretty large number and therefore the behaviour of less than a cc may often correspond to the limit $N \to \infty$. The existence of such a limit and the fact that a system with finite but large N has properties close to this limit corresponds to the physical fact that the thermodynamic quantities are proportional to the volume. But the limit for $N \to \infty$ should only be applied in a concrete physical case after it has been established that it is a good approximation.

3. BOSE-EINSTEIN CONDENSATION WITH TUNNELING

In the introduction we pointed out that there can exist no phase coherence between two separate volumes of liquid helium. In order to study this question more closely we consider two volumes separated by a potential barrier. We assume that $0 < y < L$ and $0 < z < L$. In the x-direction, however, we assume the following potential distribution

$$x > D + d \text{ and } x < -D - d \qquad V \approx \infty \;;$$

$$d < x < D + d \text{ and } -D - d < x < -d \qquad V = 0 \;;$$

$$-d < x < d \qquad V = \eta\kappa^2 \;.$$

Also we assume $D \gg d$ and $\kappa \gg 1/D$. If the potential barrier in the middle were entirely impenetrable the lowest state would be doubly degenerate. For a finite barrier this level splits up and the corresponding wavefunction is either symmetric or antisymmetric. The energy difference can easily be calculated and is found to be

$$\delta\epsilon = \left(\frac{\eta}{D^3\kappa}\right) \exp(-2\kappa d) \;,$$

the symmetric state being the lowest one. But we have

$$\alpha = |\zeta| - \epsilon_0 = \frac{f(\theta)\eta a}{L^2 D} \;.$$

Therefore if

$$\frac{D^2}{L^2} < \frac{1}{f(\theta)} \frac{1}{a\kappa} \exp(-2\kappa d) \;,$$

the bosons will condense in the lowest, i.e. in the symmetric state. In this case there is coherence through the barrier. But if this condition is not fulfilled condensation will take place indiscriminately in the two states. This means that the two volumes behave independently.

It follows from our equation that even slight tunneling is sufficient to establish phase coherence. $(D^2/L^2) \approx 10^{-8}$ is quite feasible, and if we take $\kappa a \approx 1$ we would still have coherence for $\theta = \frac{1}{2}$ and d/a as large as 10. But for larger distances, for instance $d/a \approx 100$, L/D becomes quite astronomical, and since in practice D could hardly be less than 10^{-5} cm, no coherence-tunneling takes place for any realistic system when $d/a > 100$.

Further this example shows clearly that condensation in the thermodynamical sense and long range correlation are *not* one and the same thing. The thermodynamic functions are not influenced by the height or width of the potential barrier; the long range coherence is.

4. BOSE-EINSTEIN CONDENSATION WITH EQUIDISTANT LOWEST LEVELS

In our preceding example the behaviour of the condensation phenomenon is different from that in a cube because of the special energy spectrum for the two lowest levels. We shall now discuss a case where we modify the spectrum for a narrow band of lowest levels. Suppose that the L^3/l^3 lowest levels (with $a \ll l \ll L$) do not follow a parabolic distribution law but are equidistant. The higher levels, that is those with $\epsilon > \eta/l^2$ are assumed to be as in section 2. The distance between equidistant levels is now

$$\delta\epsilon = \frac{\eta}{L^2}\frac{l}{L}.$$

Our condition for determining $\alpha = |\zeta| - \epsilon_0$ becomes

$$\sum_{n=0}^{L^3/l^3} \frac{1}{\exp[(n\delta\epsilon + \alpha)/kT] - 1} + \frac{L^3}{\eta^{\frac{3}{2}}} \int_{\eta/l^2}^{\infty} \frac{\epsilon^{\frac{1}{2}} d\epsilon}{\exp[(\epsilon + \alpha)/kT] - 1} = \frac{L^3}{a^3}.$$

Now below T_B we can again put $\alpha = 0$ in the integral and if $\eta/l^2 \ll kT$ we can even integrate from $\epsilon = 0$. Also, we can approximate the exponential in the denominator of the sum by the first term of its series expansion leading to

$$\sum_{n=0}^{L^3/l^3} \frac{kT}{n\eta l/L^3 + \alpha} = (1 - \theta^{\frac{3}{2}}) \frac{L^3}{a^3}.$$

Now we must distinguish two limiting cases. It may be that it is legitimate to neglect all terms in the series but the one with $n = 0$. Then we find

$$\alpha = f(\theta)\,\frac{\eta a}{L^3},$$

just as before. The terms from $n = 1$ to $n = L^3/l^3$ now lead to a sum (remember that $a \ll l$)

$$\sum_{n=1}^{L^3/l^3} = \theta\,\frac{L^3}{la^2} \log \frac{L^3}{l^3}.$$

This is negligible compared to the $n = 0$ term if

$$\log \frac{L^3}{l^3} \ll \frac{1}{f(\theta)} \frac{l}{a},$$

or

$$L \ll l \exp(l/3fa) .$$

The behaviour of our bosongas with modified spectrum is identical with that of a normal bosongas if this condition is fulfilled. We should now remember what we said before about the limit $L \to \infty$. Take for instance $a \approx 10^{-8}$ and $l \approx 10^{-6}$ then $L \ll 10^7$ or $\ll 100$ km. Our condition can thus hold even for quite large systems.

If on the other hand the condition does not obtain then we put $\alpha = n_0 \eta l/L^3$ and we find

$$\log \frac{L^3/l^3}{n_0} = \frac{1}{f(\theta)} \frac{l}{a},$$

or

$$n_0 = \frac{L^3}{l^3} \exp[-l/f(\theta)a] ,$$

or

$$\alpha = \frac{\eta}{l^2} \exp[-l/f(\theta)a] .$$

From a thermodynamic point of view this is essentially zero. But there is no macroscopic phase coherence over the whole of the cube. Condensation is in a narrow band, not in one single state and the width of this band is independent of the size of the cube.

It may seem a somewhat useless exercise to play around with arbitrary distributions of energy levels. The point is, that we wanted to show that a slight change in the distribution of the lowest levels does not change the thermodynamics and does still permit very long range coherence, but that this coherence does not extend beyond certain limits. Also, the spectrum we assumed may bear some resemblance to the actual spectrum for particles moving in a box with arbitrarily distributed scattering centres, l being a measure for the free path.

5. BOSE-EINSTEIN CONDENSATION IN A LONG PRISM

Consider a prism of length L and edge D. Using the analysis of section 2, we find

$$\alpha = |\zeta| - \epsilon_{000} = f(\theta)\,\frac{\eta}{D^2}\,\frac{a}{L}.$$

On the other hand the energy levels are given by

$$\epsilon = \eta\,\frac{\nu_1^2 + \nu_2^2}{D^2} + \frac{\nu_3^2}{L^2} + \epsilon_{000}$$

and therefore the lowest L-excitation lies below α when

$$L > \frac{D^2}{a}.$$

Again, we have to consider questions of limits. If L and D are increased simultaneously L/D remaining constant, then, for $L \to \infty$ we shall always have

$$\frac{L}{D} < \frac{D}{a}.$$

But if we choose for D a fixed value $\gg a$, so that a cube D^3 could be treated with good approximation as infinite, and if we then increase L we arrive at a situation where $L > D^2/a$.

We then put

$$\alpha = |\zeta| - \epsilon_{000} = (\eta/L^2)\,\nu_o^2$$

and the occupation number of a state with $\nu_1 = \nu_2 = 1$ and ν_3 arbitrary becomes

$$n(1,1;\nu_3) = \frac{\theta(\eta/a^2)}{(\eta/L^2)(\nu_3^2 + \nu_o^2)},$$

leading to

$$\sum_{\nu_3} n(1,1;\nu_3) = \frac{\theta L^2/a^2}{\nu_o}.$$

Therefore

$$\frac{\theta L^2/a^2}{\nu_0} = \frac{LD^2}{a^3}(1 - \theta^{\frac{3}{2}})$$

and

$$\nu_0 = \frac{aL}{D^2} f(\theta) , \qquad \alpha = \eta \frac{a^2}{D^4} f(\theta)^2 .$$

The bosons condense in a narrow energy band. Again the thermodynamic behaviour is not different from that of an Einstein-Bose gas in a cube, but the range of coherence is $D \cdot D/a$.

6. DISCUSSION

The preceding calculations show clearly that in Bose-Einstein condensation a difference should be drawn between the behaviour of the thermodynamic functions and the coherence phenomenon. For a body of sufficient size the thermodynamic functions will not be affected by changes of shape, or by impurities and barriers. We have also seen that coherence can prevail in the face of formidable odds: it can exist in very long thin prisms, in the presence of impurities and through high potential barriers. But we have established definite limits beyond which coherence does not exist and this confirms the ideas expressed in the introduction.

Of course a gas of non-interacting bosons is a highly artificial model but the simple and general nature of our formulae encourages us to believe that similar conditions should prevail in superfluid helium and in superconductors. This would mean that vortex quantization and flux quantization might disappear in very long thin channels or wires especially if there are obstacles or impurities and that tunneling depends on size effects. And in all cases such effects would depend on temperature.

If we admit that there is a connection between flux quantization and persisting currents then our results might lead us to believe that a loop of thin and impure superconducting wire will show traces of resistance below the transition point. Moreover the temperature at which this resistance appears should depend on the length of the loop; resistance should not be proportional to length. It seems unlikely that difficulties encountered in making large coils of hard superconductors have anything to do with this effect, but special experiments in search of such an effect might easily be designed [1].

REFERENCE

H. B. G. Casimir, Proc. Roy. Acad. Sci. Amsterdam 69 (1966) 223.

QUANTUM MECHANICS OF INFINITELY LARGE SYSTEMS

N. M. HUGENHOLTZ
Department of Theoretical Physics, University of Groningen, The Netherlands

1. MATHEMATICAL INTRODUCTION

In this introduction some general properties of C^*-algebras and their representations will be discussed [1] and [2].

1.1. C^*-*algebras*

Definition 1: Let $\mathfrak{A}$ be an algebra. An involution in $\mathfrak{A}$ is a mapping, $A \to A^*$ of $\mathfrak{A}$ in itself such that

$$(A^*)^* = A$$

$$(\lambda A + \mu B)^* = \lambda^* A^* + \mu^* B^*$$

$$(AB)^* = B^* A^* .$$

An algebra with involution is a $*$-algebra. A^* is the adjoint of A. If $A^* = A$ we say that A is selfadjoint.

Definition 2: An algebra $\mathfrak{A}$ is normed if every element A is provided with a norm $\|A\|$ such that

$$\|A\| \geqslant 0$$

$$\|A\| = 0 \text{ iff } A = 0$$

$$\|\lambda A\| = |\lambda| \; \|A\|$$

$$\|A + B\| \leqslant \|A\| + \|B\|$$

$$\|AB\| \leqslant \|A\| \cdot \|B\|$$

$$\|1\| = 1 .$$

Definition 3: $\mathfrak{A}$ is a normal *-algebra if $\mathfrak{A}$ is a normed algebra with an involution such that

$$\|A^*\| = \|A\| .$$

If in addition $\mathfrak{A}$ is complete in the norm-topology, $\mathfrak{A}$ is a Banach *-algebra.
Definition 4: A C*-algebra is a Banach *-algebra which satisfies

$$\|A^*A\| = \|A\|^2 .$$

A C*-algebra with a commutative product is called abelian.
Examples of C^*-algebras:

1. Consider the set C of all complex continuous functions $f(x)$ on the real line, which vanish for $|x| \to \infty$. With complex conjugation as involution and a norm $\|f\| = \sup|f(x)|$ the set C is an abelian C^*-algebra.

2. Consider the set of all bounded operators $\mathfrak{B}(\mathfrak{H})$ in a Hilbert space $\mathfrak{H}$. The norm $\|O\|$ of a bounded operator O is defined by

$$\|O\| = \sup_{\psi} \frac{\|O\psi\|}{\|\psi\|} .$$

Provided with this norm and with the usual definition of the adjoint $\mathfrak{B}(\mathfrak{H})$ is a C^*-algebra.

3. Any norm-closed *-subalgebra of $\mathfrak{B}(\mathfrak{H})$ is a C^*-algebra. The importance of C^*-algebras in physics follows from the fact that it can be shown that any C^*-algebra is isomorphic to a subalgebra of $\mathfrak{B}(\mathfrak{H})$.

1.2. *States*

Definition 5: A linear functional $\omega(A)$ on a C^*-algebra $\mathfrak{A}$ is called *a state* if $\omega(A^*A) \geqslant 0$ and if the norm $\|\omega\| \equiv \omega(1) = 1$. A state ω has the following properties:

1. $|\omega(A)| \leqslant \|A\|$.

2. $|\omega(A^*B)|^2 \leqslant \omega(A^*A) \cdot \omega(B^*B)$.

The states form a *convex set,* i.e. if ω_1 and ω_2 are states, each linear combination $\lambda\omega_1 + (1-\lambda)\omega_2$ with $0 \leqslant \lambda \leqslant 1$ is again a state.
Definition 6: A state ω is *pure* if there does not exist a number λ: $0 < \lambda < 1$ and two different states ω_1 and ω_2 such that

$$\omega = \lambda\omega_1 + (1-\lambda)\omega_2 .$$

A state which is not pure is called a *mixed state*.

Definition 7: An automorphism of $\mathfrak{A}$ is a linear mapping $A \in \mathfrak{A} \rightarrow A_T \in \mathfrak{A}$ such that $(AB)_T = A_T B_T$, $(A^*)_T = A_T^*$ and $\|A_T\| = \|A\|$.

An automorphism of $\mathfrak{A}$ gives rise to a transformation $\omega \rightarrow \omega_T$ of the states, defined by

$$\omega_T(A) = \omega(A_T) \text{ for all } A \in \mathfrak{A} .$$

We say that ω is invariant if $\omega_T = \omega$.

1.3. *Representations*

Definition 8: A representation of $\mathfrak{A}$ is a homomorphic mapping $A \rightarrow R(A)$ of $\mathfrak{A}$ into the bounded operators of a Hilbert-space $\mathfrak{H}$. A representation is *faithful* if the mapping is isomorphic.

It can be shown that $\|R(A)\| \leqslant \|A\|$, where the equality holds for faithful representations.

Definition 9: A representation of $\mathfrak{A}$ is *cyclic*, if there is a vector $\Omega \in \mathfrak{H}$, such that the set $\{R(A)\Omega,\ A \in \mathfrak{A}\}$ is dense in $\mathfrak{H}$; Ω is a cyclic vector.

Definition 10: Two representations R_1 in $\mathfrak{H}_1$ and R_2 in $\mathfrak{H}_2$ are *equivalent* if there is a unitary mapping U from $\mathfrak{H}_1$ on $\mathfrak{H}_2$ such that $UR_1(A)\,U^{-1} = R_2(A)$ for all $A \in \mathfrak{A}$.

We shall now see that there is an intimate connection between states and cyclic representations. A given representation defines many states $\omega_\Psi = (\Psi, R(A)\,\Psi)$, $\Psi \in \mathfrak{H}$. This is in general a subset of all states. Conversely: To every state ω one can construct a cyclic representation $\mathfrak{A} \rightarrow R(\mathfrak{A})$ with a cyclic vector Ω such that

$$\omega(A) = (\Omega,\ R(A)\,\Omega) .$$

This construction, called the *G.N.S.-construction* after Gel'fand Naimark and Segal, proceeds as follows. Let the state ω be given and let $\mathcal{I} = \{A : \omega(A^*A) = 0\}$. It is evident that $\mathcal{I}$ is a left ideal of $\mathfrak{A}$. Consider the factorspace $\mathfrak{A}/\mathcal{I}$ of residue classes ξ with respect to $\mathcal{I}$. If ξ_A denotes the residue class containing the element A, then $\xi_A = \xi_B$ if $A - B \in \mathcal{I}$. This factorspace is provided with a scalar product by

$$(\xi_A, \xi_B) = \omega(A^*B) .$$

It is easily seen that this definition fulfils all requirements of a scalar product and is independent of the choice of A and B. The ideal $\mathcal{I}$ is the zero-element in the factor space and denoted by 0. We have obtained an incomplete space, the completion of which is a Hilbert-space $\mathfrak{H}$. We shall now proceed to construct a representation of $\mathfrak{A}$ in $\mathfrak{H}$.

We define $R(A)$ by

$$R(A)\xi_B = \xi_{AB} \,. \tag{1.1}$$

This definition is again independent of the choice of $B \in \xi_B$. Equation (1.1) defines $R(A)$ on a dense set. On the other hand

$$\|R(A)\xi_B\|^2 = (\xi_{AB}, \xi_{AB}) = \omega(B^*A^*AB) \leq \omega(B^*B) \, \|A\|^2 = \|A\|^2 \, \|\xi_B\|^2 \,.$$

Hence $\|R(A)\| \leq \|A\|$ so that $R(A)$ is bounded. It is immediately seen that the mapping $A \to R(A)$ is homomorphic. Indeed, $R(AB) = R(A)\,R(B)$, $R(A)^* = R(A^*)$. Hence $R(\mathfrak{A})$ is a representation of $\mathfrak{A}$. Finally, if $\Omega = \xi_1$, then Ω is cyclic, since $R(A)\,\Omega = \xi_A$ for all $A \in \mathfrak{A}$ is dense in $\mathfrak{H}$, and $(\Omega, R(A)\,\Omega) = (\xi_1, \xi_A) = \omega(A)$, q. e. d.

1.4. *Von Neumann algebras*

We recall that a bounded operator O has a norm $\|O\|$. This norm defines a topology. We say that O is the norm limit of the sequence O_n if $\|O - O_n\| \to 0$; Notation: $O_n \overset{\text{norm}}{\to} O$. The realization $R(\mathfrak{A})$ of a C^*-algebra is closed in this topology.

We shall also need another limiting process, the so-called weak topology. We say O is the weak limit of the sequence O_n if $(\Psi, O\,\Phi) = \lim_{n\to\infty} (\Psi, O_n\,\Phi)$ for all Ψ and $\Phi \in \mathfrak{H}$; notation $O_n \overset{\text{W}}{\to} O$.
The weak topology is weaker than the norm topology in the sense that $O_n \overset{\text{norm}}{\to} O$ implies $O_n \overset{\text{W}}{\to} O$, but not conversely. In particular, a weakly closed algebra of bounded operators is also norm-closed, but not conversely.

Definition 11: A weakly closed algebra $\mathcal{R}$ of bounded operators containing the identity is a *von Neumann algebra* or a W^*-algebra.

Definition 12: The commutant $\mathcal{R}'$ of a von Neumann algebra $\mathcal{R}$ is the set of all bounded operators commuting with all operators of $\mathcal{R}$.

One can show that $\mathcal{R}'$ is also a von Neumann algebra. According to a famous theorem of von Neumann one has

$$\mathcal{R} = \mathcal{R}'' \,.$$

Definition 13: A von Neumann algebra $\mathcal{R}$ is a *factor* if the center $\mathcal{C} = \mathcal{R} \cap \mathcal{R}'$ does not contain other operators than multiples of 1.

Let $R(\mathfrak{A})$ be a representation of $\mathfrak{A}$ in $\mathfrak{H}$. As we saw $R(\mathfrak{A})$ is norm-closed. We can increase the algebra $R(\mathfrak{A})$ by adding all weak limitpoints. This leads to the von Neumann algebra $\overline{R(\mathfrak{A})}^w = \mathcal{R}$.

Definition 14: The representation $R(\mathfrak{A})$ is *irreducible* if $\mathcal{R} = \mathfrak{B}(\mathfrak{H})$; the representation is *primary* if $\mathcal{R}$ is a factor.

That "pure" and "irreducible" are almost identical concepts follows from

Theorem 1: The cyclic representation determined by a pure state is irreducible. Conversely, every vector of an irreducible representation gives rise to a pure state.

We shall end this chapter with some remarks concerning automorphisms. An automorphism $A \to A_T$ is *inner* if there exists a unitary element $Q_T \in \mathfrak{A}$ such that $A_T = Q_T A Q_T^{-1}$. The automorphisms we shall be interested in are all outer automorphisms. Let therefore $A \to A_T$ be an outer automorphism, and let $R(A)$ be a representation. Then the mapping $R(A) \to R(A_T)$ is an automorphism of $R(\;)$. We say that the automorphism is *implemented* if there exists a unitary operator U_T such that $R(A_T) = U_T R(A) U_T^{-1}$. It is clear that $U_T \notin R(\mathfrak{A})$. In some cases it happens that $U_T \in \mathcal{R} = \overline{R(\mathfrak{A})}^w$. The following theorem gives a sufficient condition for the existence of U_T.

Theorem 2: If ω is invariant for the automorphism $A \to A_T$, then the automorphism is implemented in the cyclic representation defined by ω. The unitary operator U_T is uniquely determined by the condition $U_T\Omega = \Omega$.

2. INFINITE QUANTUM SYSTEMS

2.1. *Introductory remarks*

In the many body problem and in statistical mechanics one studies systems with infinitely many degrees of freedom. Since actual systems are finite but large, it means that one studies a model which not only is mathematically simpler than the actual system, but also allows a more precise formulation of phenomena such as phase transitions, transport processes, which are typical for macroscopic systems.

How does one deal with infinitely large systems? The traditional approach has been to consider large but finite systems and to take the thermodynamical limit at the end. Although from a logical point of view this approach may seem unattractive, there was actually no choice, since a systematic classical or quantum theory of infinitely large systems was not available. In recent years progress has been made in an attempt to set up a theory (either quantum or classical) for infinitely large systems. These methods have first been developed in field theory (a field has infinitely many degrees of freedom) [3]. This series of lectures is intended to give an introduction to the quantum theory of infinitely large systems with special emphasis on systems consisting of interacting identical particles [4-8].

2.2. *Fock-space*

Our starting point will be the formulation of the problem by means of second quantization. All operators are then expressed in terms of the field operators $\psi(x)$ and $\psi^*(x)$ which annihilate and create a particle at the point x. They satisfy the canonical commutation (or anti-commutation) relations

$$[\psi(x), \psi(y)]_\pm = 0 , \qquad [\psi(x), \psi^*(y)]_\pm = \delta^3(x-y) .$$

The sign in the commutators depends on the type of particles. Since these operators are highly singular we prefer to use other operators $\psi(f)$ and $\psi^*(f)$, which are related to $\psi(x)$ and $\psi^*(x)$ by

$$\psi(f) = \int d^3x \, \psi(x) \, f(x) , \qquad \psi^*(f) = \int d^3x \, \psi^*(x) \, f(x) ,$$

where the testfunctions $f(x)$ must be chosen real and vanishing outside some bounded region. These smeared-out operators satisfy the commutation relations

$$[\psi(f), \psi(g)]_\pm = 0 , \qquad [\psi(f), \psi^*(g)]_\pm = (f, g) . \tag{2.2}$$

The Hilbert-space $\mathfrak{H}_F$ on which these operators act is called Fock-space. There is one vector φ_0, the zero particle state, which satisfies

$$\psi(f) \, \varphi_0 = 0 \text{ for all } f .$$

Fock-space is generated by repeated application of operators $\psi^*(f)$.

Among the operators in Fock-space one can distinguish local operators from global operators. Whereas local operators effect the system only in a finite region, global operators act on the system as a whole. A simple example of a local operator is $\psi(f)$ with f having support in a finite region V, while the particle number operator $N = \int d^3x \, \psi^*(x) \, \psi(x)$ and the energy operator or hamiltonian are global operators.

In contrast with the case of a finite system, where in principle all self adjoint operators are observable, it is evident that global operators are not observable. Since our measuring apparatus has finite dimensions, every observation of an infinite system is bound to be local. As we shall see later, it is possible in certain cases to compute global quantities by means of an appropriate limiting process.

We shall now define our algebra of observables. Let V be any bounded region and consider the algebra generated by the

$\psi(f)$, $\psi^*(f)$ for all functions f with support in V. By taking the weak closure we get the von Neumann algebra $\mathfrak{A}(V)$. The algebra $\mathfrak{A}_L = \bigcup_V \mathfrak{A}(V)$ is the algebra of all local operators. The algebra $\mathfrak{A}$ of observables is the closure in the norm topology of $\mathfrak{A}_L$. Hence $\mathfrak{A}$ contains not only strictly local operators but also limit-points in the norm. Such operators we shall call quasi-local. The algebra $\mathfrak{A}$ which is a C^*-algebra is a smaller set than the set of all bounded operators $\mathfrak{B}(\mathfrak{H}_F)$, which equals the weak closure of $\mathfrak{A}_L$. It is precisely this difference between norm-closure and weak closure which excludes the global operators from $\mathfrak{A}$. The definition so far was applicable only to fermions, i.e., to particles with anti-commutation relations. In the case of bosons the operators $\psi(f)$ and $\psi(f)^*$ are not bounded. The algebra $\mathfrak{A}_L$ is then an algebra of operators with infinite norm, so that the norm closure is meaningless. Since also in the case of bosons we want to have a C^*-algebra of observables we proceed as follows. We define the self-adjoint operators

$$\varphi(f) = \frac{1}{\sqrt{2}} (\psi(f) + \psi^*(f))$$

$$\pi(f) = \frac{-i}{\sqrt{2}} (\psi(f) - \psi^*(f)) ,$$

with the canonical commutation relations

$$[\varphi(f), \varphi(g)] = [\pi(f), \pi(g)] = 0$$

$$[\varphi(f), \pi(g)] = i(f, g) .$$

We then introduce unitary operators (Weyl operators):

$$U(f) = e^{i\varphi(f)} , \qquad V(f) = e^{i\pi(f)} ,$$

with the commutation relations

$$U(f)\ V(g) = V(g)\ U(f)\ e^{-i(f,g)}$$

$$U(f)\ U(g) = U(f+g) \tag{2.3}$$

$$V(f)\ V(g) = V(f+g) .$$

The operators $U(f)$ and $V(f)$ are unitary and hence bounded for all f. These operators can now be used instead of $\psi(f)$ and $\psi^*(f)$ to generate the von Neumann algebra $\mathfrak{A}(V)$ for every bounded region

V, and $\mathfrak{A}_L = \bigcup_V \mathfrak{A}(V)$. $\mathfrak{A}_L$ is now an algebra of bounded operators. Its norm-closure $\mathfrak{A}$ is the C^*-algebra of observables.

2.3. *States*

We shall find it necessary to generalize the notion of state. In ordinary quantum mechanics a pure state is characterized by a vector ψ. It allows one to find the expectation values of all observables: $\langle A \rangle = (\psi, A\psi)$ for all $A \in \mathfrak{A}$. In a somewhat more general case we have a density operator ρ with $\mathrm{Tr}\, \rho = 1$, from which we derive the average value

$$\langle A \rangle = \mathrm{Tr}\, \rho\, A\ .$$

Such states we call normal states. Vectorstates and normal states are special cases of a state as defined in 1. A state is a general prescription which assigns a number to each observable A; a positive number to a positive observable, and the dependence should be linear. One might ask the question whether there are other states than vector states and normal states. The answer depends on the algebra of observables. In the case of truly finite systems the normal states are the most general states. In our case there are other more general states as we see from the following argument. Fock-space is generated by a dense set of vectors, each of which describes a state of a finite number of particles. This has the consequence that the density of particles for every vector state and even for every normal state vanishes.

Since there are physical states with non-vanishing particle density there must be non-normal states, and all states in statistical mechanics are of this general type.

2.4. *Transformations*

A study of the automorphisms of $\mathfrak{A}$ is indispensable for a classification of states.

1. *Translations in space.* Let $f_{\boldsymbol{x}}$ be the function f shifted over a distance $\boldsymbol{x}$, then we define the shifted operator $\psi(f)_{\boldsymbol{x}}$ by the equation

$$\psi(f)_{\boldsymbol{x}} = \psi(f_{\boldsymbol{x}})\ .$$

The shifted operators $\psi(f)_{\boldsymbol{x}}$ and $\psi^*(g)_{\boldsymbol{x}}$ satisfy the commutation relations as $\psi(f)$ and $\psi^*(g)$ since $(f_{\boldsymbol{x}}, g_{\boldsymbol{x}}) = (f, g)$. All observables can be expressed in terms of ψ's and ψ^*'s. We conclude that the translations in space define a group of automorphisms of $\mathfrak{A}$.

2. *Evolution in time.* If the interparticle forces are known we

can write down the hamiltonian in Fock-space. For two particle forces we have

$$H_{\mathrm{F}} = \frac{\hbar^2}{2m} \int \mathrm{d}^3x \; (\nabla \psi^*(x)) \; (\nabla \psi(x))$$

$$+ \tfrac{1}{2} \int \mathrm{d}^3x \int \mathrm{d}^3y \; V(x-y) \; \psi^*(x) \; \psi^*(y) \; \psi(y) \; \psi(x) \; .$$

The operator for the evolution in time is $U_{\mathrm{F}}(t) = \exp(\mathrm{i}H_{\mathrm{F}}t)$. Let $A \in \mathfrak{A}$, then we define $A_t = U_{\mathrm{F}}(t) \, A \, U_{\mathrm{F}}(t)^{-1}$, where $U_{\mathrm{F}}(t)$ is a global operator, so that $U_{\mathrm{F}}(t) \notin \mathfrak{A}$, and we are faced with the question whether A_t belongs to $\mathfrak{A}$. The answer in general depends on $\mathfrak{A}$ and on H_{F}. Although a rigorous proof has not yet been given we shall assume that in all cases of physical interest time-evolution is an automorphism of $\mathfrak{A}$.

3. *Gauge transformation.* The transformation $\psi(f) \rightarrow \psi(f)\, \mathrm{e}^{\mathrm{i}\alpha}$ for all f leaves the commutation relations invariant. Gauge transformations define a one-parameter group of automorphisms. In most cases these three groups of transformations commute with each other.

Due to the quasi-local character of the observables one can prove that

$$\lim_{|x| \rightarrow \infty} \| [A_x, B] \| = 0 \; . \tag{2.4}$$

We say therefore that $\mathfrak{A}$ is asymptotically abelian for space translation. It is believed that a similar property holds with respect to time, but proofs are lacking, except in the case of free particles.

A state which is invariant with respect to translations in space and time we call respectively homogeneous and stationary. If a state is invariant for gauge transformations we say that the state has a fixed particle number.

2.5. *Distribution functions and Green's function*

We shall now exhibit the connection between our description of states with other well known methods in statistical mechanics: the reduced distribution functions and the Green's function method [9]. The connection is most easily established for the case of fermions, so that we shall limit our discussion in this section to fermions. As we have seen, every observable, i.e., every element of $\mathfrak{A}$ can be approximated in the norm by an algebraic expression in terms of $\psi(f)$ and $\psi^*(f)$ operators, where f may be any function with compact support (i.e. vanishing outside some finite region). Due to the linearity and continuity of every state ω, a state is determined as soon as its value for any finite product of ψ's and ψ^*'s is known. This means that ω is determined by an infinite set of functionals $F_{n,m}(f_1, f_2, \ldots, f_n; g_1, g_2, \ldots, g_m)$ by means of the equations

$$\omega\{\psi(f_1)\psi(f_2)\ldots\psi(f_n)\psi^*(g_1)\psi^*(g_2)\ldots\psi^*(g_n)\}$$
$$= F_{n,m}(f_1, f_2, \ldots, f_n; g_1, g_2, \ldots, g_m) ,$$

where the functions f_i and g_j have compact support. Being less rigorous one can take functions instead of functionals and write

$$\omega\{\psi(x_1)\ldots\psi(x_n)\,\psi^*(y_1)\ldots\psi^*(y_m)\} = F_{n,m}(x_1, \ldots, x_n; y_1, \ldots, y_m) .$$

These functions $F_{n,m}$ are the reduced distribution functions which are frequently used both in classical and in quantum statistical mechanics. This shows that the description of systems by means of the reduced distribution functions is equivalent to our mathematical definition of state. If ω is not stationary, one should take ω_t instead of ω to describe the evolution in time. In that case the distribution functions will become time-dependent.

In the Green's function method one takes a different time in each factor so that one obtains functions of the general form

$$G_{n,m}(x_1t_1, \ldots, x_nt_n; y_1t_1', \ldots, y_mt_m')$$
$$= \omega\{\psi(x_1t_1)\ldots\psi(x_nt_n)\,\psi^*(y_1t_1')\ldots\psi^*(y_mt_m')\} .$$

Since already the distribution functions, where all times are equal, determine the state, it is evident that the infinite set of Green's functions gives an overcomplete description of the state of the system. In the last chapter we shall once more discuss Green's functions in connection with thermal equilibrium states.

2.6. *Stationary states*

The main problem we shall be concerned with in the remaining part of these notes is to determine equilibrium states of our infinite systems and to analyse their properties. At this point we make a few preliminary remarks.

The ground state or an equilibrium state is stationary, i.e., the state is invariant for time translations. That means that in the representation-space determined by such a state ω there is a unitary operator $U(t)$ (not to be confused with the operator $U_F(t)$ in Fock-space) with the properties that $R(A_t) = U(t)\ R(A)\ U(t)^{-1}$ and $U(t)\ \Omega = \Omega$ (see 1 theorem 2). If we write $U(t) = \exp(\mathrm{i}\ Ht)$, we can call H the hamiltonian. While it is obvious that the ground state or an equilibrium state is stationary, it is less obvious that the state should be homogeneous, even though space translations commute with time translations. A well-known counterexample is a crystall.

In the next chapter we shall study the case of the ideal Bose gas. This simple model is already quite interesting in itself, and it turns out that some of the remarkable properties of this system remain true for more general systems.

3. THE IDEAL BOSE GAS

Following Araki and Woods [4] we shall in this chapter discuss properties of the ground state and of equilibrium states of the ideal Bose gas. We shall take the pragmatic point of view that the rules for dealing with a finite system in the ground state or in equilibrium are known. The corresponding states of our infinite gas are then found by taking the thermodynamical limit.

As we have seen in section 2 the algebra $\mathfrak{A}$ of observables is the norm-closure of the algebra generated by all operators $U(f)$ and $V(f)$, where the functions f have compact support ‡.

Using the linearity and the continuity of each state ω, the state is determined if we know the value ω for all finite products of operators U and V. Using the commutation relations (2.3) every such product can be reduced to a simple product of the form $U(f)\ V(g)$. Consequently the state ω is completely determined by the functional $E(f,g)$ defined by

$$E(f,g) = \omega\{U(f)V(g)\} \tag{3.5}$$

for all f and g with compact support. Our first task will be to determine this functional for the ideal Bose gas in its ground state. In section 3.2 we shall discuss the equilibrium state at finite T.

3.1. *The ground state*

For a finite system enclosed in a cubic box of volume V with periodic boundary conditions one can write

$$\psi(x) = V^{-\frac{1}{2}} \sum_k a_k \mathrm{e}^{ikx} ,$$

where a_k and a_k^* are the annihilation and creation operator for a particle of momentum k. If $f(x)$ is some arbitrary real function with support in V, we have

‡ This is not quite correct, since the definition of $\mathfrak{A}$ also involves the weak closure of each local algebra. For simplicity we shall not discuss this slight complication.

$$\psi(f) = V^{-\frac{1}{2}} \sum_k a_k \hat{f}(k)^* , \tag{3.6}$$

with

$$\hat{f}(k) = \int d^3x \, f(x) \, e^{-ikx} . \tag{3.7}$$

The ground state Φ_o of the finite Bose gas is the state where all particles are in the zero momentum state, i.e. Φ_o satisfies the relations:

$$a_k \Phi_o = 0 \text{ for } k \neq 0$$

$$a_o^* a_o \Phi_o = N \Phi_o ,$$

where N is the particle number. We first calculate expectation values of products of a_k's and a_k^*'s. These expectation values vanish unless the number of creation operators equals the number of annihilation operators. We find immediately

$$\langle a_k^* a_l \rangle = N \delta_{o,k} \delta_{o,l}$$

$$\langle a_{k_1}^* a_{k_2}^* a_{l_1} a_{l_2} \rangle = N(N-1) \quad \delta_{k_1,o} \delta_{k_2,o} \delta_{l_1,o} \delta_{l_2,o} ,$$

etc. Using (3.6) and (3.7) we get

$$\langle \psi^*(f) \, \psi(g) \rangle = \frac{N}{V} \hat{f}(0) \, \hat{g}(0)$$

$$\langle \psi^*(f_1) \, \psi^*(f_2) \, \psi(g_1) \, \psi(g_2) \rangle = \frac{N(N-1)}{V^2} \hat{f}_1(0) \, \hat{f}_2(0) \, \hat{g}_1(0) \, \hat{g}_2(0) \text{ , etc.}$$

In these expressions we can formally take the thermodynamical limit $N \to \infty$, $V \to \infty$ keeping $\rho = N/V$ finite, and we obtain for the infinite system

$$\begin{aligned} &\langle \psi^*(f) \, \psi(g) \rangle = \rho \, \hat{f}(0) \, \hat{g}(0) \\ &\langle \psi^*(f_1) \, \psi^*(f_2) \, \psi(g_1) \, \psi(g_2) \rangle = \rho^2 \, \hat{f}_1(0) \, \hat{f}_2(0) \, \hat{g}_1(0) \, \hat{g}_2(0) \text{ , etc.}^{\ddagger} \end{aligned} \tag{3.8}$$

These values for products of ψ's and ψ^*'s allow us to calculate the functional $E(f,g)$ defined in (3.5). To do this we use the commutation relations (2.3) to get the formula

‡ We have written $\langle A \rangle$ instead of $\omega(A)$, since we want to reserve the latter notation for averages of normed (bounded) observables.

$$U(f)\ V(g) = \exp\{\tfrac{i}{\sqrt{2}}\,\psi^*(f+ig)\}\cdot \exp\{\tfrac{i}{\sqrt{2}}\,\psi(f-ig)\}$$

$$\exp\{-\tfrac{1}{4}\,\|f\|^2 - \tfrac{1}{4}\|g\|^2 - \tfrac{1}{2}i(f,g)\} \tag{3.9}$$

In this expression the creation operators ψ^* occur to the left of the annihilation operators ψ. Taking the expectation value of (3.9) for the Fock-vacuum, we see that only the numerical factor remains. We, therefore, write

$$E_F(f,g) = \exp\{-\tfrac{1}{4}\|f\|^2 - \tfrac{1}{4}\|g\|^2 - \tfrac{i}{2}(f,g)\}\,.$$

The other factors in (3.9) can be expanded in a power series. Using (3.8) one finds for the ground state of the ideal Bose gas

$$\begin{aligned} E(f,g) &= E_F(f,g)\sum_n \frac{(-1)^n}{(n!)^2}\{\tfrac{1}{2}\rho(\hat{f}(0)^2+\hat{g}(0)^2)\}^n \\ &= E_F(f,g)\ J_0[\{2\rho(\hat{f}(0)^2+\hat{g}(0)^2)\}^{\frac{1}{2}}]\,. \end{aligned} \tag{3.10}$$

Having the functional $E(f,g)$, i.e. knowing the state ω, we now set out to construct the cyclic representation which is uniquely determined by ω. To simplify our calculations we shall be somewhat less rigorous and work with the unbounded $\psi(f)$ and $\psi^*(f)$ instead of with $U(f)$ and $V(f)$. That means that we shall find a cyclic representation of our algebra $\mathfrak{A}$ in a Hilbert space $\mathfrak{H}$ such that

$$(\Omega,\psi_R^*(f)\,\psi_R(g)\,\Omega) = \langle\psi^*(f)\ \psi(g)\rangle\,,\quad \text{etc.}$$

with $\psi_R(f)$ the representative of $\psi(f)$.

Let $\mathfrak{H}_F$ denote Fock space, $\psi_F(f)$, $\psi_F^*(g)$ the creation and annihilation operators in $\mathfrak{H}_F$ and φ_0 the Fock vacuum. Let moreover $\mathfrak{M}$ be the space of square integrable functions $\chi(\theta)$ on the unit circle with scalar product

$$(\chi_1,\chi_2) = \int_0^{2\pi}\frac{d\theta}{2\pi}\,\chi_1(\theta)^*\ \chi_2(\theta)\,,$$

O a bounded operator in $\mathfrak{M}$ such that

$$O\,\chi(\theta) = e^{i\theta}\,\chi(\theta)\,,$$

and χ_0 the vector in $\mathfrak{M}$ corresponding to $\chi_0(\theta)\equiv 1$. We then define the representation space $\mathfrak{H}$ as the direct product ‡ of $\mathfrak{H}_F$ and $\mathcal{M}$:

$$\mathfrak{H} = \mathfrak{H}_F \otimes \mathcal{M} . \tag{3.11}$$

The representatives $\psi_R(f)$ and $\psi_R^*(f)$ of $\psi(f)$ and $\psi^*(f)$ as operators in $\mathfrak{H}$ are defined by

$$\begin{aligned} \psi_R(f) &= \psi_F(f) \otimes 1 + 1 \otimes \rho^{\frac{1}{2}} \hat{f}(0)\, O , \\ \psi_R^*(f) &= \psi_F^*(f) \otimes 1 + 1 \otimes \rho^{\frac{1}{2}} \hat{f}(0)\, O^* , \end{aligned} \tag{3.12}$$

and the cyclic vector Ω by

$$\Omega = \varphi_0 \otimes \chi_0 \, . \tag{3.13}$$

We shall now show that equations (3.11), (3.12) and (3.13) determine the desired cyclic representation, i.e. we shall prove that

1. The commutation relations of the operators $\psi_R(f)$ and $\psi_R^*(f)$ are the same as those of $\psi(f)$ and $\psi^*(f)$.
2. $(\Omega, \psi_R^*(f)\, \psi_R(g)\Omega) = \rho\, \hat{f}(0)\, \hat{g}(0)$

   $(\Omega, \psi_R^*(f_1)\, \psi_R^*(f_2)\, \psi_R(g_1)\, \psi_R(g_2)\, \Omega) = \rho^2 \hat{f}_1(0)\, \hat{f}_2(0)\, \hat{g}_1(0)\, \hat{g}_2(0)$

   etc.
3. Ω is cyclic.

   Proof:
1. We notice that the operators O and O^* commute. Therefore

$$[\psi_R(f), \psi_R^*(g) = [\psi_F(f), \psi_F^*(g)] \otimes 1 = (f, g) \, .$$

2. $(\Omega, \psi_R^*(f)\, \psi_R(g)\Omega) = (\varphi_0, \psi_F^*(f)\, \psi_F(g)\, \varphi_0)$

$$+ \rho\, \hat{f}(0)\, \hat{g}(0)\, (\chi_0, O^*O\, \chi_0) = \rho\, \hat{f}(0)\, \hat{g}(0) \text{ etc.}$$

3. The functions $e^{im\theta}$ $(m = 0, \pm 1, \pm 2, \ldots)$ form a complete set of vectors in $\mathcal{M}$. Hence the set

$$\{\psi_F^*(f_1) \ldots \psi_F^*(f_n)\, \varphi_0 \otimes O^m\, \chi_0\}, \quad m = 0, \pm 1, \pm 2, \ldots$$

is a complete set of vectors in $\mathfrak{H}$. It is sufficient to show that the elements of this set can be approximated by applying ele-

‡ The direct product of two Hilbert-spaces is a generalization of the direct product or tensor product of finite dimensional vector spaces.

ments of $R(\mathfrak{A})$ or of its strong limitpoints to Ω. We shall make use of the following
Lemma: The operator $1 \otimes O$ is an element of the von Neumann algebra generated by $R(\mathfrak{A})$. The cyclicity of Ω follows now immediately.

The cyclicity of Ω follows now immediately. Indeed, as follows from (3.12) also $\psi_F(f) \otimes 1$ belongs to the algebra, and

$$\psi_F^*(f_1) \ldots \psi_F^*(f_n)\, \varphi_0 \otimes O^m \chi_0 = \{\psi_F^*(f_1) \otimes 1\} \ldots \{\psi_F^*(f_n) \otimes 1\}\{1 \otimes O\}^m \Omega .$$

For the proof of the lemma we refer to [4]. Here we shall present the following argument. We consider the space average of the field operator $\psi_R(x)$:

$$\mathrm{Av}\, \psi_R(x) = \lim_{V \to \infty} \frac{1}{V} \int d^3x\, \psi_R(x) .$$

If $f_V(x)$ is the characteristic function of the volume V ($f_V(x) = 1$ for x in V, $f_V(x) = 0$ otherwise) we can write

$$\mathrm{Av}\, \psi_R(x) = \lim_{V \to \infty} \frac{1}{V}\, \psi_R(f_V) =$$

$$\lim_{V \to \infty} \frac{1}{V}\, \psi_F(f_V) \otimes 1 + 1 \otimes \lim_{V \to \infty} \frac{1}{V}\, \hat{f}_V(0) \rho^{\frac{1}{2}} O = \mathrm{Av}\, \psi_F(x) \otimes 1 + 1 \otimes \rho^{\frac{1}{2}} O .$$

Consider the commutator

$$[\mathrm{Av}\, \psi_F, \psi_F^*(g)] = \lim_{V \to \infty} \frac{1}{V} (f_V, g) = \lim_{V \to \infty} \frac{1}{V} \int_V g(x)\, dx = 0 .$$

Consequently the operator $\mathrm{Av}\, \psi_F$ commutes with all $\psi_F(f)$, $\psi_F^*(g)$. Since the representation in Fock space is irreducible $\mathrm{Av}\, \psi_F$ must be a multiple of the identity (a c-number). When applied to φ_0 the operator gives zero. Hence $\mathrm{Av}\, \psi_F = 0$, and it follows that

$$1 \otimes O = \rho^{-\frac{1}{2}}\, \mathrm{Av}\, \psi_R . \tag{3.14}$$

We shall discuss next some properties of this representation. We notice first of all that our representation is reducible. Consider operators of the form $1 \otimes T$ with T a bounded operator in $\mathcal{M}$ defined by

$$T\, \chi(\theta) = t(\theta)\, \chi(\theta)$$

for each $\chi \in \mathcal{M}$. Since T commutes with O and O^*, $1 \otimes T$ commutes with $\psi_R(f)$ and $\psi_R^*(f)$ for all f, and thus $1 \otimes T \in \mathcal{R}'$. In particular O and O^* are such operators. Since $1 \otimes O$ and $1 \otimes O^*$ are also elements of $\mathcal{R}$, they must be center-elements. Consequently $\mathcal{R}$ is not a factor. Since the operators T can be expressed in terms of O and O^* (Fourier series) one can see without difficulty that the commutant $\mathcal{R}'$ is the von Neumann algebra generated by the unitary operator $1 \otimes O$ so that $\mathcal{R}'$ coincides with the center.

The physical meaning of the operator $1 \otimes O$ is found by comparing (3.6) and (3.14). One finds

$$1 \otimes O = \rho^{-\frac{1}{2}} \lim_{V \to \infty} V^{-\frac{1}{2}} a_0 \,, \tag{3.15}$$

where a_0 is the annihilation operator for zero-momentum particles. Since the system contains infinitely many such particles, the effect of a_0 or a_0^* will not be noticeable. However, since we have a ground state with a definite number of particles, a_0 and a_0^* cannot be c-numbers.

We shall now discuss some transformation properties of the state ω defined in (3.8).

Space translation. Since $\hat{f}(0)$ is invariant for translation, ω is invariant for space translation.

Gauge transformation. The non-vanishing averages in (3.8) contain an equal number of ψ's and ψ^*'s, and hence ω is invariant for gauge transformation. The unitary operator $U(\alpha)$ which implements the gauge transformation $\psi(f) \to \psi(f)\, e^{i\alpha}$ has the form

$$U(\alpha) = U_F(\alpha) \otimes \exp\left(\alpha \frac{d}{d\theta}\right) ,$$

where $U_F(\alpha)$ is the corresponding operator in Fock-space. The generator of this unitary operator is

$$N = N_F \otimes 1 + 1 \otimes -i \frac{d}{d\theta} \,.$$

Time-translation. Since we have no interaction the field operator $\psi(x, t)$ satisfies the free one-particle Schrödinger equation. Therefore

$$\psi(f, t) = \psi(f_t) \,,$$

where f_t is the solution of the Schrödinger equation for one free particle which coincides with f at $t = 0$. Since $\hat{f}(0)$ is invariant for the substitution $f \to f_t$, we conclude that ω is invariant for time translation. Since in (3.12) only the first term varies in time, we see that the unitary operator $U(t)$, which implements translation in time has the form

$$U(t) = U_{\mathrm{F}}(t) \otimes 1 .$$

$U_{\mathrm{F}}(t)$ is the time-translation operator in Fock-space as defined in 2.4. The generator H of $U(t)$ we call the hamiltonian. We have

$$H = H_{\mathrm{F}} \otimes 1 .$$

The spectrum of H is therefore the same as that of H_{F}, i.e. the positive axis including the origin. The ground state Ω satisfies

$$H\Omega = 0 ,$$

but any other vector Ω' obtained from Ω by multiplication with an element from the center also satisfies that equation. That means that there are many different vectors in $\mathfrak{H}$ which are time independent; the origin in the spectrum is highly degenerate.

Let us next construct a one-parameter family of inequivalent irreducible representations of $\mathfrak{A}$, which are physically indistinguishable from the previous one. We define:

$$\begin{aligned} \mathfrak{H} &= \mathfrak{H}_{\mathrm{F}}; \qquad \Omega = \varphi_0 ; \\ \psi_\theta(f) &= \psi_{\mathrm{F}}(f) + \rho^{\frac{1}{2}} \hat{f}(0)\, \mathrm{e}^{\mathrm{i}\theta} ; \\ \psi^*(f) &= \psi^*_{\mathrm{F}}(f) + \rho^{\frac{1}{2}} \hat{f}(0)\, \mathrm{e}^{-\mathrm{i}\theta} . \end{aligned} \tag{3.16}$$

The state defined by the vector Ω is again determined by the expectation values of products of ψ_θ and ψ^*_θ. One finds

$$\begin{aligned} \langle \psi(f) \rangle_\theta &\equiv (\Omega,\ \psi_\theta(f)\Omega) = \rho^{\frac{1}{2}} \hat{f}(0)\, \mathrm{e}^{\mathrm{i}\theta} , \\ \langle \psi^*(f)\, \psi(g) \rangle_\theta &= \rho \hat{f}(0)\, \hat{g}(0), \text{ etc.} \end{aligned} \tag{3.17}$$

If we compare (3.17) with (3.8) we see that for equal numbers of ψ's and ψ^*'s the averages are equal. For unequal numbers the averages vanish in (3.8), whereas in (3.17) they depend on the parameter θ. Clearly the state defined by (3.17) does not have a fixed particle number. One checks easily that the representation (3.16) is irreducible. Instead of (3.15) we have now

$$\rho^{-\frac{1}{2}} \lim_{V \to \infty} V^{-\frac{1}{2}} a_0 = \mathrm{e}^{\mathrm{i}\theta} ,$$

i.e., the annihilation operator a_0 and the creation operator a_0^* are now c-numbers. Since all physical observables involve equal numbers of ψ's and ψ^*'s the states (3.8) and (3.17) are physically indistinguishable. The latter state is on the one hand more simple since it is pure because it gives rise to an irreducible representation. On the other hand it lacks symmetry. In contrast to (3.8) the state (3.17) is not gauge invariant. A particle number operator N does not exist in this case. Since we have a ground state with less symmetry than the dynamical equations of the system (the Hamilton-operator H_F in Fock-space commutes with the particle number N_F) we speak of a spontaneous breaking of symmetry, a phenomenon encountered frequently in infinite systems.

The state ω_θ defined by (3.17) is invariant for translations in space and time. Accordingly, there exists a hamiltonian H, which has the form

$$H = H_F ,$$

so that the spectrum of H extends again from zero to infinity. Since $\mathfrak{H} = \mathfrak{H}_F$ the vector $\Omega = \varphi_0$ is the only vector in $\mathfrak{H}$ with eigenvalue zero. In this case the ground state is non-degenerate.

There exists a simple mathematical relation between the family of states (3.17) and the state (3.8). This relation can be expressed as follows

$$\langle \quad \rangle = \frac{1}{2\pi} \int_0^{2\pi} d\theta \langle \quad \rangle_\theta ,$$

or

$$\omega(A) = \frac{1}{2\pi} \int_0^{2\pi} d\theta\, \omega_\theta(A) .$$

It expresses the mixed state ω as an integral over the pure states ω_θ. As a result of this averaging over all angles θ the gauge invariance is restored.

3.2. *The equilibrium state at $T > T_c$*

As for the ground state we shall determine the thermal equilibrium state by the averages of products of ψ's and ψ^*'s. For a finite system the average $\langle A \rangle$ of an operator A is given by

$$\langle A \rangle = \mathrm{Tr}\, \rho\, A / \mathrm{Tr}\, \rho ,$$

with $\rho = e^{-\beta(H-\mu N)}$, where $\beta = (kT)^{-1}$ and μ is the chemical potential. We first calculate the average of products of a's and a^*'s. These averages vanish unless the numbers of a's and a^*'s are equal. We find

$$\langle a_k^* a_l \rangle = \delta_{k,l} N_k ,$$

where N_k is the average particle number of momentum k and is given by

$$N_k^{-1} = \exp\left\{\beta\left(\frac{\hbar^2 k^2}{2m} - \mu\right)\right\} - 1 . \tag{3.18}$$

Similarly

$$\langle a_{k_1}^* a_{k_2}^* a_{l_1} a_{l_2} \rangle = \delta_{k_1, l_1} \delta_{k_2, l_2} \rho_{k_1} \rho_{k_2} + \delta_{k_1, l_2} \delta_{k_2, l_1} \rho_{k_1} \rho_{k_2} ,$$

etc. Using (3.6) and (3.7) we get

$$\langle \psi^*(f)\, \psi(g) \rangle = V^{-1} \sum_k N_k \hat{f}(k)^* \hat{g}(k) . \tag{3.19}$$

Averages of four or more factors are written as sums of products of averages of pairs:

$$\langle \psi^*(f_1)\, \psi^*(f_2)\, \psi(g_1)\, \psi(g_2) \rangle = \langle \psi^*(f_1)\, \psi(g_1) \rangle \langle \psi^*(f_2)\, \psi(g_2) \rangle$$
$$+ \langle \psi^*(f_1)\, \psi(g_2) \rangle \langle \psi^*(f_2)\, \psi(g_1) \rangle , \text{ etc.}$$

If we now perform the thermodynamical limit, i.e. the limit $V \to \infty$, the particle distribution $\rho_k = (2\pi)^{-3} N_k$ becomes a continuous function of k and equation (3.19) is to be replaced by

$$\langle \psi^*(f)\, \psi(g) \rangle = \int d^3k\, \rho_k \hat{f}(k)^* \hat{g}(k) . \tag{3.20}$$

The integral in (3.20) we shall denote by $(f, \rho g)$. We then have

$$\langle \psi^*(f)\, \psi(g) \rangle = (f, \rho g) ,$$
$$\langle \psi^*(f_1)\, \psi^*(f_2)\, \psi(g_1)\, \psi(g_2) \rangle = (f_1, \rho g_1)(f_2, \rho g_2) + (f_1, \rho g_2)(f_2, \rho g_1) , \tag{3.21}$$

etc.

From (3.21) one derives the expression for $E(f, g)$. Using (3.9) and expanding the first and the second factor at the right-hand side in a power series we find

$$E(f,g) = E_F(f,g)\exp[-\tfrac{1}{2}\{(f,\rho f) + (g,\rho g)\}] \tag{3.22}$$

Like in the ground state case we shall now construct the cyclic representation which is uniquely determined by the state. As the representation space $\mathfrak{H}$ we take the direct product of two Fock-spaces:

$$\mathfrak{H} = \mathfrak{H}_F \otimes \mathfrak{H}_F \,. \tag{3.23}$$

The representatives of $\psi(f)$ and $\psi^*(f)$ which we shall denote by $\psi_R(f)$ and $\psi_R^*(f)$ are given by

$$\begin{aligned} \psi_R(f) &= \psi_F(\sqrt{1+\rho}\, f) \otimes 1 + 1 \otimes \psi_F^*(\sqrt{\rho}\, f) \,, \\ \psi_R^*(f) &= \psi_F^*(\sqrt{1+\rho}\, f) \otimes 1 + 1 \otimes \psi_F(\sqrt{\rho}\, f) \,. \end{aligned} \tag{3.24}$$

Here ψ_F and ψ_F^* are the usual annihilation and creation operators in Fock-space. The functions $\sqrt{1+\rho}\, f$ and $\sqrt{\rho}\, f$ have Fourier transforms given by $\sqrt{1+\rho_k}\, \hat{f}(k)$ and $\sqrt{\rho_k}\hat{f}(k)$. The cyclic vector Ω is defined by

$$\Omega = \varphi_0 \otimes \varphi_0 \,, \tag{3.25}$$

where φ_0 is again the Fock vacuum.

We shall show that the equations (3.23), (3.24) and (3.25) determine the desired representation, i.e. we shall prove that

1. The commutation relations of the operators $\psi_R(f)$ and $\psi_R^*(f)$ are the same as those of $\psi(f)$ and $\psi^*(f)$.
2. $(\Omega, \psi_R^*(f)\psi_R(g)\Omega) = (f, \rho g)$,

   $(\Omega, \psi_R^*(f_1)\psi_R^*(f_2)\psi_R(g_1)\psi_R(g_2)\Omega) =$

   $\qquad (f_1, \rho g_1)(f_2, \rho g_2) + (f_1, \rho g_2)(f_2, \rho g_1)$, etc.
3. Ω is cyclic.

Proof:

1. $[\psi_R(f), \psi_R^*(g)] = [\psi_F(\sqrt{1+\rho}\, f), \psi_F^*(\sqrt{1+\rho}\, g)] \otimes 1$

   $+ 1 \otimes [\psi_F^*(\sqrt{\rho}\, f), \psi_F(\sqrt{\rho}\, g)] = (f, (1+\rho)\, g) - (f, \rho g) = (f,g)$.

2. $(\Omega, \psi_R^*(f)\psi_R(g)\Omega) = (\varphi_0, \psi_F^*(\sqrt{1+\rho} f)\psi_F(\sqrt{1+\rho} g)\varphi_0)$

   $+ (\varphi_0, \psi_F^*(\sqrt{1+\rho} f)\varphi_0)(\varphi_0, \psi_F^*(\sqrt{\rho} g)\varphi_0)$

   $+ (\varphi_0, \psi_F(\sqrt{1+\rho} g)\varphi_0)(\varphi_0, \psi_F(\sqrt{\rho} f)\varphi_0) + (\varphi_0, \psi_F(\sqrt{\rho} f)\psi_F^*(\sqrt{\rho} g)\varphi_0)$.

Only the last term does not vanish. We get

$$(\Omega, \psi_R^*(f)\psi_R(g)\Omega) = (f, \rho g), \quad \text{etc.}$$

3. We must prove that the set S of vectors

$$\psi_R^*(f_1)\psi_R^*(f_2)\ldots\psi_R^*(f_n)\psi_R(g_1)\ldots\psi_R(g_m)\Omega$$

forms a complete set, which means that every vector in $\mathfrak{H}$ can be approximated by linear combinations of such vectors. Therefore we prove that any vector of the dense set S:

$$\psi_F^*(f_1)\psi_F^*(f_2)\ldots\psi_F^*(f_n)\varphi_0 \otimes \psi_F^*(g_1)\ldots\psi_F^*(g_m)\varphi_0$$

can be expressed linearly in terms of vectors of S. We show this by induction and suppose that all vectors of S' with $n \leqslant N$ and $m \leqslant M$ are linear combinations of vectors of S. We now take

$$\psi_F^*(f_1)\ldots\psi_F^*(f_{N+1})\varphi_0 \otimes \psi_F^*(g_1)\ldots\psi_F^*(g_M)\varphi_0$$

and show that also this vector can be expressed in terms of the set S. We apply the operator $\psi_R^*((1+\rho)^{-\frac{1}{2}} f_{N+1})$ to the vector

$$\psi_F^*(f_1)\ldots\psi_F^*(f_N)\varphi_0 \otimes \psi_F^*(g_1)\ldots\psi_F^*(g_M)\varphi_0 \,.$$

This gives two terms, one of which is our given vector, the other one has a smaller number of ψ^*'s and is thus expressible in terms of the set S.

We shall now discuss this representation. It differs in many respects from the ground state representation as we see from the following properties:

1. *The representation is reducible.*

Proof: consider the operators

$$\begin{aligned} \psi_S(f) &= 1 \otimes \psi_F(\sqrt{1+\rho}\, f) + \psi_F^*(\sqrt{\rho} f) \otimes 1 \\ \psi_S^*(f) &= 1 \otimes \psi_F^*(\sqrt{1+\rho}\, f) + \psi_F(\sqrt{\rho} f) \otimes 1 \,. \end{aligned} \tag{3.26}$$

We notice that these operators differ from $\psi_R(f)$ and $\psi_R^*(f)$ by interchanging the two Fock-spaces in the direct product space $\mathfrak{H}$.

The operators $\psi_S(f)$ and $\psi_S^*(f)$ commute with $\psi_R(f)$ and $\psi_R^*(f)$. Consequently $\psi_S(f)$ and $\psi_S^*(f)$ belong to the commutant $\mathcal{R}'$ of the von Neumann algebra $\mathcal{R}$ generated by $R(\mathfrak{A})$. If $\mathcal{S}$ is the von Neumann algebra generated by all $\psi_S(f)$ and $\psi_S^*(f)$ we have $\mathcal{S} \subset \mathcal{R}'$. We shall see that even $\mathcal{S} = \mathcal{R}'$. Since $\mathcal{R}$ has a non-trivial commutant the representation is reducible.

2. *The representation is primary*

Proof: By taking appropriate linear combinations of $\psi_R(f)$, $\psi_R^*(f')$,

$\psi_S(f'')$, $\psi_S^*(f''')$ one can construct the operators

$$\psi_F(g) \otimes 1, \quad 1 \otimes \psi_F(g) ,$$
$$\psi_F^*(g) \otimes 1, \quad 1 \otimes \psi_F^*(g) ,$$

which are clearly an irreducible set in the product-space $\mathfrak{H}_F \otimes \mathfrak{H}_F$. It follows that the algebra generated by all ψ_R, ψ_R^*, ψ_S, ψ_S^* is irreducible, i.e. $(\mathcal{R} \cup \mathcal{S})'' = \mathfrak{B}(\mathfrak{H})$ and thus $(\mathcal{R} \cup \mathcal{S})'$ is the set of multiples of the identity. Using the identity $(\mathcal{R} \cup \mathcal{S})' = \mathcal{R}' \cap \mathcal{S}'$ and the inequality $\mathcal{R} \subset \mathcal{S}'$ we arrive at the conclusion that the center $\mathcal{R} \cap \mathcal{R}'$ is trivial, so that $\mathcal{R}$ is a factor. Since we had $(\mathcal{R} \cup \mathcal{R}')'' = (\mathcal{R} \cup \mathcal{S})''$ we have at the same time the equality $\mathcal{S} = \mathcal{R}'$.

3. *The mapping* $\psi(f) \to \psi_S(f)$ *is an equivalent representation of* $\mathfrak{A}$ *in* $\mathfrak{H}$.

The proof is a simple repetition of the proof for $\psi_R(f)$. We have therefore, two equivalent representations of $\mathfrak{A}$, one in $\mathcal{R}$ and the other in $\mathcal{R}'$.

Like the ground state, the thermal equilibrium state, as defined by (3.21) is invariant for the following transformations:

Space translation. Since the expression $(f, \rho\, g)$ is invariant for a simultaneous translation of f and g in space, ω is invariant for translations in space.

Gauge transformation. It is evident that ω is invariant, because we always have an equal number of ψ's and ψ^*'s.

Time translation. The scalar products $(f, \rho\, g)$ occuring in (3.21) are invariant for the replacement $f \to f_t$ and $g \to g_t$; the state ω is, therefore, invariant for translation in time. This was to be expected; an equilibrium state should be stationary.

We shall now construct the unitary operators $U(\boldsymbol{x})$, $U(\alpha)$ and $U(t)$ which implement these automorphisms. We use equation (3.24) which expresses $\psi(f)$ in terms of the Fock-operators in the product space $\mathfrak{H} = \mathfrak{H}_F \otimes \mathfrak{H}_F$. Due to the very simple nature of the transformations involved, their effect on $\psi(f)$ can be expressed in terms of transformations of the testfunction f: $f \to f_{\boldsymbol{x}}$, $f \to f e^{i\alpha}$ and $f \to f_t$, respectively. In Fock-space these transformations are implemented by the known operators $U_F(\boldsymbol{x})$, $U_F(\alpha)$ and $U_F(t)$, with the momentum operator $\boldsymbol{P}_F$, the particle number operator N_F and the hamiltonian H_F as generators:

$$\begin{aligned} \psi_F(f_{\boldsymbol{x}}) &= U_F(\boldsymbol{x})\, \psi_F(f)\, U_F(\boldsymbol{x})^{-1} , \\ \psi_F(f)\, e^{i\alpha} &= U_F(\alpha)\, \psi_F(f)\, U_F(\alpha)^{-1} , \\ \psi_F(f_t) &= U_F(t)\, \psi_F(f)\, U_F(t)^{-1} . \end{aligned} \tag{3.27}$$

We note that

$$\begin{aligned} \psi_F^*(f)\, e^{i\alpha} &= U_F(\alpha)^{-1}\, \psi_F^*(f)\, U_F(\alpha) \\ \psi_F^*(f_t) &= U_F(t)^{-1}\, \psi_F^*(f)\, U_F(t)\ . \end{aligned} \tag{3.28}$$

Using (3.27), (3.28) and the fact that the operators $\sqrt{\rho}$ and $\sqrt{1+\rho}$ commute with translations in space and time we find that the unitary operators defined by

$$\begin{aligned} U(\boldsymbol{x}) &= U_F(\boldsymbol{x}) \otimes U_F(\boldsymbol{x})\ , \\ U(t) &= U_F(t) \otimes U_F(t)^*\ , \\ U(\alpha) &= U_F(\alpha) \otimes U_F(\alpha)^*\ , \end{aligned} \tag{3.29}$$

implement the automorphisms. At the same time they evidently satisfy the conditions

$$U(\boldsymbol{x})\,\Omega = U(t)\,\Omega = U(\alpha)\,\Omega = \Omega\ .$$

The generators $\boldsymbol{P}$, H, and N can be expressed in the corresponding quantities in Fock-space:

$$\begin{aligned} \boldsymbol{P} &= \boldsymbol{P}_F \otimes 1 + 1 \otimes \boldsymbol{P}_F\ , \\ H &= H_F \otimes 1 - 1 \otimes H_F\ , \\ N &= N_F \otimes 1 - 1 \otimes N_F\ . \end{aligned} \tag{3.30}$$

These generators are well-defined unbounded operators in $\mathfrak{H}$, although they do not "belong" to the von Neumann algebra $\mathcal{R}$ generated by $R(\mathfrak{A})$. We see that by noticing that also the operators in $\mathcal{R}'$ are transformed by $U(\boldsymbol{x})$, $U(t)$ and $U(\alpha)$:

$$\begin{aligned} U(\boldsymbol{x})\, \psi_S(f)\, U(\boldsymbol{x})^{-1} &= \psi_S(f_{\boldsymbol{x}})\ , \\ U(t)\, \psi_S(f)\, U(t)^{-1} &= \psi_S(f_{-t})\ , \\ U(\alpha)\, \psi_S(f)\, U(\alpha)^{-1} &= \psi_S(f)\, e^{-i\alpha}. \end{aligned}$$

Hence $\mathcal{R}'$ is transformed in a similar way as $\mathcal{R}$ with as only difference the change of sign in t and α; the representation of $\mathfrak{A}$ in $\mathcal{R}'$ runs backward in time.

Although the unitary operators and their generators cannot be expressed in terms of ψ_R and ψ_R^* alone, there must exist an expression in terms of ψ_R, ψ_R^*, ψ_S, ψ_S^* since these operators generate all operators. We shall now derive these expressions.

Let $\rho(x) = \psi^*(x)\,\psi(x)$ be the particle density operator in our original Fock-space. Represented in $\mathfrak{H}$ we have $\rho_R(x) = \psi_R^*(x)\,\psi_R(x)$. The integral

$$N_R(V) = \int_V \rho_R(x)\,\mathrm{d}^3x$$

is the operator for the number of particles in the finite volume V. This is observable. The limit of $N_R(V)$ for $V \to \infty$ does not exist, since for all states in $\mathfrak{H}$ the system has infinitely many particles. Using the symmetry between $\mathcal{R}$ and $\mathcal{R}'$ we define

$$N_S(V) = \int_V \rho_S(x)\,\mathrm{d}^3x\ ,$$

which also exists for any finite V. The difference $N_R(V) - N_S(V)$ has a finite weak limit which equals N. Hence

$$N = \int \mathrm{d}^3x[\psi_R^*(x)\,\psi_R(x) - \psi_S^*(x)\,\psi_S(x)]\ .$$

Similarly

$$H = \frac{\hbar^2}{2m}\int \mathrm{d}^3x[\nabla\psi_R^*(x)\,\nabla\psi_R^*(x) - \nabla\psi_S^*(x)\,\nabla\psi_S(x)]$$

and

$$\boldsymbol{P} = \frac{\hbar}{\mathrm{i}}\int \mathrm{d}^3x[\psi_R^*(x)\,\nabla\psi_R(x) - \psi_S^*(x)\,\nabla\psi_S(x)]\ .$$

These equations are proved by substituting (3.24) and (3.26) for ψ_R and ψ_S. The resulting expressions can be reduced to (3.30).

4. INTERACTING SYSTEMS

In the previous chapter some properties of the ground state and the equilibrium state of the ideal Bose-gas were discussed. Some of these properties are not limited to the ideal gas, but hold generally for an equilibrium state of any system. The aim of this chapter is to study some of these more general properties.

4.1. *The K.M.S. boundary condition*

In section 3.2 we proved for the ideal Bose-gas at $T \neq 0$ and, above the Bose-Einstein transition point, that there is a complete symmetry between the von Neumann algebra $\mathcal{R}$ generated by $R(\mathfrak{A})$ and its commutant $\mathcal{R}'$. It has been shown that this symmetry holds more generally for equilibrium states of any system [8]. This is a consequence of a boundary condition which was first formulated by Kubo [10] and used extensively by Martin and Schwinger [9] in their Green's function method for the description of equilibrium states.

This K. M. S. -boundary condition can be formulated as follows: Let A and B be observables and let A_t now denote the operator $\exp(\mathrm{i}H_\mathrm{F}'t)\cdot A\exp{-\mathrm{i}H_\mathrm{F}'t}$, with $H_\mathrm{F}' = H_\mathrm{F} - \mu N_\mathrm{F}$. We define the functions

$$G_{AB}^{(1)}(t) = \omega(A_tB) \text{ and } G_{AB}^{(2)}(t) = \omega(BA_t) .$$

For an equilibrium state with temperature $T = (K\beta)^{-1}$ and chemical potential μ the function $G_{AB}^{(1)}(t)$ is continuous in the strip $-\beta \leqslant \mathrm{Im}\, t \leqslant 0$ and analytic in the interior, and similarly the function $G_{AB}^{(2)}(t)$ is continuous in the strip $0 \leqslant \mathrm{Im}\, t \leqslant \beta$ and analytic in the interior, and

$$G_{AB}^{(1)}(t - \mathrm{i}\beta) = G_{AB}^{(2)}(t) \tag{4.31}$$

for all t in the strip $0 \leqslant \mathrm{Im}\, t \leqslant \beta$.

These properties are most easily shown for a finite equilibrium system, when

$$\omega(A) = \mathrm{Tr}\, \exp(-\beta H')A/\mathrm{Tr}\, \exp(-\beta H) .$$

If $z = t + \mathrm{i}\gamma$ the operator

$$A_z = \exp(\mathrm{i}H'z)\ A\exp(-\mathrm{i}H'z)$$

will, in general, not be bounded. But

$$A_z\, \mathrm{e}^{-\beta H'} = \mathrm{e}^{\mathrm{i}H't}\, \mathrm{e}^{-H'\gamma}\, A\, \mathrm{e}^{-\mathrm{i}H't}\, \mathrm{e}^{-H'(\beta - \gamma)}$$

is bounded and has a finite trace for $0 \leqslant \gamma \leqslant \beta$. Similarly $\exp(-\beta H')A_z$ is bounded and has a trace for $-\beta \leqslant \gamma \leqslant 0$. We conclude that the function

$$G_{AB}^{(1)}(t) = \mathrm{Tr}\, \mathrm{e}^{-\beta H'} A_t\, B/\mathrm{Tr}\, \mathrm{e}^{-\beta H'}$$

exists and is continuous for $-\beta \leqslant \mathrm{Im}\, t \leqslant 0$ and that the function

$$G_{AB}^{(2)}(t) = \text{Tr } e^{-\beta H'} \, B \, A_t / \text{Tr } e^{-\beta H'}$$

exists and is continuous for $0 \leqslant \text{Im } t \leqslant \beta$. The equation (4.31) is an immediate consequence of the invariance of the trace under cyclic permutations.

To prove the K. M. S. -boundary condition for the infinite system one must take the thermodynamical limit. A proof of the existence of this limit has been given only in some special cases [11]. In order to conclude to the validity of the K. M. S. -condition, some assumptions have to be made, one of which is the assumption of the existence of the thermodynamical limit. For a more detailed discussion on the conditions for the validity of the K. M. S. -boundary condition for an infinite system we refer to [8]. We shall now take the attitude that the K. M. S. -boundary condition holds for any equilibrium system.

To explore some of the consequences of this condition, we shall give another, equivalent, formulation of it. Let $\tilde{\mathfrak{A}}$ be the set of all elements A of $\mathfrak{A}$ such that A_t has a Fourier-transform $\hat{A}(\epsilon)$ of compact support. $\tilde{\mathfrak{A}}$ is a *-algebra. The elements $A \in \tilde{\mathfrak{A}}$ have the property that A_t is a bounded operator for any complex value of t. The mapping $A \to A_{ix}$ is an automorphism, but not a *-automorphism of $\tilde{\mathfrak{A}}$. The condition

$$\omega\{(A_{\frac{1}{2}i\beta})^* \, B_{\frac{1}{2}i\beta}\} = \omega(BA^*) \, , \tag{4.32}$$

for all A, $B \in \tilde{\mathfrak{A}}$ is equivalent with the K. M. S. -boundary condition. Condition (4.32) has the consequence that the left ideal of elements A for which $\omega(A^*A) = 0$ is a two-sided ideal. Since the algebras of the canonical commutation relations and of the anti-commutation relations are simple, i.e. don't have any non-trivial two-sided ideals, it follows that

$$\omega(A^*A) = 0 \text{ implies } A = 0 \, . \tag{4.33}$$

Without proof we mention some of the properties of the cyclic representation defined by a state ω satisfying conditions (4.32) and (4.33).

1. The representation is reducible.
2. There exists a conjugation operator J, i.e. an anti-unitary operator J with $J^2 = 1$, which transforms the von Neumann algebra $\mathcal{R}$ generated by $R(\mathfrak{A})$ into its commutant:

$$J\mathcal{R}J = \mathcal{R}'$$

and which has the additional properties

$$J\,C\,J = C^* ,$$

for all center-elements and

$$J\,\Omega = \Omega .$$

3. Ω is cyclic for $\mathcal{R}'$; in particular, if $S(A) = J R(A) J$, the set of vectors $S(A)\Omega$ is dense in $\mathfrak{H}$.
4. The unbounded operator T, defined by

$$T\,R(A)\,\Omega = S(A)^*\,\Omega$$

has an inverse, and

$$T\,R(A)\,T^{-1} = R(A_{\frac{1}{2}i\beta})$$

for all $A \in \mathfrak{A}$. It can be shown that

$$T^2 = e^{-\beta(H-\mu N)} .$$

The properties 1, 2, 3, 4 are well-known in the mathematical literature, where the algebra $\tilde{\mathfrak{A}}$ is called quasi-unitary, the von Neumann algebra $\mathcal{R}$ quasi-standard [1].

4.2. *General definitions and theorems*

In section 3 and section 4.1 we have explored properties of the ground state and of the thermal equilibrium states, starting from a finite system and taking the thermodynamical limit. Another and more fundamental approach would be to require some very general conditions, such as the behaviour under various groups of transformations. One would then hope that states satisfying such conditions correspond to the ground state or equilibrium states of the system. To discuss this we must first give some definitions.

As we know any automorphism of $\mathfrak{A}$ for which the state ω is invariant can be implemented by a unitary operator. For a stationary state this implies the existence of a unitary operator $U(t)$ and a hamiltonian H.

Definition: We say that the stationary state ω satisfies the *spectrum condition* if the spectrum of H is bounded from below by the origin.

That this condition is quite strong follows from the following theorem [12].

Theorem: If the stationary state ω satisfies the spectrum condition and if the cyclic vector Ω is the only invariant vector in $\mathfrak{H}$, then the representation defined by ω is irreducible; hence ω is a pure state.

This theorem tells us that the ground state of a system is pure, unless due to a symmetry breaking, the ground state is degenerate, as in (3.8).

The absence of infinitely long range correlations can be formulated by means of the cluster property.

Definition: A state ω has the *cluster property* with respect to space-translations if

$$\lim_{|x|\to\infty} [\omega(A_x B) - \omega(A_x)\omega(B)] = 0 . \tag{4.34}$$

If one defines the state by means of distribution functions $F(x_1, \ldots, x_n; y_1, \ldots, y_m)$ condition (4.34) implies that these functions factorize if one divides the points into two far apart sets, and hence (4.34) states the absence of long range correlations. The following theorem gives a sufficient condition for the cluster property.

Theorem: A factor state has the cluster property.

One checks without difficulty that the pure state defined in (3.17) and the factor states (3.21) satisfy the cluster property. On the other hand the ground state with definite particle number (3.8) does not cluster.

A somewhat weaker property is the mean cluster property [5].

Definition: A state has the *mean cluster property* with respect to space-translations if

$$\lim_{V\to\infty} \frac{1}{V} \int_V d^3x\, [\omega(A_x B) - \omega(A_x)\omega(B)] = 0 \tag{4.35}$$

Similar properties as (4.34) and (4.35) can be defined with respect to time-translations.

Another concept which proves extremely useful in this context is that of extremal invariance [5-7].

Definition: A state ω, which is invariant for a certain group of transformations is *extremal invariant* if it cannot be written as a convex sum of two other invariant states, i.e. ω is extremal invariant if $\omega = \lambda\omega_1 + (1-\lambda)\omega_2$ with $0 \leq \lambda \leq 1$ and $\omega_1 \neq \omega_2$ invariant states implies either $\lambda = 0$ or $\lambda = 1$. Extremal invariant states have a number of interesting properties, in particular if the algebra of observables $\mathfrak{A}$ is asymptotically abelian (see 2.4) with respect to the transformations considered. Since it is known that $\mathfrak{A}$ is asymptotically abelian for space-translations, we shall first consider this case. One then has the following equivalent statements:

1. The state ω is extremal invariant with respect to space-translations.

2. The state ω is invariant and the algebra generated by $R(\mathfrak{A})$ and U_x is irreducible.

3. The average

$$\lim_{V\to\infty} \frac{1}{V} \int_V \mathrm{d}^3x \; R(A_x)$$

exists, is independent of the way the limit is performed, and is a multiple of $\mathbf{1}$.

4. Ω is the only invariant vector in $\mathfrak{H}$.

5. The invariant state satisfies the mean cluster property.

We shall first comment on some of these properties. The third statement means that the space-average of any observable is a c-number. A simple example is the operator $\mathrm{Av}\psi_R$ in section 3.1. In that section we have discussed two different states ω and ω_θ both describing the ideal Bose-gas in its ground state; the state ω is a mixed state whereas ω_θ is pure. The mixed state is not extremal invariant as seen from the last equation of that section where we expressed ω as a convex sum of other invariant states ω_θ. The state ω_θ is pure and thus extremal invariant. The operator $\mathrm{Av}\psi_\theta$ is a c-number.

In more complicated cases property 3 may lead to a considerable simplification. A well-known example is the way Bogoliubov treated the imperfect Bose-gas in 1947 [13]. In that paper Bogoliubov considers the operators $V^{-\frac{1}{2}}a_0$ and $V^{-\frac{1}{2}}a_0^*$, where a_0 and a_0^* are the annihilation and creation operators for a particle of zero momentum and V is the volume of the system. He argues that in the limit where $V\to\infty$ these operators may be replaced by c-numbers. These operators can be rewritten in x-space:

$$\frac{1}{V} \int_V \mathrm{d}^3x \; \psi(x) \; , \qquad \frac{1}{V} \int_V \mathrm{d}^3x \; \psi^*(x) \; ,$$

and one sees that in the limit they are space-averages of local observables, and may be replaced by a number if the state is extremal invariant. Like in the case of the ideal gas one may select a ground state which is pure, a stronger condition than extremal invariant. The arguments presented here are due to Haag, who discussed the B.C.S.-model along these lines [14].

Statement 5, the mean cluster property, does not exclude the presence of infinitely long range correlations, as long as they have some sort of periodic behaviour, as for example in a crystal.

The importance of extremal invariant states is also based on the following [6]

Theorem: Any state ω, which is invariant with respect to translations in space may be decomposed uniquely into extremal invariant states.

This means that any invariant state can be written *uniquely* as a sum or integral over extremal invariant states. As is well-known any state can be written as a sum or integral over pure states. However, in quantum mechanics this decomposition is not in general unique.

In very recent papers extremal invariant states are called ergodic states [15]. I shall give some arguments to support that term, although the connection with the classical considerations on ergodicity is not entirely clear.

Let us suppose that the hamiltonian H_F in our original Fock-space, which determines the dynamics of the system, is such that the algebra of observables $\mathfrak{A}$ is also asymptotically abelian with respect to time-translation, i.e.,

$$\lim_{|t|\to\infty} \|[A_t, B]\| = 0 .$$

We then have for time-translations the same set of equivalent statements as for space-translations. We would in particular have the result that for an extremal invariant state for translations in time, the time-average of each observable exists, is independent of the way the average was taken, and is a c-number. That means that for any vector state $\psi \in \mathfrak{H}$ the time-average of an observable exists and is independent of ψ. If we speculate somewhat further and compare this situation with that of classical statistical mechanics it seems that our condition that $\mathfrak{A}$ is asymptotically abelian for t is similar to the condition of metrical transitivity in classical physics.

Let us now go back to our problem to characterize a thermal equilibrium state by general conditions. After what we learned about extremal invariant states for time-translations, it seems that these are natural candidates for equilibrium states consisting of one phase. A proof has not yet been given. We know that equilibrium states are characterized by a finite number of parameters: β, μ, and if the system as a whole is not at rest by a vector α. These parameters are intimately related to the automorphisms for time-translation, gauge-transformation and space-translation. One shall have to prove that all extremal invariant states are labeled by those few parameters. This problem is closely related to the following problem: what is the group of all automorphisms of $\mathfrak{A}$ that commute with time-translation. Does this group contain any other transformations than time-translation, space-translation, gauge-transformation and rotation?

REFERENCES

[1] J.Dixmier, *Les algèbres d'opérateurs dans l'espace hilbertien* (Algèbres de Von Neumann) (Gauthier-Villars, Paris 1957).
[2] J.Dixmier, *Les C*-algèbres et leurs représentations* (Gauthier-Villars, Paris 1964).
[3] R.Haag, *Lecture Notes*, Pacific Summer School, Hawaii (1965).
[4] H.Araki and E.J.Woods, J. Math. Phys. 4 (1963) 637.
[5] S.Doplicher, D.Kastler and D.W.Robinson, Commun. Math. Phys. 3 (1966) 1.
[6] D.Ruelle, Commun. Math. Phys. 3 (1966) 133.
[7] D.Kastler and D.W.Robinson, Commun. Math. Phys. 3 (1966) 151.
[8] R.Haag, N.M.Hugenholtz and M.Winnink, Commun. Math. Phys. 5 (1967) 215.
[9] P.C.Martin and J.Schwinger, Phys. Rev. 115 (1959) 1342.
[10] R.Kubo, J. Phys. Soc. Japan 12 (1957) 570.
[11] J.Ginibre, J. Math. Phys. 6 (1965) 238; 6 (1965) 252.
[12] H.Araki, *thesis*, Princeton University (1960).
[13] N.N.Bogoliubov, J. Phys. 11 (1947) 23.
[14] R.Haag, Nuovo Cimento 25 (1962) 287.
[15] S.Doplicher and D.Kastler, *Ergodic states*, preprint.

THE KINETIC THEORY OF DENSE GASES

E. G. D. COHEN
The Rockefeller University, New York, USA

1. INTRODUCTION

In these lectures I will give a brief survey of the present status of the kinetic theory of dense gases. For details on many points, the reader is referred to previous surveys of a similar nature [1-3].

The aim is to understand a number of observed macroscopic phenomena of gases viz.

I) Why does a gas which is originally not in thermal equilibrium always go to thermal equilibrium;

II) How does the gas approach to thermal equilibrium; in particular:

IIa) How fast does the gas approach thermal equilibrium.

If one wants an explanation on the basis of the microscopic properties of the molecules out of which the gas consists, then one has to know:

1) What is the appropriate macroscopic description of the gas; in particular:

1a) What are the appropriate variables to describe the gas;

2) What are the equations these variables obey and can one on the basis of them explain I) - IIa)?

In order to understand I) some generalization of the H-theorem seems necessary; for II) the validity of the hydrodynamical equations must a.o. be established; IIa) is related to the transport coefficients; in particular one would like to know their equation of state, i.e. their dependence on the density and temperature of the gas.

We shall illustrate the above on the case of a dilute gas, where the connection between the microscopic properties of the molecules and the macroscopic properties of the gas is fairly well understood (at least compared to the case of a dense gas or a liquid!).

2. DILUTE GAS [1,3,4]

The appropriate variable for a macroscopic description of the dilute gas is $f(\boldsymbol{r}\,\boldsymbol{v}\,t)$, the average number of molecules at the position $\boldsymbol{r}$ with velocity $\boldsymbol{v}$ at time t.

The equation f satisfies is *the Boltzmann equation:*

$$\frac{\partial f(\boldsymbol{r}\,\boldsymbol{v}\,t)}{\partial t} + \boldsymbol{v}\cdot\frac{\partial f}{\partial \boldsymbol{r}} = J(ff) =$$

$$= \iiint [f(\boldsymbol{r}\,\boldsymbol{v}'\,t)\, f(\boldsymbol{r}\,\boldsymbol{v}_1'\,t) - f(\boldsymbol{r}\,\boldsymbol{v}\,t)\, f(\boldsymbol{r}\,\boldsymbol{v}_1\,t)]\, gb\, \mathrm{d}b\, \mathrm{d}\psi\, \mathrm{d}\boldsymbol{v}_1. \qquad (2.1)$$

Here $g = |\boldsymbol{g}| = |\boldsymbol{v}_1 - \boldsymbol{v}|$ is the absolute value of the relative velocity $\boldsymbol{g}$ of a molecule with velocity $\boldsymbol{v}$ and a molecule with velocity $\boldsymbol{v}_1$; b is the impact parameter and ψ the azimuth of the plane of the relative motion of $\boldsymbol{v}_1$ with respect to $\boldsymbol{v}$; $\boldsymbol{v}'$ and $\boldsymbol{v}_1'$ are the velocities of 2 molecules after a collision, in which they had the velocities $\boldsymbol{v}$ and $\boldsymbol{v}_1$ respectively before and where the geometry of the collision is characterized by b and ψ (cf. fig. 1). They depend for given $\boldsymbol{v}$, $\boldsymbol{v}_1$, b and ψ on the intermolecular potential field $\phi(r)$, which is always assumed in these lectures to be spherically symmetric.

The point I) can be answered on the basis of the eq. (2.1) by the proof of the H-theorem.

To answer point II) one notes the existence of 2 basic relaxation times in the gas: the average time to traverse a mean free path, t_{mi}, and the average time to traverse a macroscopic length (for instance of the order of the size of the container), t_{ma}. As for a dilute gas one has $t_{\mathrm{mi}} \ll t_{\mathrm{ma}}$ one can distinguish, for a large class of initial conditions, two stages in the approach of such a gas to thermal equilibrium:

1) if $t \leq t_{\mathrm{mi}}$, a *kinetic stage*, where f is needed to describe the gas adequately and where the Boltzmann equation (2.1) is the equation that f obeys;

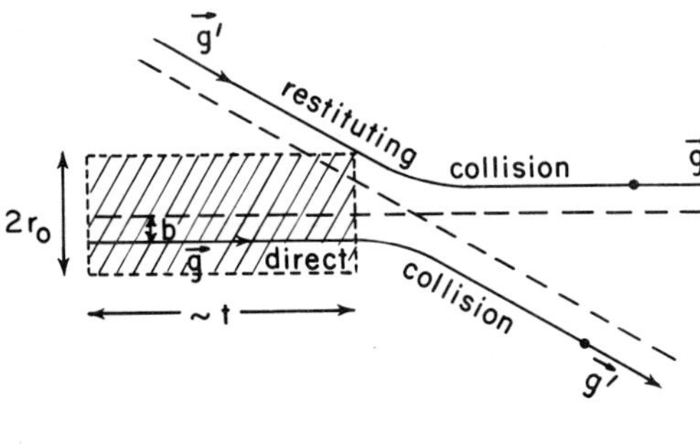

Fig. 1. Binary collision. $\boldsymbol{g} = \boldsymbol{v}_1 - \boldsymbol{v}$; $\boldsymbol{g}' = \boldsymbol{v}_1' - \boldsymbol{v}'$; $g = |\boldsymbol{g}| = |\boldsymbol{g}'|$; shaded area: cross-section of collision cylinder.

2) if $t \gg t_{mi}$, a *hydrodynamical stage*, where the first five moments of f, the hydrodynamical variables: the local density $n(\boldsymbol{r}, t)$, the local velocity $\boldsymbol{u}(\boldsymbol{r}, t)$ and the local temperature $T(\boldsymbol{r}, t)$, are sufficient to describe the gas adequately *.

The equations which these five variables satisfy can be *derived* on the basis of the Chapman-Enskog method to solve the Boltzmann equation. Then one looks for a special solution of the Boltzmann equation, where the whole time dependence of f is through n, $\boldsymbol{u}$ and T and one tries to find such a solution by an expansion in terms of the non-uniformity of the system:

$$f(\boldsymbol{r}\,\boldsymbol{v}\,|\,n\boldsymbol{u}\,T) = f_0(\boldsymbol{r}\,\boldsymbol{v}\,|\,n\boldsymbol{u}\,T) + \mu f_1(\boldsymbol{r}\,\boldsymbol{v}\,|\,n\boldsymbol{u}\,T) + \mu^2 f_2(\boldsymbol{r}\,\boldsymbol{v}\,|\,n\boldsymbol{u}\,T) + \ldots \quad (2.2)$$

Here the expansion parameter μ is physically the relative variation of the hydrodynamical variables over a mean free path l viz. for instance:

$$\mu = \frac{l}{L}\frac{\Delta T}{T} = \frac{l/\bar{v}}{L/\bar{v}}\frac{\Delta T}{T} \sim t_{mi}/t_{ma} ,$$

where ΔT is the temperature difference over a macroscopic length L in the system; $\bar{v}$ is an average velocity characteristic for the molecular motions, for instance the velocity of sound.

Solving then successively for $f_0, f_1, f_2, \ldots$, one can derive from the Boltzmann equation the Euler-, Navier- Stokes-, Burnett-.... hydrodynamical equations, which correspond to taking into account f_0, $f_0 + f_1$, $f_0 + f_1 + f_2, \ldots$ respectively.

In particular the validity of the Navier-Stokes equations proves the validity of the linear laws between fluxes and forces (gradients): Newton's law for viscous flow and Fourier's law of heat conduction, which read, in the case that the gradients are in the z-direction:

$$P_{xz} = -\eta \frac{\partial u_x}{\partial z} , \quad (2.3)$$

$$J_z = - \frac{\partial T}{\partial z} . \quad (2.4)$$

Here P_{xz} is the xz-component of the pressure tensor of the gas and J_z is the z-component of the heat flux vector of the gas; η is the viscosity (coefficient) and λ is the heat conductivity (coefficient).

* After a time $t \gg t_{mi}$, n, $\boldsymbol{u}$ and T suffice to describe the gas, if they are slowly varying quantities for times $t \leqslant t_{mi}$, i.e. in the kinetic stage. This will in general be so if $t_{mi}/t_{ma} \ll 1$, i.e. when $\mu \ll 1$ (see below).

The equation of state of η and λ also follows from the Chapman-Enskog solution of the Boltzmann equation:

$$\eta = \eta_0(T) \,, \tag{2.5}$$

$$\lambda = \lambda_0(T) \,, \tag{2.6}$$

where $\eta_0(T)$ and $\lambda_0(T)$ are well-defined functions of the temperature T, which agree for a realistic intermolecular potential field (for instance a 12-6 Lennard-Jones potential) very well with experiment over a wide range of temperature.

We remark that:

1) η and λ are independent of n;
2) the bulk viscosity $\kappa = 0$;
3) the validity of the laws of irreversible thermodynamics - in particular the validity of the Onsager reciprocal relations in dilute gas mixtures - can be proved on the basis of the Chapman-Enskog solution of the eq. (2.1) or its generalization to gas mixtures.
4) All results quoted so far, with the exception of the proof of the H-theorem, only depend on the linearized Boltzmann equation and constitute therefore no check on the validity of the full non-linear eq. (2.1) for f.

In the next section we shall discuss Bogoliubov's attempt to generalize the Boltzmann-Chapman-Enskog ideas from dilute to dense gases.

3. BOGOLIUBOV'S THEORY [1,3,5]

The first question that arises is: what are the appropriate variables to describe a dense gas and what are the equations they obey? Bogoliubov assumes that $f(\boldsymbol{r}\,\boldsymbol{v}\,t)$ can still be used and obeys a generalized Boltzmann equation of the form:

$$\frac{\partial f}{\partial t} + \boldsymbol{v} \cdot \frac{\partial f}{\partial \boldsymbol{r}} = J(ff) + K(fff) + L(ffff) + \dots \,. \tag{3.1}$$

Here $J, K, L, \dots$ are time independent operators, which depend on the solution of the mechanical 2-, 3-, 4-, ... body problem respectively.

The equation (3.1) should result in some way from a virial expansion (i.e. density expansion) of the Liouville equation or from the hierarchy of equations which can be derived from the Liouville equation by integrating over an increasing number of particles.

To derive this hierarchy one starts with the Liouville equation for $D_N(x_1 \ldots x_N;t)$, the probability density to find the molecules $1,2,\ldots,N$ of the gas of N molecules in volume V, in the phases $x_1 \equiv \boldsymbol{r}_1\boldsymbol{p}_1$, $x_2 \equiv \boldsymbol{r}_2\boldsymbol{p}_2, \ldots, x_N \equiv \boldsymbol{r}_N\boldsymbol{p}_N$ respectively at time t:

$$\frac{\partial D_N(x_1 \ldots x_N;t,V)}{\partial t} = \left\{H_N^W, D_N\right\} = -\mathscr{H}_N^W(x_1 \ldots x_N)D_N\ , \tag{3.2}$$

where the Hamilton function H_N^W is given by $(r_{ij} = (|\boldsymbol{r}_i - \boldsymbol{r}_j|)$:

$$H_N^W = H_N^W(x_1 \ldots x_N) = \sum_{i=1}^{N} \frac{\boldsymbol{p}_i^2}{2m} + \sum\sum_{\substack{i<j\\1}}^{N} \phi(r_{ij}) + \sum_{i=1}^{N} \phi^W(\boldsymbol{r}_i) \tag{3.3}$$

and the Hamilton operator $\mathscr{H}_N^W(x_1 \ldots x_N)$ results from working out the Poisson brackets $\{\ \ \}$:

$$\mathscr{H}_N^W(x_1 \ldots x_N) = \sum_{i=1}^{N} \frac{\boldsymbol{p}_i}{m} \cdot \frac{\partial}{\partial \boldsymbol{r}_i} - \sum\sum_{\substack{i<j\\1}}^{N} \theta_{ij} - \sum_{i=1}^{N} \theta_i^W\ , \tag{3.4}$$

with

$$\theta_{ij} = \frac{\partial \phi(r_{ij})}{\partial \boldsymbol{r}_i} \cdot \frac{\partial}{\partial \boldsymbol{p}_i} + \frac{\partial \phi(r_{ij})}{\partial \boldsymbol{r}_j} \cdot \frac{\partial}{\partial \boldsymbol{p}_j} \tag{3.5}$$

and similarly for θ_i^W.

Here $\phi^W(\boldsymbol{r}_i)$ is the potential due to the wall of the container. In (3.4) it has been assumed that the intermolecular potential is additive.

Introducing reduced distribution functions:

$$F_s(x_1 \ldots x_s;t;N,V) = V^s \int dx_{s+1} \ldots \int dx_N \frac{D_N(x_1 \ldots x_N;t;V)}{V^N}\ , \tag{3.6}$$

one can obtain from the eq. (3.2), by multiplying both sides with $V^{-(N-s)}$ and using that D_N vanishes for $\boldsymbol{p}_i \to \infty$ as well as for $\boldsymbol{r}_i \to \infty$ (because of the wall potential $\phi^W(\boldsymbol{r}_i)$), the following equation for F_s:

$$\frac{\partial F_s(x_1\ldots x_s;t;N,V)}{\partial t} + \mathcal{H}_s^W(x_1\ldots x_s)F_s$$
$$= \frac{N-s}{V}\int \mathrm{d}x_{s+1}\left[\sum_{i=1}^{s}\theta_{i,\,s+1} + \sum_{i=1}^{s}\theta_i^W\right] F_{s+1}(x_1\ldots x_{s+1};t;N,V) \tag{3.7}$$
$$(s = 1,2,\ldots,N-1)\;.$$

We are interested in the bulk properties of molecules deep in the interior of the container and therefore we will consider, rather than the equations (3.7), the corresponding equations for an infinite system, where $N, V \to \infty$ at $N/V = n$ = constant and $\phi_i^W = 0$:

$$\frac{\partial F_s(x_1\ldots x_s;t)}{\partial t} + \mathcal{H}_s(x_1\ldots x_s)\,F_s$$
$$= n\int \mathrm{d}x_{s+1}\sum_{i=1}^{s}\theta_{i,\,s+1}\,F_{s+1}(x_1\ldots x_{s+1};t)\;,\quad (s = 1,2,\ldots) \tag{3.8}$$

where

$$F_s(x_1\ldots x_s;t) = \lim_{\substack{N,V\to\infty\\ N/V = n}} F_s(x_1\ldots x_s;t;N,V) \tag{3.9}$$

and

$$\mathcal{H}_s(x_1\ldots x_s) = \sum_{i=1}^{s}\frac{\boldsymbol{p}_i}{m}\cdot\frac{\partial}{\partial \boldsymbol{q}_i} - \sum_{\substack{i<j\\1}}^{s}\sum\theta_{ij}\;. \tag{3.10}$$

The infinite set of coupled equation (3.8) is usually called the B-B-K-G-Y- hierarchy.

In order to avoid extra complications we shall restrict ourselves, with Bogoliubov, to the case that the intermolecular potential is monotonically repulsive (so that no molecular binding is possible) and of finite range r_0. In that case all properties of interest of the gas can be expressed in terms of F_1 and F_2 alone, where F_2 is only needed for intermolecular distances $r_{12} \leqslant r_0$!

The Bogoliubov assumption, in this context, is now that not only a dilute gas but also a dense gas can be adequately described by $F_1(t)$ alone, which implies that F_2 is determined by F_1. More precisely Bogoliubov asserts that in the approach of a dense gas to equilibrium one can distinguish, in addition to the kinetic- and hydrodynamical stages, mentioned before already for the dilute gas, an *initial stage* which is related to the existence of a third basic relaxation time $t_c = r_0/\bar{v}$, the average duration of a collision. As $t_c \ll t_{mi} \ll t_{ma}$, one can distinguish for a large class of initial conditions, three stages in the approach of a dense gas to thermal equilibrium:

1) if $t \leqslant t_c$, an *initial stage*, where one needs $D_N(x_1 \ldots x_N; t)$ (or $F_1, F_2, F_3, \ldots$) to describe the gas adequately and where the Liouville equation (or the B-B-G-K-Y-hierarchy) is the equation that D_N (or $F_1, F_2, F_3, \ldots$) obeys:

2) if $t_c \ll t < t_{mi}$, a *kinetic stage*, where $nF_1 = f$ is needed to describe the gas adequately* and where the generalized Boltzmann equation (3.1) is the equation that f obeys;

3) if $t_{mi} \ll t < t_{ma}$, a *hydrodynamical stage*, where n, $\boldsymbol{u}$ and T are needed to describe the gas adequately and where the hydrodynamical equations are the equations that n, $\boldsymbol{u}$ and T obey.

Finally after $t \gg t_{ma}$ thermal equilibrium is established, where only n and T are needed to describe the system adequately: a progressing simplification of the description of the gas, accompanies therefore the approach to thermal equilibrium!

The equation that F_1 (or f) satisfies was *derived* by Bogoliubov from the hierarchy (3.8) by generalizing the Chapman-Enskog method to solve the Boltzmann equation, i.e. he looked for a special solution of the hierarchy equations, where the *whole* time dependence of F_2, F_3, $F_4, \ldots$ is through F_1. He tried to find this solution by an expansion of F_2, $F_3, \ldots$ in terms of the density of the system:

$$F_s(x_1 \ldots x_s | F_1) = F_s^{(0)}(x_1 \ldots x_s | F_1)$$

$$+ \rho F_s^{(1)}(x_1 \ldots x_s | F_1) + \rho^2 F_s^{(2)}(x_1 \ldots x_s | F_1) + \ldots \quad (s \geqslant 2). \quad (3.11)$$

Here the expansion parameter ρ is the reduced density of the gas:

$$\rho = nr_0^3 \sim \frac{t_c}{t_{mi}}, \qquad (3.12)$$

where it has been used that $t_{mi} \sim l \sim 1/(nr_0^2)$.

* After a time $t \gg t_c$, $nF_1 = f$ suffices to describe the gas, if f is a slowly varying quantity for times $t \leqslant t_c$, i.e. in the initial stage. This will be so if $t_c/t_{mi} \ll 1$, i.e. if $\rho \ll 1$ (see (3.12)). For the hydrodynamical stage the same argument applies as before.

Solving then successively for $F_s^{(0)}, F_s^{(1)}, F_s^{(2)}, \ldots (s \geq 2)$, one can derive from the hierarchy the generalized Boltzmann equation (3.1), where $J, K, L, \ldots$ correspond to taking into account $F_s^{(0)}(|F_1)$, $F_s^{(0)} + F_s^{(1)}$, $F_s^{(0)} + F_s^{(1)} + F_s^{(2)}, \ldots$ respectively.

In the hydrodynamical stage, one can solve the generalized Boltzmann equation à la Chapman-Enskog by expanding f in terms of μ (cf. eq. (2.2)). Then the hydrodynamical equations can be derived, where now the transport coefficients are given by a virial expansion in ρ. In particular in the Navier-Stokes equations, the equations of state for the viscosity and heat conductivity are respectively:

$$\eta(\rho, T) = \eta_0(T) + \rho\eta_1(T) + \rho^2\eta_2(T) + \ldots , \tag{3.13}$$

$$\lambda(\rho, T) = \lambda_0(T) + \rho\lambda_1(T) + \rho^2\lambda_2(T) + \ldots . \tag{3.14}$$

Sengers has computed η_1 and λ_1 for the case of hard spheres. We remark further that:

1) the bulk viscosity κ is given by:

$$\kappa(\rho, T) = \kappa_2(T)\rho^2 + \kappa_3(T)\rho^3 + \ldots .$$

2) the validity of the laws of irreversible thermodynamics has been proved on the basis of the Chapman-Enskog solution of the eq. (3.1) with only J and K included [3]. However, no proof has been given, as yet, for the validity of the Onsager reciprocal on the basis of the eq. (3.1);

3) No H-theorem has been proved on the basis of the eq. (3.1);

4) No computations have been made as yet for a realistic intermolecular potential, since it is not clear that an equation of the form (3.1) still holds in that case, because of the attractive part of the intermolecular potential.

4. CLUSTER EXPANSIONS [2, 3, 6]

Bogoliubov's results were obtained on the basis of a number of well-defined, but unproven, approximations. In order to ascertain the validity of these approximations, we shall try to derive the same results in a different way, using cluster expansions similar to those employed in equilibrium statistical mechanics.

We first introduce *streaming operators*, which will be used extensively throughout the rest of these lectures.

A formal solution of the Liouville equation (3.2) is

$$D_N(x_1 \ldots x_N; t, V) = S^W_{-t}(x_1 \ldots x_N) D_N(x_1 \ldots x_N; 0, V)$$
$$= e^{-t \mathcal{H}^W_N(x_1 \ldots x_N)} D_N(x_1 \ldots x_N; 0, V) . \tag{4.1}$$

Here the streaming operator

$$S^W_{-t}(x_1 \ldots x_N) = \exp\{-t\mathcal{H}^W_N(x_1 \ldots x_N)\} \tag{4.2}$$

transforms the phases $x_1 \ldots x_N$ of the particles $1, \ldots, N$ at time 0 into the phases $S^W_{-t}(x_1 \ldots x_N)x_1, \ldots, S^W_{-t}(x_1 \ldots x_N)x_N$, which they had a time t earlier. As we are interested in the behavior of a large number particles deep inside the vessel, disregarding collisions with the wall, we introduce in addition to the streaming operators defined by (4.2), streaming operators

$$S_{-t}(x_1 \ldots x_N) = \exp\{-t\mathcal{H}_N(x_1 \ldots x_N)\} , \tag{4.3}$$

which transform the phases $x_1 \ldots x_N$ of the particles $1 \ldots N$ at time $t = 0$ into these a time t earlier if the particles move in *infinite space* under the influence of their mutual interaction only.

Properties of the S-operators (4.3) are given elsewhere [1]. We now make cluster expansions of $S^W_{-t}(x_1 \ldots x_N)$ and $S_{-t}(x_1 \ldots x_N)$, which, when substituted into (3.6) for $s = 1$ and $s = 2$, yield cluster expansions of $F_1(x_1; t; N, V)$ and $F_2(x_1 x_2; t; N, V)$ and with (3.9) of $F_1(x_1; t)$ and $F_2(x_1 x_2; t)$. We shall only write down explicitly the expansions for $S_{-t}(x_1 \ldots x_N)$, as those for $S^W_{-t}(x_1 \ldots x_N)$ follow immediately from these through replacing $\mathcal{H}_s(x_1 \ldots x_s)$ by $\mathcal{H}^W_s(x_1 \ldots x_s)$ $(s = 1, 2, \ldots, N)$.

For the cluster expansion of F_1 we define to the S-operators the following set of U-operators:

$$S_{-t}(x_1) = U_t(x_1) ,$$

$$S_{-t}(x_1 x_2) = U_t(x_1) S_{-t}(x_2) + U_t(x_1 | x_2) ,$$

$$S_{-t}(x_1 x_2 x_3) = U_t(x_1) S_{-t}(x_2 x_3) + U_t(x_1 | x_2) S_{-t}(x_3)$$
$$+ U_t(x_1 | x_3) S_{-t}(x_2) + U_t(x_1 | x_2 x_3) ,$$

$$S_{-t}(x_1 \ldots x_s) = U_t(x_1) S_{-t}(x_2 \ldots x_s)$$
$$+ \sum_{r=2}^{s} U_t(x_1 | x_r) S_{-t}(x_1 \ldots x_{r-1} x_{r+1} \ldots x_s) + \ldots + U_t(x_1 | x_2 \ldots x_s) . \tag{4.4}$$

The equations (4.4) define the $U_t(x_1|x_2\ldots x_s)$ $(s = 1,2,\ldots)$, as each line introduces one new U-operator. In fact, solving the eqs. (4.4) successively for the $U_t(x_1|x_2\ldots x_s)$ yields:

$$U_t(x_1) = S_{-t}(x_1)\ ,$$

$$U_t(x_1|x_2) = S_{-t}(x_1x_2) - S_{-t}(x_1)S_{-t}(x_2)\ ,$$

$$U_t(x_1|x_2x_3) = S_{-t}(x_1x_2x_3) - S_{-t}(x_1x_2)S_{-t}(x_3) - S_{-t}(x_1x_3)S_t(x_2)$$

$$- S_{-t}(x_2x_3)S_{-t}(x_1) + 2S_{-t}(x_1)S_{-t}(x_2)S_{-t}(x_3)\ . \qquad (4.4a)$$

In the corresponding expansions of $S^W_{-t}(x_1\ldots x_s)$ s will not exceed N, of course. Substituting the expansion for $S^W_{-t}(x_1\ldots x_N)$ into the right hand side (r.h.s.) of eq. (3.6), taking then the limit of an infinite system according to (3.9) and *assuming* that this limit can be taken term by term, one obtains the following expansion for $F_1(x_1;t)$:

$$F_1(x_1;t) = U_t(x_1)F_1(x_1;0) + n\int \mathrm{d}x_2\ U_t(x_1|x_2)F_2(x_1x_2;0)$$

$$+ \frac{n^2}{2!}\int \mathrm{d}x_2 \int \mathrm{d}x_3\ U_t(x_1|x_2x_3)F_3(x_1x_2x_3;0) + \ldots\ . \qquad (4.5)$$

Proof
From (3.6), (4.1) and (4.4), one has:

$$F_1(x_1;t;N,V) = V\int \mathrm{d}x_2\ldots\int \mathrm{d}x_N \frac{D_N(x_1\ldots x_N;t,V)}{V^N}$$

$$= V\int \mathrm{d}x_2\ldots\int \mathrm{d}x_N\ S^W_{-t}(x_1\ldots x_N)\frac{D_N(x_1\ldots x_N;0,V)}{V^N}$$

$$= V\int \mathrm{d}x_2\ldots\int \mathrm{d}x_N \prod_{i=1}^{N} U^W_t(x_i)\frac{D_N(x_1\ldots x_N;0,V)}{V^N}$$

$$+ V\int \mathrm{d}x_2\ldots\int \mathrm{d}x_N \sum_{r=2}^{N} U^W_t(x_1|x_r)\sum_{i\neq r}^{N} U^W_t(x_i)\frac{D_N(x_1\ldots x_N;t,V)}{V^N} + \ldots$$

$$= U^W_t(x_1)F_1(x_1;0;N,V) + \frac{(N-1)}{V}\int \mathrm{d}x_2\ U^W_t(x_1|x_2)F_2(x_1x_2;0;N,V)$$

$$+ \frac{(N-1)(N-2)}{2V^2}\int \mathrm{d}x_2\int \mathrm{d}x_3\ U^W_t(x_1|x_2x_3)F_3(x_1x_2x_3;0;N,V) + \ldots,$$

where one has used in the last equality the identity of all the particles and that $\int \mathrm{d}x_2 = \int \mathrm{d}x_2\, S_{-t}(x_2)$. The superscript W on the U_t-operators indicates that H^W_s-operators are used instead of H_s-operators. Taking now the limit $N, V \to \infty$ at $N/V = n =$ constant *term by term*, one obtains with eq. (3.9) the expansion (4.5).

A cluster expansion for F_2 can be obtained in a similar fashion. Thereto we introduce to the S-operators the following set of U-operators:

$$\begin{aligned} S_{-t}(x_1x_2) &= U_t(x_1x_2)\,, \\ S_{-t}(x_1x_2x_3) &= U_t(x_1x_2)S_{-t}(x_3) + U_t(x_1x_2|x_3)\,, \\ S_{-t}(x_1x_2x_3x_4) &= U_t(x_1x_2)S_{-t}(x_3x_4) + U_t(x_1x_2|x_3)S_{-t}(x_4) \\ &\quad + U_t(x_1x_2|x_4)S_{-t}(x_3) + U_t(x_1x_2|x_3x_4)\,, \\ S_{-t}(x_1x_2\ldots x_s) &= U_t(x_1x_2)S_{-t}(x_3\ldots x_s) \\ &\quad + \sum_{r=3}^{s} U_t(x_1x_2|x_r)S_{-t}(x_3\ldots x_{r-1}x_{r+1}\ldots x_s) + \ldots \\ &\quad + U_t(x_1x_2|x_3\ldots x_s)\,. \end{aligned} \tag{4.6}$$

Solving the equation (4.6) for the $U_t(x_1x_2|\,x_3\ldots x_s)$ yields:

$$\begin{aligned} U_t(x_1x_2) &= S_{-t}(x_1x_2)\,, \\ U_t(x_1x_2|x_3) &= S_{-t}(x_1x_2x_3) - S_{-t}(x_1x_2)S_{-t}(x_3)\,, \\ U_t(x_1x_2|x_3x_4) &= S_{-t}(x_1x_2x_3x_4) - S_{-t}(x_1x_2x_3)S_{-t}(x_4) \\ &\quad - S_{-t}(x_1x_2x_4)S_{-t}(x_3) - S_{-t}(x_1x_2)S_{-t}(x_3x_4) \\ &\quad + 2S_{-t}(x_1)S_{-t}(x_2)S_{-t}(x_3)S_{-t}(x_4)\,. \end{aligned} \tag{4.6a}$$

Proceeding in a similar way as before, one derives from (3.6), (4.1), (4.6) and (3.9):

$$\begin{aligned} F_2(x_1x_2;t) &= U_t(x_1x_2)F_2(x_1x_2;0) + n\int \mathrm{d}x_3\; U_t(x_1x_2|x_3)F_3(x_1x_2x_3;0) \\ &\quad + \frac{n^2}{2!}\int \mathrm{d}x_3 \int \mathrm{d}x_4\; U_t(x_1x_2|x_3x_4)F_4(x_1x_2x_3x_4;0) + \ldots\,. \end{aligned} \tag{4.7}$$

The expansions for $F_1(x_1;t;N,V)$ and $F_2(x_1x_2;t;N,V)$ that correspond to (4.5) and (4.7) certainly represent these functions for every finite time t as they contain a finite number of terms N, each of which exists. One would hope that the same is true for the expansions (4.5) and (4.7) for $F_1(x_1;t)$ and $F_2(x_1x_2;t)$ respectively, in spite of the term by term limit taken in their derivation. In fact if we assume that the $F_s(x_1 \dots x_s;0)$ effectively vanish as soon as the velocity of any of the particles exceeds a certain value v_M, then all the terms on the r.h.s. of (4.5) and (4.7) *exist* for any finite t. We know nothing, however, about the convergence of the series.

What we are really interested in, is to use the expansion (4.5) and (4.7) to compute F_1 and F_2 for times $t \gg t_{mi} \gg t_c$. Clearly for this purpose the expansions (4.5) and (4.7) are only then of *practical* interest if one can restrict oneself for such a computation to the first few terms of these expansions only and does not have to sum the whole infinite series.

In order to investigate this point, we shall study the behavior of the terms on the r.h.s. of (4.5) and (4.7) for large t, i.e. for $t \gg t_c$. The occurrence of the S-operator makes an analysis in terms of the dynamics of isolated groups of particles in infinite space necessary.

We note that due to the disappearance of the wall potential, the only length occurring in the U-operators is r_0, so that the characteristic time associated with the S_t- or U_t-operators is t_c.

Study of $F_1(x_1;t)$. The second term on the r.h.s. of (4.5) behaves for large $t \sim t/t_c$ as the phase space associated with particle 2, for which the integrand does not vanish grows $\sim t/t_c$. This is so because the integrand does not vanish whenever the (initial) phase of particle 2 is such that a binary collision takes place of 1 and 2 between 0 and t (fig. 2a). For this to happen, particle 2 has to be in the collision cylinder of particle 1, which grows $\sim t/t_c$ (cf. fig. 1).

Similarly, the third term on the r.h.s. of (4.5) behaves for large $t \sim (t/t_c)^2$, as the combined phase space of the particles 2 and 3 for two successive binary collisions to occur between 0 and t grows $\sim (t/t_c)^2$ (cf. fig. 2b).

In general one expects that the lth term on the r.h.s. of (4.5) grows $\sim (t/t_c)^{l-1}$ due to the occurrence of $(l-1)$ uncorrelated successive binary collisions (cf. fig. 2c).

Study of $F_2(x_1x_2;t)$. On the basis of similar arguments as presented above, one expects that the lth term on the r.h.s. of (4.7) grows for $r_{12} \leq r_0$, $\sim (t/t_c)^{l-1}$ (cf. fig. 3).

On the basis of the above analysis it is clear that the expansions (4.5) and (4.7) cannot be used to compute F_1 and F_2 for times $t \gg t_c$, if one wants to restrict oneself to the first few terms in these expansions only.

On the contrary for such times the most important contributions seem to come from higher terms!

We would like to stress that the analysis of dynamical events

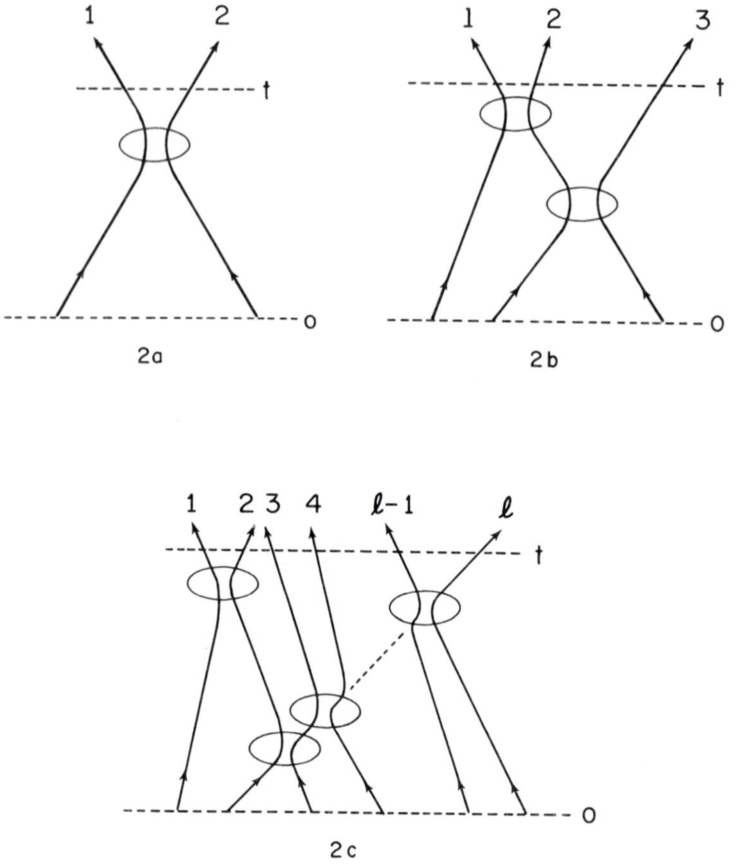

Fig. 2. Two-, three- and l-particle events leading to secular contribution to $F_1(x_1;t)$ of (4.5). (a): $\approx t/t_c$; (b): $\sim (t/t_c)^2$; (c): $\sim (t/t_c)^{l-1}$.

which lead to this conclusion - plausible as it may seem - can only be put on a firm basis if a classification of all possible dynamical events between a (finite) number of particles is known. If such a classification were known then one could hope *to prove* that the sequence of (l-1) or l successive binary collisions indeed determine the dominant behavior of the lth term of the r.h.s. of (4.5) and (4.7) respectively.

These sequences of dynamical events are represented by diagrams which could be called "connected". A diagram is connected if it is impossible to separate the diagram into two or more (sub) diagrams such that there is no collision joining a particle of one (sub) diagram to that of another (sub) diagram. In addition they

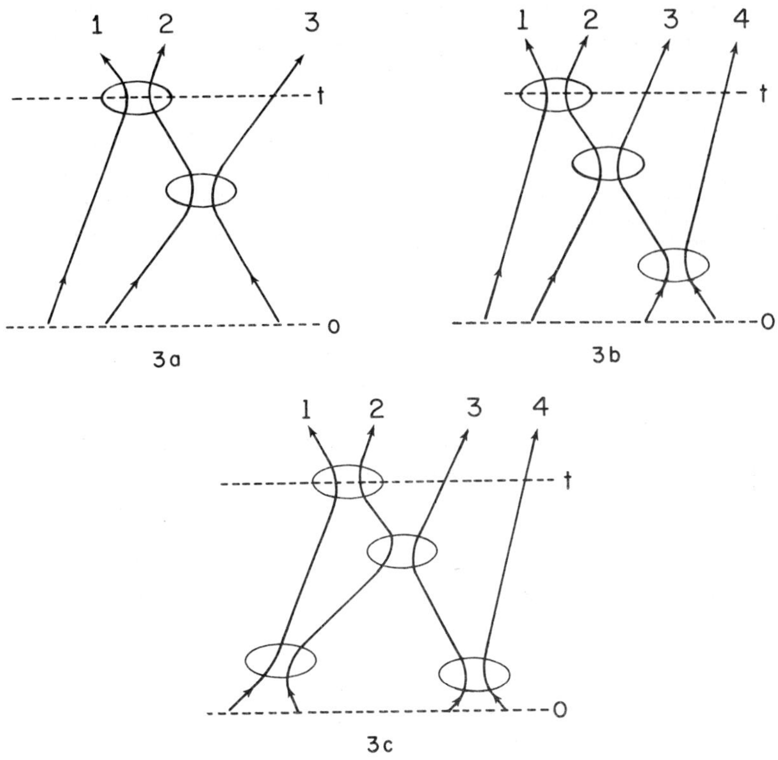

Fig. 3. Three- and four-particle events leading to secular contributions to $F_2(x_1x_2;t)$ of (4.7), (a): $\sim t/t_c$; (b): $\sim (t/t_c)^2$; (c): $\sim (t/t_c)\ln(t/t_c)$.

could be called "completely reducible" as they are so loosely connected that if one cuts the diagram between *any* two successive binary collisions, the diagram is not anymore connected (fig. 2b, c; 3a, b).

To be sure, there are many more sequences of dynamical events, that contribute to the r.h.s. of (4.5) and (4.7) and lead to contributions that grow with time, than those represented by the completely reducible diagrams (cf. fig. 3c; 4a, b, c). The claim, however, is that the completely reducible ones *dominate* the long time behavior of each term in the expansion of F_1 and F_2. The contributions of the "completely reducible events" are the *most secular* contributions to each term on the r.h.s. of (4.5) and (4.7), as they grow with the highest power of t. One also sometimes refers to this by saying that they are the most "divergent" contributions, classifying then according to their behavior for $t \to \infty$.

Physically, the origin of the secular contributions to F_1 and F_2 is the fact that the first few terms of the expansions (4.5) and (4.7) will suffice for a computation of F_1 and F_2 only if a few successive binary collisions can take place between particles during the time t. Clearly this will not be so if $t \gg t_c$.

In order to obtain expansions for F_1 and F_2 which *can* be used for large t, a rearrangement of the expansions (4.5) and (4.7) has to be carried out such that not only the most secular contributions, but in fact *all* secular contributions are eliminated. We shall attempt such a rearrangement for the special case that the initial state is such that all $F_s(x_1 \ldots x_s;0)$ factorize, viz.:

$$F_s(x_1 \ldots x_s;0) = \prod_{i=1}^{s} F_1(x_i;0) \ . \tag{4.8}$$

Clearly the eq. (4.8) is unrealistic. The justification of the results obtained on the basis of it must come from the independence of the state of the gas at a time $t \gg t_c$ from the precise initial state at $t = 0$. In how far one has been able to do this has been discussed elsewhere [7, 3]. Using (4.8) in (4.7), one can say that the elimination of the most secular contributions to each term on the r.h.s. of (4.7) can be achieved by:

1) Replacing with (4.5) in each term of (4.7) the $F_1(x_i;0)$ by $F_1(x_i;t)$;
2) Replacing the U_t-operators by $\mathcal{T}_t$-operators, which vanish for sequences of collisions represented by completely reducible diagrams.

By "upgrading" $F_1(x_i;0)$ to $F_1(x_i;t)$ the origin of the secular contributions, viz. the occurrence for $t \gg t_c$ of too many successive binary collisions, seems to be eliminated.

One obtains then the following expansion for $F_2(x_1 x_2;t)$:

$$\begin{aligned} F_2(x_1x_2;t) &= \mathcal{T}_t(x_1x_2)F_1(x_1;t)F_1(x_2;t) \\ &\quad + n \int \mathrm{d}x_3\ \mathcal{T}_t(x_1x_2|x_3) \prod_{i=1}^{3} F_1(x_i;t) \\ &\quad + \frac{n^2}{2!} \int \mathrm{d}x_3 \int \mathrm{d}x_4\ \mathcal{T}_t(x_1x_2|x_3x_4) \prod_{i=1}^{4} F_1(x_i;t) + \ldots \ , \end{aligned} \tag{4.9}$$

where

$$\mathcal{T}_t(x_1x_2) = \mathcal{S}_t(x_1x_2) ,$$

$$\mathcal{T}_t(x_1x_2|x_3) = \mathcal{S}_t(x_1x_2x_3) - \mathcal{S}_t(x_1x_2)\mathcal{S}_t(x_1x_3) - \mathcal{S}_t(x_1x_2)\mathcal{S}_t(x_2x_3) + \mathcal{S}_t(x_1x_2) , \tag{4.10}$$

with

$$\mathcal{S}_t(x_1 \ldots x_s) = S_{-t}(x_1 \ldots x_s) \prod_{i=1}^{s} S_t(x_i) . \tag{4.11}$$

For a derivation of (4.9) from (4.5) and (4.7) which has been discussed several times before already, we refer to the literature [3,6].

That for instance the operator $\mathcal{T}_t(x_1x_2|x_3)$ vanishes for 2 successive binary collisions sketched in fig. 3a is due to 1) the factorization of $\mathcal{S}_t(x_1x_2x_3)$ into $\mathcal{S}_t(x_1x_2)$ $\mathcal{S}_t(x_2x_3)$ and 2) that $\mathcal{S}_t(x_1x_3) = 1$ for such a sequence of collisions.

The replacement of $F_1(x_i;0)$ by $F_1(x_i;t)$ has eliminated not only the contributions of all sequences of collisions represented by completely reducible diagrams, but many more (for instance those represented by a diagram as in fig. 3c).

The expansion (4.9) of F_2 in terms of F_1 defines one relation between F_2 and F_1. A second relation is furnished by the first hierarchy equation:

$$\frac{\partial F_1(x_1;t)}{\partial t} + \frac{\boldsymbol{p}_1}{m} \cdot \frac{\partial F_1}{\partial \boldsymbol{r}_1} = n \int \mathrm{d}x_2 \, \theta_{12} \, F_2(x_1x_2;t) , \tag{4.12}$$

which is obtained from (3.8) for $s = 1$. The equation (4.9) can serve as a starting point for investigating the validity of the results obtained by Bogoliubov, discussed in the previous section.

In fact replacing $\mathcal{T}_t(x_1x_2)$ and $\mathcal{T}_t(x_1x_2|x_3)$ by $\mathcal{T}_\infty(x_1x_2)$ and $\mathcal{T}_\infty(x_1x_2|x_3)$ respectively, one can show that the first two terms on the r.h.s. of eq. (4.9) do reduce to Bogoliubov's $F_2^{(0)}(x_1x_2|F_1)$ and $F_2^{(1)}(x_1x_2|F_1)$ respectively [1,8].

Although it would seem that the replacement of $\mathcal{T}_t$ by $\mathcal{T}_\infty$ would also be allowed in the higher terms in the expansion (4.9), a closer analysis has revealed that this is *not* so. The reason is that although the replacement of $F_1(x_i;0)$ by $F_1(x_i;t)$ has eliminated all most secular contributions to the terms in the expansion (4.7), it has eliminated *all* secular contributions *only* in the second term of the expansion (4.7) (cf. section 5). The sequences of collisions that lead to these remaining secular contributions lead to a divergence of all the terms $F_2^{(l)}(F_1)$ with $l \geq 2$ in the expansion (3.11) of F_2 as well as of all the "virial" coefficients $\eta_l(T)$ and $\lambda_l(T)$ with $l \geq 2$ in the expansions (3.13) and (3.14) given in the previous section*. Therefore a kinetic stage in the sense of Bogoliubov - and in particular a generalized Boltzmann equation of the form (3.1) - can only be given a meaning if one can restrict oneself in the time evolution of a gas to contributions of binary and triple collisions. Within this limitation $F_1(x;t)$ can be called a "regularizing function" as it eliminates all secular contributions to the first two terms in the expansion (4.7) and all most secular contributions to all higher terms.

5. SECULAR CONTRIBUTIONS TO F_2 [2, 3, 7]

As already mentioned in the previous section, the rearrangement of the expansion (4.7) for F_2 with the help of the expansion (4.5) for F_1 into a new expansion (4.9) of F_2 does not eliminate all secular contributions in the higher terms of the expansion (4.9). Careful study reveals that for a repulsive intermolecular potential of finite range r_0 and for general F_1:

1) the most secular contributions to the higher terms of the expansion (4.9) come from sequences of successive binary collisions (cf. fig. 4);

2) the second term on the r.h.s. of (4.9), i.e. the four-body term, grows for $t \gg t_c \sim \ln(t/t_c)$, as the combined phase space associated with the particles 3 and 4 leading to sequences of 4 or more binary collisions grows in this way (in 3 dimensions) (fig. 4b);

3) the l-body term on the r.h.s. of (4.9) grows for $t \gg t_c \sim (t/t_c)^{l-4}$ (in 3 dimensions) (cf. fig. 4c).

As these statements have been discussed before extensively several times, we refer for details to the literature [2, 3, 7].

Here we shall illustrate the three points on the 2-dimensional Lorentz model which is particularly instructive.

We consider then an infinite number of independent point particles which move with constant (magnitude of their) velocity v through the plane and collide with an infinite number of infinitely heavy scatterers, whose centers are randomly distributed through

* It should be remarked, however, that for the special case that F_1 is the equilibrium Maxwell distribution function, the replacement of $\mathcal{T}_t$ by $\mathcal{T}_\infty$ *is* allowed, as then the r.h.s. of (4.9) approaches term by term to the equilibrium virial expansion of F_2! The approach of the expansion (4.9) to the equilibrium expansion seems therefore to be highly non-uniform.

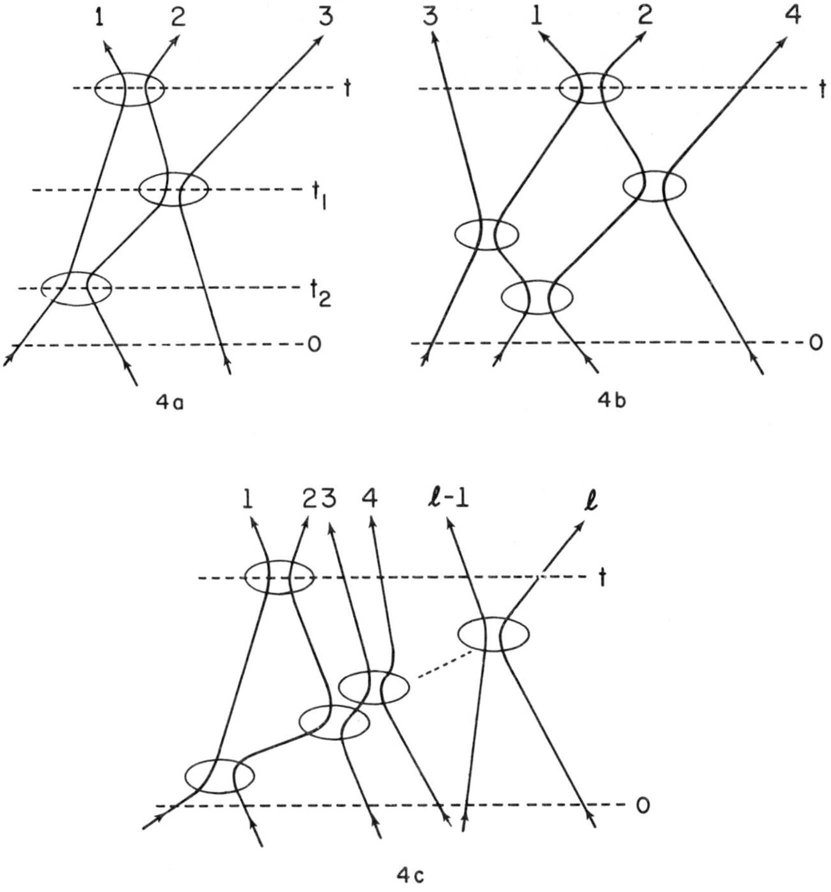

Fig. 4. Three-, four- and l-particle events contributing to $F_2(x_1x_2;t)$ of (4.9) and $\psi_2(x_1x_2;t)$ of (7.21). (a): $\sim (t_c/t)$; (b): $\sim \ln(t/t_c)$; (c): $\sim (t/t_c)^{l-4}$.

the plane *. We consider two cases:

a) the scatterers are oriented squares with diameter $2a$** [10] (fig. 5);

b) the scatterers are circles with diameter $2a$ (fig. 6) [11].

* Random could be taken to mean here that every configuration of the scatterers is weighed with the canonical distribution function for instance.

** This special case of the Lorentz model was studied by Ehrenfest and is often referred to as the "wind-tree" model [9].

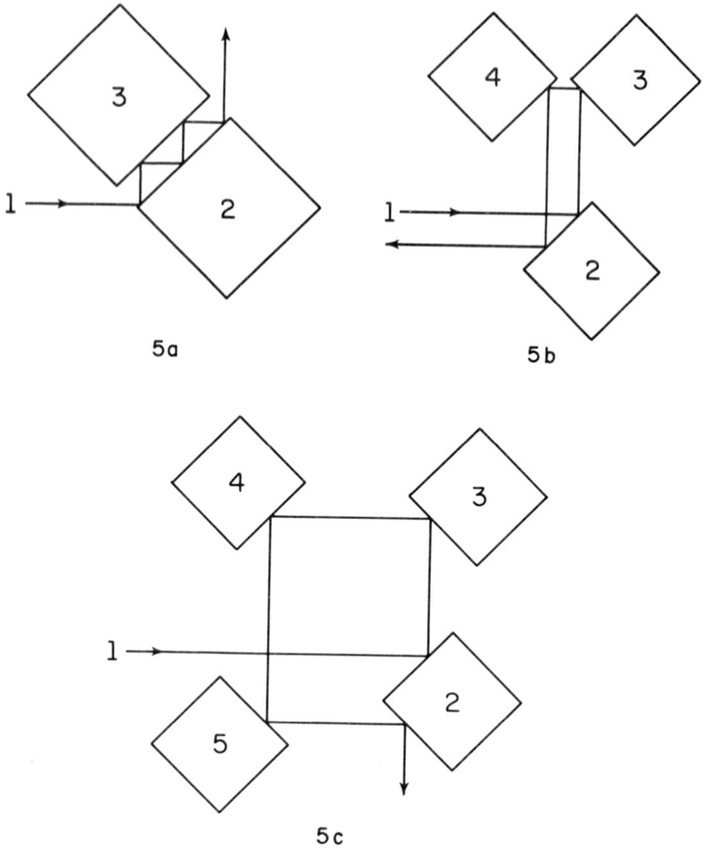

Fig. 5. Three-, four- and five- particle events contribution to $F_2(x_1x_2;t)$ of (4.7) for Lorentz model with oriented hard squares as scatterers (wind-tree model). (a): constant : (b) $\sim (t/t_c)$; (c): $\sim (t/t_c)^2$.

As the point particles are independent it suffices 1) to consider only one point particle and 2) to restrict the discussion of l-body term to the case of *one* point particle and $(l-1)$ scatterers.

Furthermore in these models the diagrams that represent the sequences of collisions of the point particle with the scatterers and for which the $\mathcal{T}_t$ do not vanish, can be characterized very simply. In fact, they are such that the scatterers remain connected by the path of the point particle if one cuts the path between any two successive collisions. We will say that the $\mathcal{T}_t$ do not vanish for irreducible paths of the point particle.

Case a) We allow the point particle only to move in the 4 directions along the $\pm$ x-axis and the $\pm$ y-axis, i.e. $\rightarrow, \leftarrow, \uparrow, \downarrow$. Due to

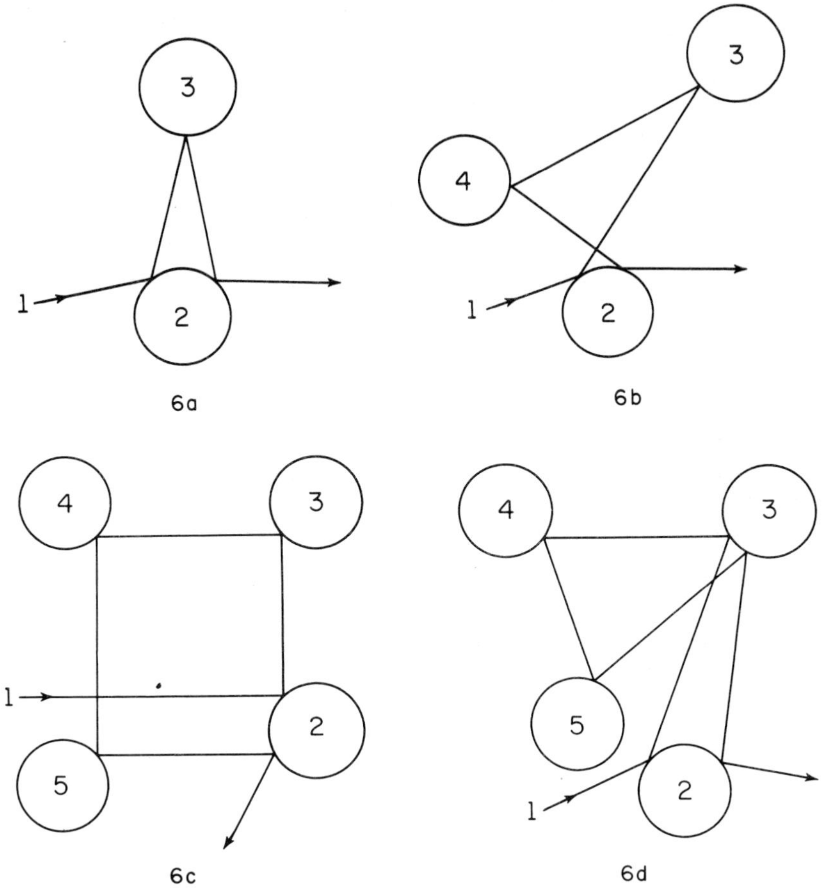

Fig. 6. Three-, four- and five-particle events contributing to $F_2(x_1x_2;t)$ of (4.7) for Lorentz model with hard circles as scatterers. (a): $\sim \ln(t/t_c)$; (b): $\sim (t/t_c)$; (c): $\sim (t/t_c)^2$; (d): $\sim (t/t_c)\ln(t/t_c)$.

the perfect orientational alignment of all the squares this discrete velocity space for the point particle is maintained for all times if one starts the particle in it at $t = 0$.

The 3-body term does *not* contain secular contributions because the condition of irreducibility of the path of point particle 1 restricts the position of square 3 to the neighborhood of square 2 (cf. fig. 5a); the 4-body term contains for $t \gg t_c = a/v$, contributions that grow linearly with time, as the (vertical) distance of the squares 3 and 4 and the square 2 can grow $\sim (t/t_c)$ (cf. fig. 5b); the 5-body term contains for $t \gg t_c$ contributions that grow $\sim (t/t_c)^2$

as the (horizontal) distance between the squares 2 and 5 (or 3 and 4) *as well as* the (vertical) distance between the squares 2 and 3 (or 5 and 4) can grow $\sim (t/t_c)$ (cf. fig. 5c).

In general one expects (no rigorous proof exists !) that the l-body term grows $\sim (t/t_c)^{l-3}$, i.e. that all most secular contributions grow with powers of t and that no logarithmic time dependence appears *.

Case b) In this case the point particle will move in all directions, if one starts it at $t = 0$ in any particular direction.

The 3-body term contains a secular contribution $\sim \ln(t/t_c)$ (cf. fig. 6a); the 4-body term contains a most secular contribution $\sim (t/t_c)$ (cf. fig. 6b); the 5-body term contains a most secular contribution $\sim (t/t_c)^2$ (cf. fig. 6c); a less secular contribution $\sim (t/t_c) \ln (t/t_c)$ is for instance given in fig. 6d. In general one expects (no proof!) secular contributions that grow with powers of t and $\ln t$. Whether this is all, or whether there are, for instance secular contributions that grow $\sim (t/t_c)^r [\ln \ln (t/t_c)]^s$ (r, s positive integers) nobody knows. This can be decided again only by a classification of all paths, which is not available at the moment.

The nature of the secular behavior of the terms in the expansion (4.9) is not of academic interest as this determines directly the nature of the density dependence of the pair distribution function F_2 and the transport coefficients (see below)!

Comparing the cases a) (hard square) and b) (hard circle) one sees that the nature of the secular behavior of the terms in the expansion (4.9) depends critically on the nature of the intermolecular potential field. (Compare for instance the secular behavior of the sequence of collisions sketched in fig. 5a and the corresponding one for the circles sketched in fig. 6a).

In particular the appearance of contributions which contain a logarithmic time dependence seem to be absent in the case the velocity space of the point particle is discrete and to occur only if all velocity directions are allowed.

6. RESUMMATION

The secular contributions to the higher terms in the expansion (4.9) make this expansion unsuitable for the computation of F_2 for times $t \gg t_{mi} \gg t_c$, if one wants to restrict oneself to a few terms only.

The secular contributions are a consequence of the whole set up

* One expects this still to be true if one replaces the squares by (regular) polygons, oriented with corresponding diagonals parallel.

of the expansion viz. as an expansion of F_2 in terms of the dynamics of *isolated* groups of particles, as incorporated in the S-operators.

This allows for instance in the three-particle term on the r.h.s. of (4.9) the occurrence of "free" paths of the particles 1, 2 and 3, which are much larger than the mean free path l, because of the absence of collisions with other particles. This in turn leads to contributions to $F_2(x_1 x_2;t)$ from particles which are much further away from 1 and 2 than l, which is unphysical. Clearly one has somehow to introduce the mean free path to "cut-off" "free" paths at the mean free path. However, the mean free path is a property of *all* the particles of the gas, not of a small group of particles. Therefore, to introduce the mean free path, one has to make a *second* rearrangement of the expansion of F_2.

Hopefully, then all secular contributions to the terms in the expansion (4.9) could be eliminated by combining them in such a way that in each term a damping is introduced which effectively cuts-off all paths at the mean free path. Such a resummation was carried out for the first time by K. Kawasaki and I. Oppenheim for the self diffusion coefficient (see section 7) [12].

Very recently Dorfman and Cohen summed all secular contributions to the 4-body term and all most secular contributions to the higher terms in the expansion (4.9). However, not all of these secular contributions could be combined to a damping in the three-particle term. In fact all the secular contributions mentioned could be classified into 2 classes (cf. section 8):

a) those that do combine to a damping in the 3-body term*;

b) those that lead to a "degrading" of $F_1(x_i; t)$ to $F_1(x_i;t_1)$, where $t_1 < t$ is the time of an intermediate collision (cf. fig. 3). *After* this resummation, it would seem that all the remaining secular contributions to the 5-body term as well as all the then most secular contributions to the higher terms can be eliminated by summing them up to a damping and F_1-"degrading" in the 4-particle term etc.

Continuing in this fashion, one could hope that the expansion (4.9) of F_2 could be rearranged in such a way that *all* secular terms in *all* terms would be absorbed in a damping and F_1-degrading so that then an expansion free of secular contributions would be obtained. It will be appreciated that

1) In order to be sure that *all* secular contributions in the l-particle term are resummed, a classification of all dynamical events between l particles and their secular behavior should be made;

* As the 3-body term is free of secular contributions in 3 dimensions such a damping is not needed for the 3-body term itself. In 2-dimensions, however, where the 3-body term does contain secular contributions, this damping is necessary.

2) Due to the resummation, the F_2 - and also $\partial F_1/\partial t$ - at time t depend not only on F_1 at the same time t (as in the eq. (3.1)) but on F_1 over a time *interval* of the order of t_{mi}: on F_1 from t to $t - t_{mi}$.

3) The resummation would not lead to a new regularizing function instead of F_1.

As pointed out before, the secular contributions to $F_2(x_1x_2;t)$ lead to divergent contributions to the virial coefficients $\eta_2, \lambda_2, \ldots$ of the transport coefficients. In particular η_2 and λ_2 contain contributions from the 4-particle term of F_2 which grow with time $\sim \ln(t/t_c)$. One would expect that after a resummation which introduces a damping and makes the 4-particle term in F_2 free of secular contributions, the η_2 and λ_2 would contain contributions $\sim \ln(t_{mi}/t_c)$, i.e. $\sim \ln\rho$.

Therefore, instead of the virial expansions (3.13) and (3.14) for η and λ, one expects, after resummation, the following expansions for η and λ respectively:

$$\eta(\rho,T) = \eta_0(T) + \rho\eta_1(T) + \rho^2\eta_2'(T)\ln\rho + \rho^2\eta_2(T) + \ldots\,, \tag{6.1}$$

$$\lambda(\rho,T) = \lambda_0(T) + \rho\lambda_1(T) + \rho^2\lambda_2'(T)\ln\rho + \rho^2\lambda_2(T) + \ldots\,. \tag{6.2}$$

It seems likely - but has not been proved - that $\eta_2'(T)$ and $\lambda_2'(T)$ are identical with the coefficients of the secular contributions $\sim \ln t/t_c$, so that *all* that the resummation does, as far as these terms is concerned, is to replace t/t_c by t_{mi}/t_c [15,19].

Similarly, for a computation of the transport coefficients up to $O(\rho^l)$, the $(l+2)$-particle term should be free of secular contributions.

7. TIME CORRELATION FUNCTIONS

One might think that if one is only interested in a computation of the transport coefficients of a dense gas, the difficulties discussed in the previous sections, which were largely connected with the general kinetic theory of dense gases, could be avoided.

In particular one might hope that if one would start from the general expressions for transport coefficients in terms of time correlation functions, these difficulties would not arise.

In this section we will show that this is *not* so.

We will discuss as an example the self-diffusion coefficient (for other transport coefficients see [13]):

$$D = \frac{1}{3} \lim_{T\to\infty} \lim_{\substack{N,V\to\infty \\ N/V=n}} \int_0^T dt \, \langle \boldsymbol{v}_1(0)\cdot \boldsymbol{v}_1(t)\rangle_{eq} \, . \tag{7.1}$$

Here D is given as the time integral over the velocity auto correlation function $\langle \boldsymbol{v}_1(0)\cdot \boldsymbol{v}_1(t)\rangle_{eq}$ of a particle 1 of the gas if the gas is in thermal equilibrium. The average $\langle\ \rangle_{eq}$ is over a canonical ensemble. We remark that in (7.1) the transport coefficient is *identified* with a double limit: first the thermodynamic limit of the system, then the limit of infinite time.

By taking the limits in this order;

1) the duration of the Poincaré cycle has become infinite, so that the time integral does not necessarily diverge;

2) the effects of the walls can be eliminated as the effective duration of a velocity correlation is of $O(t_{mi})$ and the effective time interval over which the time integration is performed is therefore also of $O(t_{mi})$.

A derivation of (7.1) proceeds as follows. If the particle 1 is at time $t = 0$ at $\boldsymbol{r}_1 = 0$, then at time t one has:

$$\frac{d}{dt}[\boldsymbol{r}_1(t)]^2 = 2\boldsymbol{r}_1(t)\cdot \boldsymbol{v}_1(t) = 2\int_0^t dt' \, \boldsymbol{v}_1(t')\cdot \boldsymbol{v}_1(t) = 2\int_0^t d\tau \, \boldsymbol{v}_1(t-\tau)\cdot \boldsymbol{v}_1(t) \tag{7.2}$$

by a simple change of variable $(t - t' = \tau)$. Averaged over a canonical ensemble one has:

$$\frac{d}{dt}\langle \boldsymbol{r}_1^2(t)\rangle_{eq} = 2\int_0^t d\tau \, \langle \boldsymbol{v}_1(0)\cdot \boldsymbol{v}_1(\tau)\rangle_{eq} \, , \tag{7.3}$$

as then only the time difference τ but not t is relevant.

If the limit $t \to \infty$ on the r.h.s. exists * and does not vanish, then one can write asymptotically:

$$\frac{d}{dt}\langle \boldsymbol{r}_1^2(t)\rangle_{eq} \sim 2\int_0^\infty d\tau \, \langle \boldsymbol{v}_1(0)\cdot \boldsymbol{v}_1(\tau)\rangle_{eq} \, , \tag{7.4}$$

or

$$\langle \boldsymbol{r}_1^2(t)\rangle_{eq} \sim 6Dt \, , \tag{7.5}$$

where

$$D = \frac{1}{3}\int_0^\infty d\tau \, \langle \boldsymbol{v}_1(0)\cdot \boldsymbol{v}_1(\tau)\rangle_{eq} \, . \tag{7.6}$$

* For this to be true, the thermodynamic limit has to be taken first. We do not indicate this explicitly here.

We remark that:

1) The same expression holds in the Lorentz model for the diffusion coefficient of a point particle 1 through stationary (spherically symmetric) scatterers;
2) In that case as well as in the case of self diffusion, one has to connect the D as defined by the r.h.s. of the eq. (7.1) with the diffusion coefficient occurring in the diffusion equation, to prove that the D of eq. (7.1) is indeed the true diffusion coefficient for the processes considered.

An evaluation of (7.1) for a dilute and moderately dense gas has been performed on the basis of two methods:

1) One writes D in the form:

$$D = \tfrac{1}{3} \lim_{T\to\infty} \int_0^T \mathrm{d}t \int \mathrm{d}\boldsymbol{p}_1 \frac{\boldsymbol{p}_1}{m} \cdot \boldsymbol{\Phi}_1(\boldsymbol{p}_1;t) , \tag{7.7}$$

where

$$\boldsymbol{\Phi}_1(\boldsymbol{p}_1;t) = \lim_{\substack{N,V\to\infty \\ N/V=n}} \frac{V}{Z_N} \int \mathrm{d}x_2 \ldots \int \mathrm{d}x_N \exp(-\beta H_N) S_{-t}(x_1 \ldots x_N) \frac{\boldsymbol{p}_1}{m} \tag{7.8}$$

and observes that:

$$D = \tfrac{1}{3} \lim_{\epsilon\to 0} \int \mathrm{d}\boldsymbol{p}_1 \frac{\boldsymbol{p}_1}{m} \hat{\boldsymbol{\Phi}}_1(\boldsymbol{p}_1;\epsilon) , \tag{7.9}$$

if $\hat{\boldsymbol{\Phi}}_1(\boldsymbol{p}_1;\epsilon)$ is the Laplace transform of $\boldsymbol{\Phi}_1(\boldsymbol{p}_1;t)$:

$$\hat{\boldsymbol{\Phi}}_1(\boldsymbol{p}_1;\epsilon) = \int_0^\infty \mathrm{d}t \exp(-\epsilon t)\, \boldsymbol{\Phi}_1(\boldsymbol{p}_1;t) . \tag{7.10}$$

In (7.8) Z_N is the partition function of the system and $\beta = 1/kT$, where k is Boltzmann's constant and T the absolute temperature.

The computation of D based on eq. (7.9) has been reviewed extensively elsewhere [16,17] (see also section 8).

2) One writes D in the form:

$$D = \tfrac{1}{3} \lim_{t\to\infty} \int \mathrm{d}\boldsymbol{p}_1 \frac{\boldsymbol{p}_1}{m} \boldsymbol{\Psi}_1(\boldsymbol{p}_1;t) , \tag{7.11}$$

where

$$\boldsymbol{\Psi}_1(\boldsymbol{p}_1;t) = \psi_1(x_1;t) - \psi_1(x_1;0) , \tag{7.12}$$

with

$$\psi_1(x_1;t) = \lim_{\substack{N,V\to\infty \\ N/V=n}} \frac{V}{Z_N} \int dx_2 \ldots \int dx_N \exp(-\beta H_N) S_{-t}(x_1 \ldots x_N)(-\boldsymbol{r}_1) . \tag{7.13}$$

The eq. (7.11) can be derived from the eq. (7.7) as follows. We first show that

$$\boldsymbol{\Phi}_1(\boldsymbol{p}_1,t) = \frac{\partial \psi_1(x_1;t)}{\partial t} .$$

Indeed, using (7.13) and (4.3) and (3.10) one has:

$$\frac{\partial \psi_1(x_1;t)}{\partial t} = \lim_{\substack{N,V\to\infty \\ N/V=n}} \frac{V}{Z_N} \int dx_2 \ldots \int dx_N \exp(-\beta H_N) S_{-t}(x_1 \ldots x_N) \mathcal{H}_N(x_1 \ldots x_N)\, \boldsymbol{r}_1$$

$$= \lim_{\substack{N,V\to\infty \\ N/V=n}} \frac{V}{Z_N} \int dx_2 \ldots \int dx_N \exp(-\beta H_N) S_{-t}(x_1 \ldots x_N) \frac{\boldsymbol{p}_1}{m} = \boldsymbol{\Phi}_1(\boldsymbol{p}_1;t) . \tag{7.14}$$

Using this in (7.7) and carrying out the time integration, immediately leads with (7.12) to (7.11).

We remark that:

1) $\boldsymbol{\Phi}_1$ and $\boldsymbol{\Psi}_1$ only depend on $\boldsymbol{p}_1$, while ψ_1 depends on $\boldsymbol{r}_1$ and $\boldsymbol{p}_1$;

2)
$$\psi_1(x_1;0) = -\boldsymbol{r}_1 F_1^0(p_1) , \tag{7.15}$$

where $F_1^0(p_1)$ is the equilibrium distribution function:

$$F_1^0(p_1) = \left(\frac{2\pi m}{\beta}\right)^{-\frac{3}{2}} \exp\left(-\frac{\beta p_1^2}{2m}\right) . \tag{7.16}$$

From (7.11) and (7.12) follows that D can be computed if the long time behavior of $\psi_1(x_1;t)$ is known. Thus the computation of the transport coefficient D from a time correlation function ψ_1 entails the same problem as discussed before for the distribution functions and the same difficulties from secular contributions will be encountered if such a computation is performed by the same cluster expansions that were employed in that case.

We illustrate this briefly. For further details of the computation of D on the basis of ψ_1 as well as for a comparison of the two methods of computation of D, we refer to elsewhere [6,16,18,19].

Introducing correlation functions ψ_s by:

$$\psi_s(x_1\dots x_s;t)$$

$$= \lim_{\substack{N,V\to\infty \\ N/V=n}} \frac{V^s}{Z_N} \int dx_{s+1}\dots\int dx_N \exp(-\beta H_N) S_{-t}(x_1\dots x_N)(-r_1), \quad (7.17)$$

one can expand ψ_1 and ψ_2 - as F_1 and F_2 before - using the equation (7.13), the equation (7.17) for $s = 2$ and the cluster expansions (4.4) and (4.6) respectively for $S_{-t}(x_1\dots x_N)$:

$$\psi_1(x_1;t) = U_t(x_1)\psi_1(x_1;0) + n\int dx_2\, U_t(x_1|x_2)\psi_2(x_1x_2;0)$$

$$+\frac{n^2}{2!}\int dx_2\int dx_3\, U_t(x_1|x_2x_3)\psi_3(x_1x_2x_3;0) + \dots, \quad (7.18)$$

$$\psi_2(x_1x_2;t)$$

$$= U_t(x_1x_2)\psi_2(x_1x_2;0) + n\int dx_3\, U_t(x_1x_2|x_3)\psi_3(x_1x_2x_3;0) + \dots \quad (7.19)$$

The long time behavior of the terms in these expansions can be analyzed in the same fashion as those in the corresponding expansions of F_1 and F_2 in section 4. In fact, since this analysis was confined to a study of the U_t-operators and was independent of the functions behind the operators, the identical analysis applies here. Thus one concludes that:

1) the terms in the expansions (7.18) and (7.19) contain secular contributions from the same sequences of dynamical events as the corresponding terms in the expansions (4.5) and (4.7);

2) the most secular contributions to ψ_2 can be eliminated with the help of the expansion for ψ_1. Assuming that for times $t \gg t_{\mathrm{mi}} \gg t_{\mathrm{c}}$ the correlations present in the $\psi_s(x_1\dots x_s;0)$ can be neglected [6, 7] and setting therefore (cf. eq. (4.8)):

$$\psi_s(x_1\dots x_s;0) = \psi_1(x_1;0)\prod_{i=2}^{s} F_1^0(p_i), \quad (7.20)$$

one obtains the following expansion for ψ_2 in terms of ψ_1 (cf. eq. (4.9)):

$$\psi_2(x_1x_2;t) = \mathcal{T}_t(x_1x_2)\psi_1(x_1;t)F_1^0(p_2)$$

$$+\, n\int dx_3\ \mathring{\mathcal{T}}_t(x_1x_2|x_3)\psi_1(x_1;t)\prod_{i=2}^{3} F_1^0(p_i)+\dots \tag{7.21}$$

Just as in the expansion (4.9), a dynamical analysis of the $\mathcal{T}_t$-operators reveals that the 4-particle term and all higher terms in the expansion (7.21) still contain secular contributions (cf. fig. 4). This leads to the same difficulties in the evaluation of the D as noticed before for η and λ (cf. (3.13) and (3.14)) viz. that the virial coefficients in the density expansion of D all become infinite from D_2 on:

$$\rho D = D_0 + \rho D_1 + \rho^2 D_2 + \dots\ . \tag{7.22}$$

One expects that all secular contributions to the 4-particle term as well as all most secular contributions to each of the higher terms in (7.21) can be eliminated by combining them a.o. to a damping in the 3-particle term.

In the next sections I will discuss such a resummation for $\Phi_2(x_1x_2;t)$.

8. RESUMMATION

For the resummation of $\Phi_2(x_1x_2;t)$ in terms of $\Phi_1(p_1;t)$, which will be carried out in this section, we start from the following expansions for Φ_1 and Φ_2, which can be obtained in precisely the same manner as the corresponding expansions (4.5) and (4.7) for F_1 and F_2 in section 4, or (7.18) and (7.19) for ψ_1 and ψ_2 in the previous section:

$$\Phi_1(p_1;t) = U_t(x_1)F_1^0(p_1)\frac{p_1}{m} + n\int dx_2\ U_t(x_1|x_2)F_2^0(x_1x_2)\frac{p_1}{m}$$

$$+\frac{n^2}{2!}\int dx_2\int dx_3\ U_t(x_1|x_2x_3)F_3^0(x_1x_2x_3)\frac{p_1}{m}+\dots\ , \tag{8.1}$$

$$\Phi_2(x_1x_2;t) = U_t(x_1x_2)F_2^0(x_1x_2)\frac{p_1}{m} + n\int dx_3\ U_t(x_1x_2|x_3)F_3^0(x_1x_2x_3)\frac{p_1}{m}$$

$$+\frac{n^2}{2!}\int dx_3\int dx_4\ U_t(x_1x_2|x_3x_4)F_3^0(x_1x_2x_3x_4)\frac{p_1}{m}+\dots\ . \tag{8.2}$$

Here one has used that for Φ_s defined by:

$$\Phi_s(x_1 \ldots x_s;t)$$

$$= \lim_{\substack{N,V\to\infty\\ N/V=n}} \frac{V^s}{Z_N} \int dx_{s+1} \ldots \int dx_N \exp(-\beta H_N) S_{-t}(x_1 \ldots x_N) \frac{p_1}{m}, \tag{8.3}$$

one has:

$$\Phi_s(x_1 \ldots x_s;0) = \frac{p_1}{m} F_s^0(x_1 \ldots x_s) , \tag{8.4}$$

where F_s^0 is the equilibrium s-particle distribution function:

$$F_s^0(x_1 \ldots x_s) = \lim_{\substack{N,V\to\infty\\ N/V=n}} \frac{V^s}{Z_N} \int dx_{s+1} \ldots \int dx_N \exp(-\beta H_N) . \tag{8.5}$$

We do *not* proceed now as in section 7 and eliminate secular contributions in the Φ_2 expansion by using the Φ_1 expansion, to obtain an expansion of Φ_2 in terms of Φ_1 analogous to (7.21), as such an expansion still contains secular contributions and we do not know at present how to eliminate those.

It has proved more convenient to use the expansions (8.1) and (8.2) directly. The reason is that a classification of all secular contributions to (8.1) and (8.2) - and in particular of the most secular contributions to each term - can be made most easily for the Laplace transforms $\hat{\Phi}_1(p_1;\epsilon)$ and $\hat{\Phi}_2(x_1 x_2;\epsilon)$ which can directly be obtained from (8.1) and (8.2) *. For this classification the binary collision expansion (BCE) of the U-operators will be used. The advantage of this expansion, which is most conveniently carried out using Laplace transforms, is that the secular contributions to each term in (8.1) and (8.2) are easily identifiable. This is because most of these contributions - and in particular the most secular ones - are due to sequences of successive binary collisions. The disadvantage of the expansion is that contributions such as for instance a genuine triple collision event is decomposed into an infinite series of binary collisions.

The Laplace transform (L.T.) of (8.1) and of (8.2) read, using (4.3), (4.4a) and (4.6a):

* It will be appreciated that the Laplace transform of the expansion of Φ_2 in terms of Φ_1, that corresponds to (7.21), is *not* easy, as the time dependence of the S-operators and the Φ_1, does not appear in the form of a time convolution.

$$\hat{\Phi}_1(p_1;\epsilon) = \hat{G}_1(x_1)F_1^0(p_1)\frac{p_1}{m} + n\int dx_2\ \hat{G}_2(x_1|x_2)F_2^0(x_1x_2)\frac{p_1}{m}$$

$$+\frac{n^2}{2!}\int dx_2 \int dx_3\ \hat{G}_3(x_1|x_2x_3)F_3^0(x_1x_2x_3)\frac{p_1}{m}+\ldots\ , \qquad (8.6)$$

$$\hat{\Phi}_2(x_1x_2;\epsilon) = \hat{G}_2(x_1x_2)F_2^0(x_1x_2)\frac{p_1}{m}$$

$$+n\int dx_3\ \hat{G}_3(x_1x_2|x_3)F_3^0(x_1x_2x_3)\frac{p_1}{m}$$

$$+\frac{n^2}{2!}\int dx_3 \int dx_4\ \hat{G}_4(x_1x_2|x_3x_4)F_4^0(x_1x_2x_3x_4)\frac{p_1}{m}+\ldots\ . \qquad (8.7)$$

Here *

$$\hat{G}_2(x_1|x_2) = \hat{G}_2(x_1x_2) - \hat{G}_2(x_1;x_2)\ , \qquad (8.8)$$

$$\hat{G}_3(x_1|x_2x_3) = \hat{G}_3(x_1x_2x_3) - \hat{G}_3(x_1x_2;x_3) - \hat{G}_3(x_1x_3;x_2) - \hat{G}_3(x_2x_3;x_1)$$
$$+ 2\hat{G}_3(x_1;x_2;x_3)\ , \qquad (8.9)$$

while

$$\hat{G}_3(x_1x_2|x_3) = \hat{G}_3(x_1x_2x_3) - \hat{G}_3(x_1x_2;x_3)\ , \qquad (8.10)$$

$$\hat{G}_4(x_1x_2|x_3x_4) = \hat{G}_4(x_1x_2x_3x_4) - \hat{G}_4(x_1x_2x_3;x_4) - \hat{G}_4(x_1x_2x_4;x_3)$$
$$- \hat{G}_4(x_1x_2;x_3x_4) + 2\hat{G}_4(x_1x_2;x_3;x_4)\ , \qquad (8.11)$$

with

$$G_N(x_1\ldots x_s;x_{s+1}\ldots x_{s+r};x_{s+r+1};\ldots;x_N)$$

$$= \{\epsilon + \mathcal{H}_s(x_1\ldots x_s) + \mathcal{H}_r(x_{s+1}\ldots x_{s+r}) + \sum_{i=s+r+1}^{N} \mathcal{H}_1(x_i)\}^{-1}\ , \qquad (8.12)$$

$$^*\int_0^\infty dt\ \exp(-\epsilon t)\ S_{-t}(x_1\ldots x_s) = \int_0^\infty dt\ \exp(-\epsilon t)\exp\{-t\mathcal{H}_s(x_1\ldots x_s)\}$$

$$= \{\epsilon + \mathcal{H}_s(x_1\ldots x_s)\}^{-1} = \hat{G}_s(x_1\ldots x_s)\ \text{(cf. eq. (8.12))}$$

so that for instance:

$$\hat{G}_3(x_1x_2;x_3) = \{\epsilon + \mathcal{H}_2(x_1x_2) + \mathcal{H}_1(x_3)\}^{-1} , \tag{8.13a}$$

$$\hat{G}_4(x_1x_2;x_3x_4) = \{\epsilon + \mathcal{H}_2(x_1x_2) + \mathcal{H}_2(x_3x_4)\}^{-1} , \text{ etc.} \tag{8.13b}$$

In (8.12) only the interaction of the s particles $1,\ldots,s$ amongst each other and of the r particles $s+1,\ldots,s+r$ amongst each other is taken into account, whereas all other interactions between the particles $1,\ldots,N$ are neglected*.

BCE of $\hat{\Phi}_1(p_1;\epsilon)$. For our purpose it is sufficient to obtain a BCE of the 3 terms on the r.h.s. of (8.6) only. One obtains:

$$\begin{aligned}\hat{\Phi}_1(p_1;\epsilon) = {} & \frac{1}{\epsilon}F_1^0(p_1)\frac{p_1}{m} + \frac{n}{\epsilon}\int \mathrm{d}x_2\ \hat{C}_2(x_1x_2)F_2^0(x_1x_2)\frac{p_1}{m} \\ & + \frac{n^2}{2\epsilon}\int \mathrm{d}x_2 \int \mathrm{d}x_3 [\{\hat{C}_3(x_1x_2)\hat{C}_3(x_1x_3) + \hat{C}_3(x_1x_3)\hat{C}_3(x_1x_2) + \ldots\} \\ & + \{\hat{C}_3(x_1x_2)\hat{\Lambda}_3(x_1x_2|x_3)\hat{C}_3(x_1x_2) \\ & + \hat{C}_3(x_1x_3)\hat{\Lambda}_3(x_1x_3|x_2)\hat{C}_3(x_1x_3) + \ldots\} + \ldots] \\ & \cdot F_3^0(x_1x_2x_3)\frac{p_1}{m} + \ldots .\end{aligned} \tag{8.14}$$

Here

$$\hat{C}_3(x_1x_2) = \theta_{12}\hat{G}_3(x_1x_2;x_3) \tag{8.15}$$

and

$$\hat{\Lambda}_3(x_1x_2|x_3) = \hat{C}_3(x_1x_3) + \hat{C}_3(x_2x_3)(1+P_{23}) , \tag{8.16}$$

* One could also say that the $\hat{G}_N$ of (8.12) is the Laplace transform of a streaming operator

$$S_{-t}(x_1\ldots x_s;x_{s+1}\ldots x_{s+r};x_{s+r+1};\ldots;x_N)$$

$$= \exp(-t\mathcal{H}_s(x_1\ldots x_s))\exp(-t\mathcal{H}_r(x_{s+1}\ldots x_{s+r}))\prod_{i=s+r+1}^{N}\exp(-t\mathcal{H}_1(x_i)) ,$$

where in the streaming of the N particles $1,\ldots,N$ only the interaction of the s particles $1,\ldots,s$ and of the r particles $s+1,\ldots,s+r$ is taken into account.

where P_{23} is the permutation operator acting on the particles 2 and 3, *if and only if they occur in a* $\hat{C}$*-operator:* $P_{23}\hat{C}_3(x_1x_2) = \hat{C}_3(x_1x_3)$ etc.

The P_{23} operator does *not* affect therefore particles 2 or 3 for instance, if they occur in an F^0-function (for instance the 3 in $F_1^0(p_3)$).

To obtain (8.14) one has further used:

a) $\{\epsilon + \mathcal{H}_s(x_1 \ldots x_s)\}^{-1} = \{\epsilon + \mathcal{H}_s^0(x_1 \ldots x_s) - \sum_\alpha \theta_\alpha\}^{-1} = \hat{G}_s(x_1 \ldots x_s)$

$$= \hat{G}_s^0(x_1 \ldots x_s) + \sum_\alpha \hat{G}_s^0 \theta_\alpha \hat{G}_s^0 + \sum_\alpha\sum_\beta \hat{G}_s^0 \theta_\alpha \hat{G}_s^0 \theta_\beta \hat{G}_s^0 + \ldots$$

$$= \hat{G}_s^0 + \sum_\alpha \hat{G}_s^0 \hat{T}_s(\alpha) \hat{G}_s^0 + \sum_\alpha \sum_{\beta \neq \alpha} \hat{G}_s^0 \hat{T}_s(\alpha) \hat{G}_s^0 T_s(\beta) \hat{G}_s^0 + \ldots$$

$$= \hat{G}_s^0[1 + \sum_\alpha \hat{C}_s(\alpha) + \sum_\alpha \sum_{\beta\neq\alpha} \hat{C}_s(\alpha)\hat{C}_s(\beta) + \ldots] . \tag{8.17}$$

Here

$$\hat{G}_s^0 = \hat{G}_s^0(x_1 \ldots x_s) = \hat{G}_s(x_1;x_2;\ldots;x_s) = \frac{1}{\epsilon + \mathcal{H}_s^0(x_1 \ldots x_s)} , \tag{8.18}$$

where

$$\mathcal{H}_s^0(x_1 \ldots x_s) = \sum_{i=1}^{s} \mathcal{H}_1(x_i) . \tag{8.19}$$

A Greek character α or $\beta, \ldots$ indicates a pair out of the set $1, 2, \ldots, s$; $\sum_\alpha$ means a summation over all pairs; $\sum_{\beta\neq\alpha}$ a summation over all pairs but α etc.

$$T_s(\alpha) = \theta_\alpha + \theta_\alpha G_s^0 \theta_\alpha + \theta_\alpha G_s^0 \theta_\alpha G_s^0 \theta_\alpha + \ldots \tag{8.20}$$

$$\hat{C}_s(\alpha) = \hat{T}_s(\alpha)\hat{G}_s^0(x_1 \ldots x_s) . \tag{8.21}$$

If $\alpha = 12$, one has for instance *

$$\hat{C}_s(12) = \hat{C}_s(x_1x_2) = \hat{T}_s(12)\hat{G}_s^0(x_1 \ldots x_s) . \tag{8.22}$$

See also eq. (8.31) for a more useful expression for $\hat{C}_s(\alpha)$.

The expansion (8.17) is used to obtain expansions for $\hat{G}_2(x_1x_2)$ and $\hat{G}_3(x_1x_2x_3)$.

b) In a completely similar fashion an expansion for

$$\hat{G}_s(x_1 \ldots x_r;x_{r+1};\ldots;x_s) = \{\epsilon + \mathcal{H}_s^0(x_1 \ldots x_s) - \sum_\alpha{}' \theta_\alpha\}^{-1}$$

* $\hat{C}_s(12)$ as well as $\hat{C}_s(x_1x_2)$ are abbreviations for $\hat{C}_s(x_1x_2;x_3;x_4;\ldots;x_s)$ which would perhaps be a more consistent notation when compared with that previously used for the $\hat{G}_s$.

can be obtained, where in $\sum_\alpha'$ the prime indicates that the summation extends only over pairs α out of the subset $1,2,\ldots,r$ of the set $1,2,\ldots,s$.

c)
$$\hat{G}_1(x_1)F_1^0(p_1)\frac{\boldsymbol{p}_1}{m} = \int_0^\infty dt\, \exp(-\epsilon t)\exp\{-t\mathcal{H}_1(x_1)\}\, F_1^0(p_1)\frac{\boldsymbol{p}_1}{m}$$

$$= \int_0^\infty dt\, \exp(-\epsilon t)F_1^0(p_1)\frac{\boldsymbol{p}_1}{m} = \frac{1}{\epsilon}F_1^0(p_1)\frac{\boldsymbol{p}_1}{m}, \qquad (8.23)$$

d)
$$\int dx_2\, \hat{G}_2^0(x_1x_2)\ldots = \int_0^\infty dt\, \exp(-\epsilon t) \int dx_2\, \exp\{-t\mathcal{H}_1(x_2)\}\exp\{-t\mathcal{H}_1(x_1)\}\ldots$$

$$= \int_0^\infty dt\, \exp(-\epsilon t)\exp\{-t\mathcal{H}_1(x_1)\} \int dx_2\, \exp\{-t\mathcal{H}_1(x_2)\}\ldots$$

$$= \int_0^\infty dt\, \exp(-\epsilon t) \int dx_2 \ldots = \frac{1}{\epsilon}\int dx_2\ldots, \qquad (8.24)$$

as $\int dx_2\ldots$ only depends on $\boldsymbol{p}_1$. Similarly:

$$\int dx_2 \int dx_3\, \hat{G}_3^0(x_1x_2x_3)\ldots = \frac{1}{\epsilon}\int dx_2 \int dx_3\ldots\,. \qquad (8.25)$$

The following property of the $\hat{C}_s(\alpha)$ has been used:

e) $\int dx_2 \int dx_3\, \hat{C}_3(x_2x_3)\ldots F_3^0(x_1x_2x_3)$

$$= \int_0^\infty dt\, \exp(-\epsilon t) \int dx_2 \int dx_3\, \theta_{23}\, S_{-t}(x_2x_3)\ldots F_3^0(x_1x_2x_3) = 0, \text{ using (3.5)}\,. \qquad (8.26)$$

BCE of $\theta_{12}\hat{\Phi}_2(x_1x_2;\epsilon)$.

We are interested in the BCE of $\theta_{12}\hat{\Phi}_2(x_1x_2;\epsilon)$ rather than in that of $\hat{\Phi}_2(x_1x_2;\epsilon)$ itself*.

We obtain

* The reason is the following. $\Phi_1(\boldsymbol{p}_1;t)$ obeys the first hierarchy equation:

$$\partial\Phi_1/\partial t = n\int dx_2\, \theta_{12}\,\Phi_2(x_1x_2;t) \text{ (cf. (4.12))}.$$

Taking the L.T. on both sides, yields the equation:

$$\epsilon\hat{\Phi}_1(\boldsymbol{p}_1;\epsilon) = \Phi_1(\boldsymbol{p}_1;0) + n\int dx_2\, \theta_{12}\,\hat{\Phi}_2(x_1x_2;\epsilon)\,.$$

To obtain therefore an equation for $\hat{\Phi}_1(\boldsymbol{p}_1;\epsilon)$, $\theta_{12}\hat{\Phi}_2(x_1x_2;\epsilon)$ suffices.

$$\theta_{12}\hat{\Phi}_2(x_1x_2;\epsilon) = \hat{C}_2(x_1x_2)F_2^0(x_1x_2)\frac{p_1}{m} + n\int dx_3\,[\{\hat{C}_3(x_1x_2)\hat{C}_3(x_1x_3)$$

$$+\ldots\} + \{\hat{C}_3(x_1x_2)\hat{\Lambda}_3(x_1x_2|x_3)\hat{C}_3(x_1x_2)+\ldots\}+\ldots]F_3^0(x_1x_2x_3)\frac{p_1}{m}$$

$$+\frac{n^2}{2!}\int dx_3\int dx_4\,[\{\hat{C}_4(x_1x_2)\hat{C}_4(x_1x_3)\hat{C}_4(x_1x_4)$$

$$+\hat{C}_4(x_1x_2)\hat{C}_4(x_1x_4)\hat{C}_4(x_1x_3)+\ldots\}$$

$$+\{\hat{C}_4(x_1x_2)\Big(\hat{\Lambda}_4(x_1x_2|x_3)\hat{\Lambda}_4(x_1x_2|x_4)$$

$$+\hat{\Lambda}_4(x_1x_2|x_4)\hat{\Lambda}_4(x_1x_2|x_3)\Big)\hat{C}_4(x_1x_2)$$

$$+\hat{C}_4(x_1x_2)\Big(\hat{C}_4(x_1x_3)\hat{\Lambda}_4(x_1x_3|x_4)+\hat{\Lambda}_4(x_1x_2|x_4)\hat{C}_4(x_1x_2)\Big)$$

$$\cdot\hat{C}_4(x_1x_3)+\hat{C}_4(x_1x_2)\Big(\hat{C}_4(x_1x_4)\hat{\Lambda}_4(x_1x_4|x_3)$$

$$+\hat{\Lambda}_4(x_1x_2|x_3)\hat{C}_4(x_1x_2)\Big)\hat{C}_4(x_1x_4)+\ldots\}+\ldots]F_4^0(x_1x_2x_3x_4)\frac{p_1}{m}+\ldots\,. \tag{8.27}$$

To obtain (8.27) one has further used:

a) $$\{\epsilon+\mathcal{H}_s^0(x_1\ldots x_s)-\sum_\alpha\theta_\alpha\}^{-1} = \hat{G}_s(x_1\ldots x_s) = \{\epsilon+\mathcal{H}_s^0(x_1\ldots x_s)-\theta_\alpha-\sum_{\beta\neq\alpha}\theta_\beta\}^{-1}$$

$$=\hat{G}_s(\alpha)+\hat{G}_s(\alpha)\sum_{\beta\neq\alpha}\theta_\beta\hat{G}_s(x_1\ldots x_s) = \hat{G}_s(\alpha)+\hat{G}_s(\alpha)\sum_{\beta\neq\alpha}\theta_\beta\hat{G}_s(\beta)$$

$$+\hat{G}_s(\alpha)\sum_{\beta\neq\alpha}\theta_\beta\hat{G}_s(\beta)\sum_{\gamma\neq\beta}\theta_\gamma\hat{G}_s(\gamma)+\ldots\,, \tag{8.28}$$

so that with (8.20)-(8.22) and $\alpha = 12$, one has:

$$\theta_{12}\hat{G}_s(x_1\ldots x_s) = \hat{C}_s(x_1x_2)\,[1+\sum_{\alpha\neq 12}\hat{C}_s(\alpha)+\sum_{\alpha\neq 12}\sum_{\beta\neq\alpha}\hat{C}_s(\alpha)\hat{C}_s(\beta)$$

$$+\sum_{\alpha\neq 12}\sum_{\beta\neq\alpha}\sum_{\gamma\neq\beta}\hat{C}_s(\alpha)\hat{C}_s(\beta)\hat{C}_s(\gamma)+\ldots]\,. \tag{8.29}$$

In (8.28):

$$\hat{G}_s(\alpha) = \frac{1}{\epsilon+\mathcal{H}_s^0(x_1\ldots x_s)-\theta_\alpha}, \tag{8.30}$$

in terms of which one has, with (8.20) and (8.21):

$$\hat{C}_s(\alpha) = \hat{T}_s(\alpha)\hat{G}_s^0 = \theta_\alpha \hat{G}_s(\alpha) . \tag{8.31}$$

b) that by taking the sum $\sum_\alpha \theta_\alpha$ not over *all* pairs of the set $1,2,\ldots,s$ but only over the pairs of one or more subsets $1\ldots r_1; r_1+1\ldots r_2$, for instance, similar expansions in terms of $\hat{G}_s(\alpha)$ can be obtained for

$$\theta_{12}\,\hat{G}_s(x_1\ldots x_{r_1}; x_{r_1+1}\ldots x_{r_2}; x_{r_2+1};\ldots;x_s) .$$

In this way for instance the expansion of $\theta_{12}\,\hat{G}_4(x_1x_2;x_3x_4)$ is derived, which was used to obtain (8.27).

In each order of n in these expansions of $\hat{\Phi}_1$ and $\hat{\Phi}_2$, products of the same number of $\hat{C}$-operators are grouped together between curly brackets { }. In each such group, only those products of $\hat{C}$-operators have been written down, which give the most secular contributions to $\hat{\Phi}_1$ or $\hat{\Phi}_2$ *.

Thus for instance in $\hat{\Phi}_1$, in the $O(n^2)$: $\hat{C}_3(x_1x_2)\hat{C}_3(x_1x_3)$ leads to a more secular contribution than $\hat{C}_3(x_1x_2)\hat{C}_3(x_2x_3)$ and $\hat{C}_3(x_1x_2)\hat{C}_3(x_1x_3)\hat{C}_3(x_1x_2)$ leads to a more secular contribution than $\hat{C}_3(x_1x_2)\hat{C}_3(x_1x_3)\hat{C}_3(x_2x_3)$, which in fact leads to a non-secular contribution. Similar considerations hold for $\hat{\Phi}_2$ in $O(n)$. In $\hat{\Phi}_2$ in the $O(n^2)$: $\hat{C}_4(x_1x_2)\hat{C}_4(x_1x_3)\hat{C}_4(x_1x_4)$ and $\hat{C}_4(x_1x_2)\hat{C}_4(x_1x_3)\hat{C}_4(x_1x_4)\hat{C}_4(x_1x_3)$ lead to more secular contributions than $\hat{C}_4(x_1x_2)\hat{C}_4(x_1x_3)\hat{C}_4(x_2x_4)$ and $\hat{C}_4(x_1x_2)\hat{C}_4(x_1x_3)\cdot\hat{C}_4(x_1x_4)\hat{C}_4(x_3x_4)$ respectively. That in $\hat{\Phi}_2$ in the 3-particle term, a term like

$$\int dx_3\; \hat{C}_3(x_1x_2)\hat{C}_3(x_1x_3)\hat{C}_3(x_2x_3)F_3^0(x_1x_2x_3)\,\frac{p_1}{m}$$

and in the four particle term a term like

$$\int dx_3 \int dx_4\; \hat{C}_4(x_1x_2)\hat{C}_4(x_1x_3)\hat{C}_4(x_1x_4)\hat{C}_4(x_3x_4)F_4^0(x_1x_2x_3x_4)\,\frac{p_1}{m}$$

are non-secular and stay finite in the limit $\epsilon \to 0$ can be demonstrated by taking the inverse L.T. of these terms and showing that the integrals thus obtained behave for large $t \sim 1/t^2$ and $\sim 1/t$ respectively, i.e. like "initial condition correction terms" of F_2 or ψ_2 as considered by Dorfman and Cohen [6, 7]. The proof uses: 1) that in each case the last $\hat{C}$-operator only acts on F_3^0 and not on p_1/m; 2) the product property of the F_s^0.

The products of $\hat{C}$-operators occurring in the various terms of (8.14) and (8.27) can be represented by similar diagrams as used before in section 4 (cf. fig. 7). This is particularly clear if one transforms from the ϵ- to the t-language, by taking the inverse L.T. of the various terms in the expansions (8.14) and (8.27) (cf. the expansions (8.32) and (8.33)).

We remark that

1) All diagrams representing the contributions to $\hat{\Phi}_1$ and $\hat{\Phi}_2$ in (8.14) and (8.27) respectively are connected;

* The product $\hat{C}_3(x_1x_2)\hat{\Lambda}_3(x_1x_2|x_3)\hat{C}_3(x_1x_2)$ in the term of $O(n)$ in $\hat{\Phi}_2$, which leads to a non-secular contribution to $\hat{\Phi}_2$, has been written down for later use (see ad B).

2) the diagrams representing the $\hat{C}$-operator products resulting from the terms with a $\hat{\Lambda}$-operator contain *ring diagrams*, i.e. diagrams which connect 2 collisions of two particles - for instance of 1 and 2 - by a sequence of uncorrelated binary collisions, i.e. each intermediate binary collision of the particle 1 or 2 is with a *different* particle 3, 4, ... (cf. fig. 7c, e, f, g).

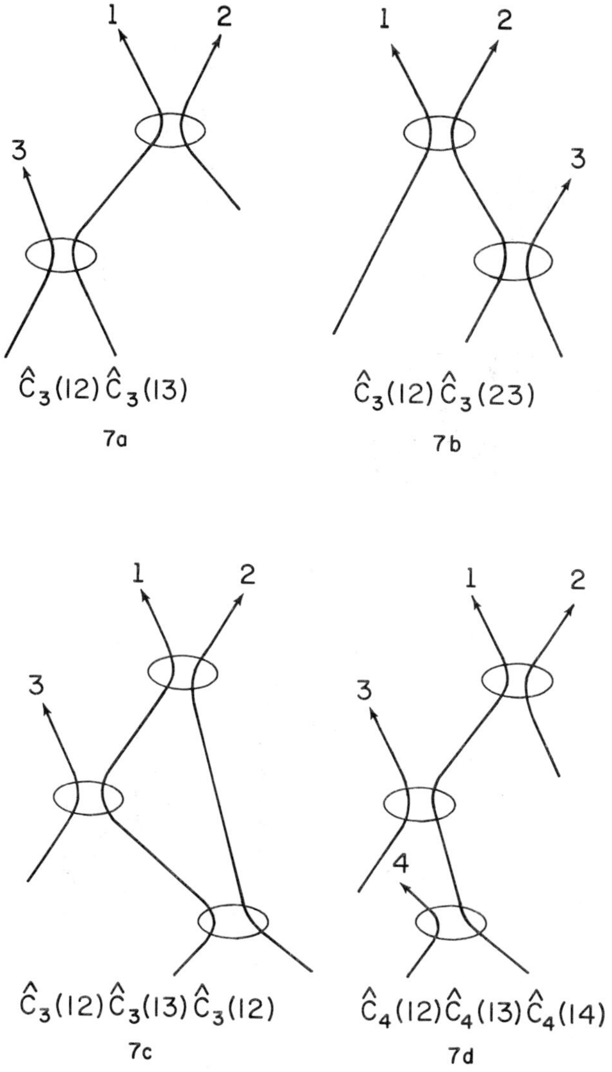

Fig. 7. Diagrams representing products of $\hat{C}$-operators.

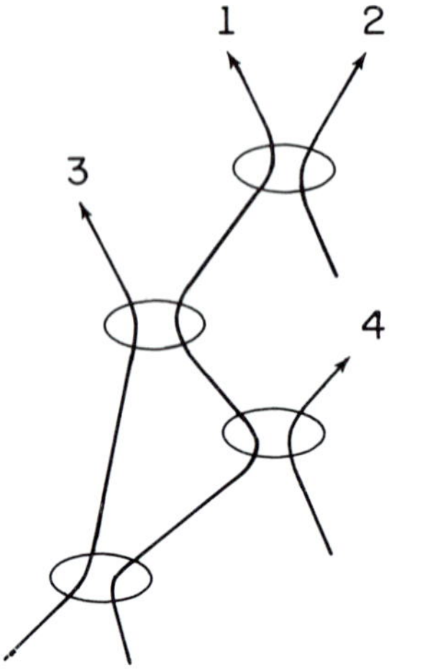

$\hat{C}_4(12)\hat{C}_4(13)\hat{C}_4(14)\hat{C}_4(13)$

7e

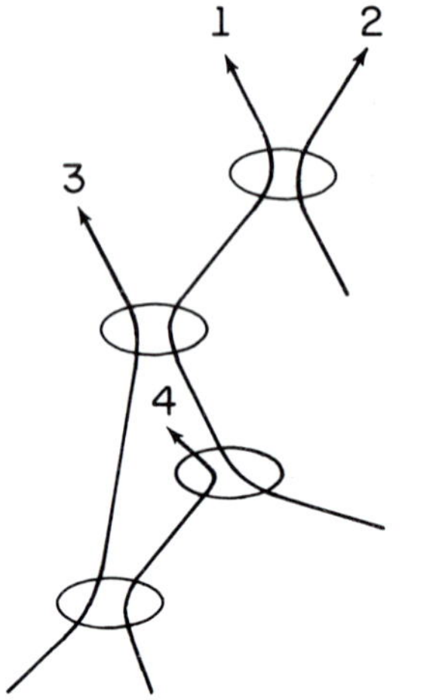

$\hat{C}_4(12)\hat{C}_4(13)\hat{C}_4(14)\hat{C}_4(34)$

7f

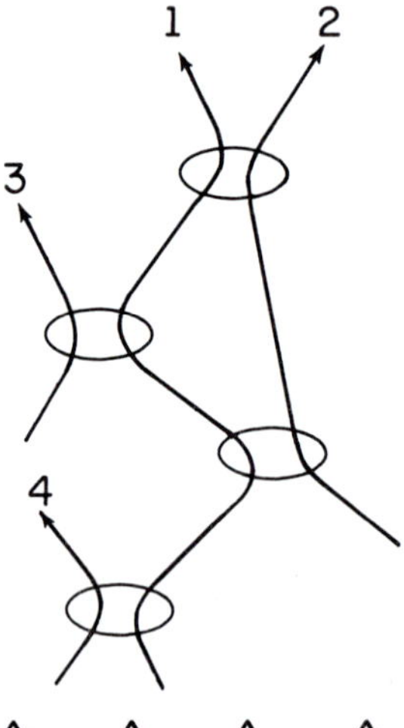

$\hat{C}_4(12)\hat{C}_4(13)\hat{C}_4(12)\hat{C}_4(14)$

7g

Fig. 7. Diagrams representing products of $\hat{C}$-operators.

A similar analysis as before (cf. section 5) reveals that in all orders of n - except in $O(n^0)$, i.e. in the first term - $\hat{\Phi}_1$ and $\hat{\Phi}_2$ contain secular contributions, i.e. contributions which grow with ϵ when $\epsilon \rightarrow 0$.

In fact by taking for instance the inverse L.T., a product of $\hat{C}$-operators in $\hat{\Phi}_1$ or $\hat{\Phi}_2$, which leads to a secular contribution $\sim (1/\epsilon)^l$ for $\epsilon \rightarrow 0$ and is represented by a certain diagram, can after Laplace transformation be represented by the same diagram and leads to a secular contribution $\sim (t/t_c)^{l-1}$.

The elimination of these secular contributions can be discussed entirely in the ϵ-language with the expansions (8.14) and (8.27) [12, 14,19]. We shall not do this here but instead transform back from the ϵ- to the t-language. This procedure has the following advantages:

1) the physical significance of the result becomes very apparent;

2) it can serve as an introduction to the non-linear case discussed in section 6 where this procedure *has* to be followed, it seems, if one wants to obtain an expansion for F_2 whose terms are free of secular contributions (see also section 8 sub 6).

In the following we shall restrict ourselves to an elimination to $O(n^2)$ of the most secular contributions in each group of $\hat{C}$-operator products in $\hat{\Phi}_2$, i.e. to an elimination of the secular contributions of those products of $\hat{C}$-operators, which have been written down explicitly on the r.h.s. of eq. (8.27). All these secular contributions to the three- and four-particle terms in the expansion (8.27) of $\theta_{12}\hat{\Phi}_2$ can be eliminated by:

A) a rearrangement of (8.27) using (8.14), leading to an "upgrading" of

$$F_1^0(p_1)\frac{p_1}{m} = \Phi_1(p_1;0)$$

to a Φ_1 at a later time;

B) a resummation, leading to a "damping" in the three-particle term.

Ad A. In order to achieve A):

1) We take the inverse L.T. of (8.14) and (8.27) and obtain:

$$\Phi_1(p_1;t) = F_1^0(p_1)\frac{p_1}{m} + n\int dx_2 \int_0^t dt_1\, \theta_{12} S_{-t_1}(x_1x_2)F_2^0(x_1x_2)\frac{p_1}{m}$$

$$+\frac{n^2}{2!}\int dx_2\int dx_3 [\int_0^t dt_1 \int_0^{t_1} dt_2\, \theta_{12}\, S_{-(t_1-t_2)}(x_1x_2;x_3)$$

$$\cdot\, \theta_{13}\, S_{-t_2}(x_1 x_3; x_2) + \ldots]F_3^0(x_1 x_2 x_3)\, \frac{\boldsymbol{p}_1}{m} + \ldots \tag{8.32}$$

and

$$\theta_{12}\, \boldsymbol{\Phi}_2(x_1 x_2; t) = \theta_{12} S_{-t}(x_1 x_2) F_2^0(x_1 x_2)\, \frac{\boldsymbol{p}_1}{m} + n \int \mathrm{d}x_3 [\int_0^t \mathrm{d}t_1\, \theta_{12}$$

$$\cdot\, S_{-(t-t_1)}(x_1 x_2; x_3) \cdot \theta_{13}\, S_{-t_1}(x_1 x_3; x_2) + \ldots]F_3^0(x_1 x_2 x_3)\, \frac{\boldsymbol{p}_1}{m}$$

$$+ \frac{n^2}{2!} \int \mathrm{d}x_3 \int \mathrm{d}x_4 [\int_0^t \mathrm{d}t_1 \int_0^{t_1} \mathrm{d}t_2\, \theta_{12}\, S_{-(t-t_1)}(x_1 x_2; x_3; x_4)$$

$$\cdot\, \{\theta_{13}\, S_{-(t_1-t_2)}(x_1 x_3; x_2; x_4) \cdot \theta_{14}\, S_{-t_2}(x_1 x_4; x_2; x_3)$$

$$+ \theta_{14}\, S_{-(t_1-t_2)}(x_1 x_4; x_2; x_3) \cdot \theta_{13}\, S_{-t_2}(x_1 x_3; x_2; x_4)\} + \ldots]$$

$$\cdot\, F_4^0(x_1 x_2 x_3 x_4)\, \frac{\boldsymbol{p}_1}{m} + \ldots\,, \tag{8.33}$$

where (cf. footnote page 258):

$$S_{-t}(x_1 \ldots x_r; x_{r+1}; \ldots ; x_s) = S_{-t}(x_1 \ldots x_r) \prod_{i=r+1}^{s} S_{-t}(x_i)\,. \tag{8.34}$$

To obtain (8.32) and (8.33) we have used that:
a) $\hat{C}_s(x_1 x_2)$ is the L.T. of $\theta_{12} S_{-t}(x_1 x_2; x_3; \ldots ; x_s)$;
b) $\frac{1}{\epsilon}\, \hat{C}_s(x_1 x_2)$ is the L. T. of $\int_0^t \mathrm{d}t_1\, \theta_{12} S_{-t}(x_1 x_2; x_3; \ldots ; x_s)$;
c) Products of $\hat{C}$ are the L.T. of time convolutions.

2) We *assume* that it is allowed in the computation of the long time behavior of $\boldsymbol{\Phi}_1$ and $\boldsymbol{\Phi}_2$, to factorize all $F_s^0(x_1 \ldots x_s)$ in the expansions (8.32) and (8.33), i.e. to set (cf. point 2) at the end of this section):

$$F_s^0(x_1 \ldots x_s) = \prod_{i=1}^{s} F_1^0(p_i)\,. \tag{8.35}$$

3) We obtain then from (8.32) for $F_1^0(p_1)\, \frac{\boldsymbol{p}_1}{m}$:

$$F_1^0(p_1)\, \frac{\boldsymbol{p}_1}{m} = \boldsymbol{\Phi}_1(\boldsymbol{p}_1; t) - n \int \mathrm{d}x_2 \int_0^t \mathrm{d}t_1\, \theta_{12}\, S_{-t_1}(x_1 x_2) F_1^0(p_2) F_1^0(p_1)\, \frac{\boldsymbol{p}_1}{m}$$

$$-\frac{n^2}{2!}\int dx_2 \int dx_3 \int_0^t dt_1 \int_0^{t_1} dt_2 \, [\theta_{12} S_{-(t_1-t_2)}(x_1x_2;x_3)$$

$$\cdot \theta_{13} S_{-t_2}(x_1x_3;x_2) + \ldots] F_1^0(p_3)F_1^0(p_2)F_1^0(p_1)\frac{p_1}{m} \ldots . \tag{8.36}$$

$O(n)$: To eliminate the most secular contributions to the term of $O(n)$ in $\theta_{12}\Phi_2$ we replace with (8.35), $F_1^0(p_1)\frac{p_1}{m} = \Phi_1(p_1;0)$

in the first term on the r.h.s. of (8.33), by the first two terms on the r.h.s. of (8.36):

$$\theta_{12}\Phi_2(x_1x_2;t) = \theta_{12}S_{-t}(x_1x_2)F_1^0(p_2)[\Phi_1(p_1;t)$$

$$-n\int dx_3 \int_0^t dt_1 \, \theta_{13}S_{-t_1}(x_1x_3)F_1^0(p_3)F_1^0(p_1)\frac{p_1}{m}]$$

$$+n\int dx_3 \int_0^t dt_1 \, [\theta_{12}S_{-(t-t_1)}(x_1x_2;x_3)\cdot\theta_{13}S_{-t_1}(x_1x_3;x_2) + \ldots]$$

$$\cdot F_1^0(p_3)F_1^0(p_1)\frac{p_1}{m}, \tag{8.37}$$

or

$$\theta_{12}\Phi_2(x_1x_2;t) = \theta_{12}S_{-t}(x_1x_2)F_1^0(p_2)\Phi(p_1;t)$$

$$-n\,\theta_{12}\int dx_3 \int_0^t dt_1 \, [S_{-(t-t_1)}(x_1x_2;x_3) - S_{-t}(x_1x_2;x_3)]$$

$$\cdot \theta_{13}S_{-t_1}(x_1x_3)F_1^0(p_1)F_1^0(p_2)F_1^0(p_3)\frac{p_1}{m} + \ldots . \tag{8.38}$$

One easily convinces oneself that the second term on the r.h.s. of (8.38) does not contain a secular contribution to $\theta_{12}\Phi_2$ anymore, as $t-t_1$ has to be of $O(t_c)$ now, if the integrand is not to vanish.

Thus the "upgrading" of $\Phi_1(p_1;0)$ to $\Phi_1(p_1;t)$ in the first term as the r.h.s. of (8.33) has indeed eliminated all most secular contributions of $O(n)$! In general one would expect that the replacement of $\Phi_1(p_1;0)$ by $\Phi_1(p_1;t)$ in the first term on the r.h.s. of

(8.33) eliminates the most secular contributions in *all* terms on the r.h.s. of (8.33).

$O(n^2)$: a) In a similar fashion taking into account the most secular contributions to the 3-body term on the r.h.s. of (8.32) reduces the most secular contributions to the 4-body term on the r.h.s. of (8.33) from $\sim (t/t_c)^2$ to $\sim (t/t_c)$.

One can make the 4-body term also free of *these* secular contributions, by upgrading the $\Phi_1(0)$ in the $O(n)$ term on the r.h.s. of (8.38), through replacing

$$F_1^0(p_1)\frac{p_1}{m},$$

using the first two terms on the r.h.s. of (8.36).

In fact in doing this, a secular contribution $\sim t/t_c$ such as for instance

$$\int dx_3 \int dx_4 \int_0^t dt_1 \int_0^{t_1} dt_2\, \theta_{12}\, [S_{-(t-t_1)}(x_1x_2;x_3;x_4)$$

$$- S_{-t}(x_1x_2;x_3;x_4)]\, \theta_{13}\, S_{-(t_1-t_2)}(x_1x_2;x_3;x_4)$$

$$\cdot\, \theta_{14}\, S_{-t_2}(x_1x_4;x_2;x_3) \prod_{i=1}^{4} F_1^0(p_i)\frac{p_1}{m},$$

- which was obtained by replacing $\Phi_1(0)$ by $\Phi_1(t)$ in the first term of (8.33) - can be transformed by replacing in addition $\Phi_1(0)$ by $\Phi_1(t_1)$ in the second term of (8.38), into:

$$\int dx_3 \int dx_4 \int_0^t dt_1 \int_0^{t_1} dt_2 \theta_{12}\, [\, S_{-(t-t_1)}(x_1x_2;x_3;x_4)$$

$$- S_{-t}(x_1x_2;x_3;x_4)] \cdot \theta_{13}\{S_{-(t_1-t_2)}(x_1x_3;x_2;x_4)$$

$$- S_{-t_1}(x_1x_3;x_2;x_4)\} \cdot \theta_{14}\, S_{-t_2}(x_1x_4;x_2;x_3)] \prod_{i=1}^{4} F_1^0(p_i)\frac{p_1}{m}.$$

which is free of secular contributions as both $t - t_1$ and $t_1 - t_2$ are of $O(t_c)$.

The expansion of $\theta_{12}\, \Phi_2(x_1x_2;t)$ reads now to $O(n)$:

$$\theta_{12}\, \Phi_2(x_1x_2;t) = \theta_{12}\, S_{-t}(x_1x_2)F_1^0(p_2)\Phi_1(p_1;t)$$

$$+ n \int dx_3 \int_0^t dt_1\, \theta_{12}[S_{-(t-t_1)}(x_1x_2;x_3) - S_{-t}(x_1x_2;x_3)]$$

$$\cdot\, \theta_{13} S_{-t_1}(x_1x_3;x_2)F_1^0(p_2)F_1^0(p_3)\boldsymbol{\Phi}_1(\boldsymbol{p}_1;t_1)\,. \quad (8.39)$$

b) The secular contributions associated with

$$\hat{C}_4(x_1x_2)\{\hat{C}_4(x_1x_3)\hat{\Lambda}(x_1x_3|x_4)\hat{C}_4(x_1x_3)$$

$$+\hat{C}_4(x_1x_4)\hat{\Lambda}_4(x_1x_4|x_3)\hat{C}_4(x_1x_4)\}\prod_{i=1}^{4} F_1^0(p_i)\frac{\boldsymbol{p}_1}{m}$$

can be eliminated by considering the two $\hat{\Lambda}_3$-terms in (8.32) - which have not been written down explicitly - in the replacement of $F_1^0(p_1)\,\boldsymbol{p}_1/m$ by $\boldsymbol{\Phi}_1(\boldsymbol{p}_1;t)$* in the first term on the r.h.s. of $\theta_{12}\boldsymbol{\Phi}_2(x_1x_2;t)$.

Ad B. The only most secular contributions left to eliminate from the 4-body term in the expansion (8.33) are the L.T. of the following terms in (8.27):

$$\hat{C}_4(x_1x_2)\{\hat{\Lambda}_4(x_1x_2|x_3)\hat{\Lambda}_4(x_1x_2|x_4)+\hat{\Lambda}_4(x_1x_2|x_4)\hat{\Lambda}_4(x_1x_2|x_3)\}\hat{C}_4(x_1x_2)$$

$$+\hat{C}_4(x_1x_2)\{\hat{\Lambda}_4(x_1x_2|x_4)\hat{C}_4(x_1x_2)\hat{C}_4(x_1x_3)+\hat{\Lambda}_4(x_1x_2|x_3)\hat{C}_4(x_1x_2)$$

$$\cdot\,\hat{C}_4(x_1x_4)\}\prod_{i=1}^{4} F_1^0(p_i)\frac{\boldsymbol{p}_1}{m}\,. \quad (8.40)$$

We shall eliminate these secular contributions in two steps.

1) First we sum the terms containing *two* $\hat{\Lambda}_4$ to: the term containing $\hat{\Lambda}_3$ in the 3-body term, the terms containing three $\hat{\Lambda}_5$ in the 5-body term etc. As this is most easily done in the L. T., we carry this out in the original expansion (8. 27). Only *after* the resummation, we shall take the inverse L.T. to interpret the result.

We obtain using (8.35):

$$n\int dx_3\,\hat{C}_3(x_1x_2)\hat{\Lambda}_3(x_1x_2|x_3)\prod_{i=1}^{3} F_1^0(p_i)\frac{\boldsymbol{p}_1}{m}+n^2\int dx_3\int dx_4\,\hat{C}_4(x_1x_2)$$

$$\cdot\,\hat{\Lambda}_4(x_1x_2|x_3)\hat{\Lambda}_4(x_1x_2|x_4)\hat{C}_4(x_1x_2)\prod_{i=1}^{4} F_1^0(p_i)\frac{\boldsymbol{p}_1}{m}$$

$$+n^3\int dx_3\int dx_4\int dx_5\,\hat{C}_5(x_1x_2)\hat{\Lambda}_5(x_1x_2|x_3)\hat{\Lambda}_5(x_1x_2|x_4)\hat{\Lambda}_5(x_1x_2|x_5)$$

* I.e. on the r.h.s. of (8.36).

$$\cdot \prod_{i=1}^{5} F_1^0(p_i)\, \frac{\boldsymbol{p}_1}{m} + \ldots$$

$$= n\,\hat{C}_2(x_1x_2)[\int \mathrm{d}x_3\, \hat{\Lambda}_3(x_1x_2|x_3)F_1^0(p_3) + n\{\int \mathrm{d}x_3\, \hat{\Lambda}_3(x_1x_2|x_3)F_1^0(p_3)\}^2$$

$$+ n^2\{\int \mathrm{d}x_3\, \hat{\Lambda}_3(x_1x_2|x_3)F_1^0(p_3)\}^3 + \ldots]\hat{C}_2(x_1x_2)F_1^0(p_1)F_1^0(p_2)\,\frac{\boldsymbol{p}_1}{m}$$

$$= n\,\hat{C}_2(x_1x_2)\,\frac{\hat{\mathcal{L}}(x_1x_2)}{1-n\hat{\mathcal{L}}(x_1x_2)}\,\hat{C}_2(x_1x_2)F_1^0(p_1)F_1^0(p_2)\,\frac{\boldsymbol{p}_1}{m}\,, \tag{8.41}$$

where the operator $\hat{\mathcal{L}}$ is defined by:

$$\hat{\mathcal{L}}(x_1x_2) = \int \mathrm{d}x_3\, \hat{\Lambda}_3(x_1x_2|x_3)F_1^0(p_3)\,. \tag{8.42}$$

To obtain (8.41) we have used that for instance

$$\int \mathrm{d}x_3\, \hat{C}_3(x_1x_2)\hat{\Lambda}_3(x_1x_2|x_3)\hat{C}_3(x_1x_2) = \int \mathrm{d}x_3 \int_0^\infty \mathrm{d}t\, \exp(-\epsilon t)\, S_{-t}(x_1x_2;x_3)$$

$$\times\, \hat{\Lambda}_3(x_1x_2|x_3)\hat{C}_3(x_1x_2) = \hat{C}_2(x_1x_2)\int \mathrm{d}x_3 \hat{\Lambda}_3(x_1x_2|x_3)\hat{C}_2(x_1x_2) \tag{8.43}$$

and

$$\int \mathrm{d}x_3 \int \mathrm{d}x_4\, \hat{C}_4(x_1x_2)\hat{\Lambda}_4(x_1x_2|x_3)\hat{\Lambda}_4(x_1x_2|x_4)\hat{C}_4(x_1x_2)$$

$$= \hat{C}_2(x_1x_2)\int \mathrm{d}x_3 \int \mathrm{d}x_4\, \hat{\Lambda}_4(x_1x_2|x_3)\hat{\Lambda}_4(x_1x_2|x_4)\hat{C}_2(x_1x_2)$$

$$= \hat{C}_2(x_1x_2)\int \mathrm{d}x_3\, \hat{\Lambda}_3(x_1x_2|x_3)\cdot \int \mathrm{d}x_4\, \hat{\Lambda}_3(x_1x_2|x_4)\cdot \hat{C}_2(x_1x_2)\,, \tag{8.44}$$

etc.

2) Similarly the terms containing *one* $\hat{\Lambda}_4$ can be combined with secular contributions in higher terms of the expansion (8.27) to give:

$$n^2\hat{C}_2(x_1x_2)\,\frac{\hat{\mathcal{L}}(x_1x_2)}{1-n\hat{\mathcal{L}}(x_1x_2)}\,\hat{C}_2(x_1x_2)\cdot \int \mathrm{d}x_4 \hat{C}_4(x_1x_4)F_1^0(p_4)\, F_1^0(p_2)F_1^0(p_1)\,\frac{\boldsymbol{p}_1}{m}\,. \tag{8.45}$$

We now transform the operator

$$\frac{\hat{\mathcal{L}}(x_1x_2)}{1-n\hat{\mathcal{L}}(x_1x_2)}\,,$$

so that the inverse L.T. can be taken.

a) We write

$$\hat{\mathscr{L}}(x_1x_2) = \hat{L}(x_1x_2) + \hat{O}(x_1x_2) , \tag{8.46a}$$

where

$$\hat{L}(x_1x_2) = \int dx_3[\theta_{13}\mathscr{S}_\infty(x_1x_3)G_3^0(x_1x_2x_3)$$

$$+ \theta_{23}\mathscr{S}_\infty(x_2x_3)G_3^0(x_1x_2x_3)(1+P_{23})]F_1^0(p_3) \tag{8.46b}$$

and where $\hat{O}(x_1x_2)$ will be neglected.

For, using (8.42), (8.16), (8.31) and (8.13a) one has:

$$\hat{\mathscr{L}}(x_1x_2) = \int dx_3\, \hat{\Lambda}_3(x_1x_2|x_3)F_1^0(p_3) = \int dx_3[\theta_{13}\,\hat{G}_3(x_1x_3;x_2)$$

$$+ \theta_{23}\,\hat{G}_3(x_2x_3;x_1)(1+P_{23})]F_1^0(p_3) = \hat{L}(x_1x_2) + \int_0^\infty dt\exp(-\epsilon t) \int dx_3[\theta_{13}\{\mathscr{S}_t(x_1x_3)$$

$$- \mathscr{S}_\infty(x_1x_3)\} + \theta_{23}\{\mathscr{S}_t(x_2x_3) - \mathscr{S}_\infty(x_2x_3)\}(1+P_{23})]\prod_{i=1}^{3} S_{-t}(x_i)F_1^0(p_3) = \hat{L}(x_1x_2) + \hat{O}(x_1x_2). \tag{8.47}$$

The ratio of the second to the first term on the r.h.s. of (8.47) is $\sim t_c/t$ or t_c/t_{mi} i.e. $\sim \rho$.

$\hat{O}(x_1x_2)$ is therefore of higher order in ρ than $\hat{L}(x_1x_2)$ and can for our purpose be neglected.

b) One has:

$$\frac{1}{1-n\hat{L}(x_1x_2)} = [\hat{G}_2^0(x_1x_2)]^{-1}\cdot[(1-n\hat{L}(x_1x_2))(\hat{G}_2^0(x_1x_2))^{-1}]^{-1}$$

$$= [\epsilon + \mathscr{H}_2^0(x_1x_2)]\{[\epsilon + \mathscr{H}_2^0(x_1x_2)] - n\hat{L}(x_1x_2)[\epsilon + \mathscr{H}_2^0(x_1x_2)]\}^{-1} . \tag{8.48}$$

Therefore, using (8.46) and (8.48), one can write:

c) $$\frac{n\hat{\mathscr{L}}(x_1x_2)}{1-n\hat{\mathscr{L}}(x_1x_2)} \cong n\{\hat{L}(x_1x_2)\cdot[\epsilon + \mathscr{H}_2^0(x_1x_2)]\}$$

$$\times \frac{1}{[\epsilon + \mathscr{H}_2^0(x_1x_2)] - n\{\hat{L}(x_1x_2)\cdot[\epsilon + \mathscr{H}_2^0(x_1x_2)]\}}$$

$$= n\lambda(x_1x_2)\cdot\frac{1}{[\epsilon + \mathcal{H}_2^0(x_1x_2) - n\lambda(x_1x_2)]}\,, \tag{8.49}$$

where

$$\lambda(x_1x_2) = \int dx_3[\theta_{13}\mathcal{S}_\infty(x_1x_3) + \theta_{23}\mathcal{S}_\infty(x_2x_3)(1 + P_{23})]F_1^0(p_3)\,. \tag{8.50}$$

Here it has been used that $S_{-t}(x_3)$ can be replaced by 1, as it only acts in (8.41) and (8.45) on $F_1^0(p_3)$.

Using (8.49) - which suffices for a computation of $\theta_{12}\boldsymbol{\Phi}_2$ to $O(n^2)$ - the inverse L.T. of (8.41) and (8.45) can easily be taken. One obtains adding the two contributions:

$$n\int_0^t dt_1\int_0^{t_1} dt_2\, \theta_{12}S_{-(t-t_1)}(x_1x_2)\lambda(x_1x_2)$$

$$\cdot \exp\{-(t_1 - t_2)[\mathcal{H}_2^0(x_1x_2) - n\lambda(x_1x_2)]\}$$

$$\cdot\, \theta_{12}S_{-t_2}(x_1x_2)F_1^0(p_2)F_1^0(p_1)\frac{\boldsymbol{p}_1}{m}$$

$$+ n^2\int_0^t dt_1\int_0^{t_1} dt_2\int_0^{t_2} dt_3\, \theta_{12}S_{-(t-t_1)}(x_1x_2)\lambda(x_1x_2)$$

$$\cdot \exp\{-(t_1 - t_2)[\mathcal{H}_2^0(x_1x_2) - n\lambda(x_1x_2)]\}$$

$$\cdot\, \theta_{12}S_{-(t_2-t_3)}(x_1x_2)\cdot\int dx_4\, \theta_{14}S_{-t_3}(x_1x_4)F_1^0(p_4)F_1^0(p_2)F_1^0(p_1)\frac{\boldsymbol{p}_1}{m}\,, \tag{8.51}$$

which, with (8.36), can be rewritten in the form:

$$n\int_0^t dt_1\int_0^{t_1} dt_2\, \theta_{12}S_{-(t-t_1)}(x_1x_2)\lambda(x_1x_2)$$

$$\cdot \exp\{-(t_1 - t_2)[\mathcal{H}_2^0(x_1x_2) - n\lambda(x_1x_2)]\}$$

$$\times\, \theta_{12}S_{-t_2}(x_1x_2)F_1^0(p_2)\boldsymbol{\Phi}_1(\boldsymbol{p}_1;t_2)$$

$$+ n^2 \int_0^t \mathrm{d}t_1 \int_0^{t_1} \mathrm{d}t_2 \int_0^{t_2} \mathrm{d}t_3 \, \theta_{12} \, S_{-(t-t_1)}(x_1 x_2) \lambda(x_1 x_2)$$

$$\cdot \exp\{(t_1 - t_2)[\mathcal{H}_2^0(x_1 x_2) - n\lambda(x_1 x_2)]\}$$

$$\cdot \theta_{12}\{S_{-(t_2-t_3)}(x_1 x_2) - S_{-t_2}(x_1 x_2)\}$$

$$\cdot \int \mathrm{d}x_4 \, S_{-t_3}(x_1 x_4) F_1^0(p_4) F_1^0(p_2) F_1^0(p_1) \frac{p_1}{m} \, . \tag{8.52}$$

Both terms in (8.52) seem free of secular contributions. For, although no proof is available, it seems likely that the factor:

$$\exp\{-(t_1 - t_2)[\mathcal{H}_2^0(x_1 x_2) - n\lambda(x_1 x_2)]\}$$

$$= \exp\{-(t_1 - t_2)[\mathcal{H}_1(x_1) - n \int \mathrm{d}x_3 \, \theta_{13} \mathcal{S}_\infty(x_1 x_3) F_1^0(p_3)]\}$$

$$\times \exp\{-t_1 - t_2)[\mathcal{H}_1(x_2) - n \int \mathrm{d}x_3 \mathcal{S}_\infty(x_2 x_3)(1 + P_{23}) F_1^0(p_3)]\} \tag{8.53}$$

in (8.52) represents a "*damping*" of the "free" paths of the particles 1 and 2 between t_1 and t_2, leading to a cut-off over a few mean free paths of these paths (cf. fig. 4a). This is suggested by 1) the appearance of the operators $\mathcal{H}_1(x_1) - n \int \mathrm{d}x_3 \, \theta_{13} \mathcal{S}_\infty(x_1 x_3) F_1^0(p_3)$ and $\mathcal{H}_1(x_2) - n \int \mathrm{d}x_3 \, \theta_{23} \mathcal{S}_\infty(x_2 x_3) F_1^0(p_3)$ in the exponent and 2) that but for the replacement of $\mathcal{S}_\infty(x_1 x_3)$ and $\mathcal{S}_\infty(x_2 x_3)$ by $S_{-\infty}(x_1 x_3)$ and $S_{-\infty}(x_2 x_3)$ respectively, these operators are identical with the linearized inhomogeneous Boltzmann operator*.

4) Using (8.4) and (8.52) the following expression for $\theta_{12} \Phi_2(x_1 x_2)$, with a "damped" 3-particle term and a 4-particle term that is free of the most secular contributions present in (8.27), is obtained:

$$\theta_{12} \Phi_2(x_1 x_2; t) = \theta_{12} \, S_{-t}(x_1 x_2) F_1^0(p_2) \Phi_1(p_1; t)$$

$$+ n \int \mathrm{d}x_3 \int_0^t \mathrm{d}t_1 \, \theta_{12} [S_{-(t-t_1)}(x_1 x_2; x_3) - S_{-t}(x_1 x_2; x_3)]$$

* The linearized inhomogeneous Boltzmann equation reads [1, 3, 4]

$$\partial F_1(x_1; t)/\partial t = -\mathcal{H}_1(x_1) F_1(x_1; t) + n \int \mathrm{d}x_2 \, \theta_{12} \, S_{-\infty}(x_1 x_2) F_1^0(p_2) F_1(x_1; t) \, . \tag{8.54}$$

$$\cdot\, \theta_{13}\, S_{-t_1}(x_1 x_3; x_2) F_1^0(p_2) F_1^0(p_3) \Phi_1(p_1;t)$$

$$+ n\, [\int_0^t \mathrm{d}t_1 \int_0^{t_1} \mathrm{d}t_2\, \theta_{12}\, S_{-(t-t_1)}(x_1 x_2) \lambda(x_1 x_2)$$

$$\cdot \exp\{-(t_1 - t_2)\,[\mathcal{H}_2^0(x_1 x_2) - n\lambda(x_1 x_2)]\}$$

$$\cdot\, \theta_{12}\, S_{-t_2}(x_1 x_2) F_1^0(p_2) \Phi_1(p_1; t_2) + \ldots]\,. \tag{8.55}$$

We conclude with a few remarks:

1) A general discussion of the elimination of the less secular contributions of $O(n)$ and $O(n^2)$ to Φ_2 - which have not been written down in (8.27) - should still be given. In contrast to the non-linear case, the situation with regard to these contributions is not entirely clear at the moment.

2) A justification of the neglect of the initial correlations as expressed in the eq. (8.35) has only been given in the context of the expansion (7.21) to $O(n^2)$ [7]. A general discussion of this point, which would be applicable to our treatment here, is not available*.

3) One could hope to accomplish in a similar fashion as presented here, the elimination of the most secular contributions in the spatially inhomogeneous case (for instance for $\psi_2(x_1 x_2;t)$);

4) It is not clear whether *all* secular contributions to *all* terms in the expansion (8.27) of $\theta_{12}\hat{\Phi}_2$ can be successively eliminated *only* by upgrading $\Phi_1(p_1;0)$ and by introducing "damped" paths of particles;

5) With a 4-body term free of secular contributions one should be able to show that the logarithmically divergent contributions to the virial coefficient D_2 in (7.22) have been eliminated and are replaced by a convergent contribution $D_2' \ln \rho$ (cf. (3.13), (3.14) and (6.1) and (6.2) respectively), where the coefficient D_2' is *identical* with the coefficient of the logarithmic divergence.

6) In the *non-linear* case, not only $F_1(x_1;0)$ but *all* $F_1(x_i;0)$ $(i = 1,2,3)$ in the 2- and 3- body term have to be upgraded. Only then can all secular contributions to the 4-body term be eliminated. However, in that case, the result cannot be written anymore in the form of a time convolution (as in (8.52)) and the transformation from ϵ- to t- language seems unavoidable.

* The reason is that the existing proof - which suffices to $O(n^2)$ - only uses phase space arguments, while one would expect that the mean free path would play an essential role in the forgetting of the initial state for large times.

REFERENCES

[1] E. G. D. Cohen, The Boltzmann equation and its generalization to higher densities, in: *Fundamental Problems in Statistical Mechanics* (ed. E. G. D. Cohen), North-Holland Publishing Company, Amsterdam, 1962.
[2] E. G. D. Cohen, On the statistical mechanics of moderately dense gases not in equilibrium, in: *Lectures in Theoretical Physics*, Vol. VIII A (ed. W. E. Brittin), University of Colorado Press, 1966, p. 145.
[3] E. G. D. Cohen, Kinetic theory of dense gases, in: *Lectures in Theoretical Physics*, Vol. IX C (ed. W. E. Brittin), University of Colorado Press, 1967, p. 279.
[4] G. E. Uhlenbeck and G. W. Ford, Lectures in Statistical Mechanics, in: *Lectures in Applied Mathematics*, Vol. I, American Mathematical Society, Providence, Rhode Island, 1963, chapters 4 and 6.
[5] Ref. 4 chapter 7.
[6] E. G. D. Cohen, Transport phenomena in gases from distribution functions and correlation functions, in: *Statistical Mechanics of Equilibrium and Non-equilibrium* (ed. J. Meixner), North-Holland Publishing Company, Amsterdam, 1965.
[7] J. R. Dorfman and E. G. D. Cohen, J. Math. Phys. 8 (1967) 282.
[8] E. G. D. Cohen, Physica 28 (1962) 1045.
[9] P. T. Ehrenfest, Begriffliche Grundlagen der Statistischen Auffassung in der Mechanik, in: *Collected Scientific Papers*, North-Holland Publising Company, Amsterdam, 1959, p. 229;
English translation: *The Conceptual Foundations of the Statistical approach in Mechanics*, Cornell University Press, 1959, p. 54.
[10] E. H. Hauge and E. G. D. Cohen, Phys. Letters 25A (1967) 78.
[11] J. M. J. van Leeuwen and A. Weyland, The density expansion of the diffusion coefficient of a Lorentz gas, in: *Statistical Mechanics* (ed. Th. A. Bak), Benjamin, New York (1967) p. 335; Physica 36 (1967) 457.
[12] K. Kawasaki and I. Oppenheim, Phys. Rev. 139 (1965) A1763; first of ref. [11] p. 313.
[13] L. K. Haines, J. R. Dorfman and M. H. Ernst, Phys. Rev. 144 (1966) 207.
[14] J. R. Dorfman, in: ref. [3] p. 443.
[15] For the 2-dimensional case see for instance: J. V. Sengers, Phys. of Fluids 9 (1966) 1685; ref. [3] p. 335; see also ref. [12].
[16] M. H. Ernst, in ref. [3] p. 417.
[17] K. Kawasaki and I. Oppenheim, Phys. Rev. 136 (1964) A1519.
[18] M. H. Ernst, J. R. Dorfman and E. G. D. Cohen, Physica 31 (1965) 493; M. H. Ernst, *Thesis*, University of Amsterdam (1965).
[19] L. K. Haines, *Thesis*, University of Maryland (1966).

KINETIC THEORY OF DILUTE GASES WITH INTERNAL MOLECULAR DEGREES OF FREEDOM *

L. WALDMANN
Institut für Theoretische Physik der Universität Erlangen-Nürnberg, Erlangen, Germany

INTRODUCTION

The kinetic theory of dilute polyatomic gases, as understood in this article, is not subject to the considerable difficulties which are met by the theory of dense monoatomic gases. The binary molecular collisions will be treated as local, point-like events. This enables one to formulate kinetic equations which are still local and quadratic in the one particle distribution function like the classical Boltzmann equation for the monoatomic gas. Thus, a lucid conceptual basis is retained. It would be lost if one thought of higher and higher states of internal molecular excitation with bigger and bigger molecules near and beyond dissociation; the dense gas and chemical reaction aspects then would come into play. But this case will be excluded in the following.

Even then, polyatomic gases show several new and interesting features, compared with monoatomic gases. Some of these features are already apparent with non-rotating spherical molecules or, more precisely, with molecules having several non-degenerate eigen-states of internal energy. This group of phenomena will be briefly discussed in the first part of this lecture, starting from the Wang Chang-Uhlenbeck kinetic equation. Other phenomena occur only with rotating, non-spherical molecules or - more simple - particles with spin. A gas composed of such particles in general shows correlations between the molecular velocity and spin orientation in a non-equilibrium situation, e.g. if a heat flow is going on. These correlations are changed by an external magnetic field which, via the molecular magnetic moment, causes the spins to precess. This change in correlations manifests itself in macro-

* Amplified version of a seminar given at the Noordwijk-Summer Session, June 1967.

scopic phenomena. The heat conductivity e.g., which without magnetic field is a scalar, becomes a tensor with magnetic field (Senftleben effect in heat conduction, transverse heat flow). Boundary conditions may be changed and thermal stresses too will become field-dependent. This apparently explains the magneto-thermal effect experimentally observed by Scott, Sturner and Williamson (torque on a heated cylinder surrounded by a polyatomic gas in a magnetic field). Phenomena linked with rotating molecules, especially particles with spin, will be discussed in the second part of this lecture. Completeness can't be the aim, however it is hoped that the basic principles become clear.

Part I. GAS COMPOSED OF MOLECULES WITH SEVERAL NON-DEGENERATE STATES OF INTERNAL ENERGY

The kinetic equation of Wang Chang-Uhlenbeck [1] will be taken as more or less evident from physical intuition, from a Stosszahlansatz. Only its range of validity and the meaning and symmetry properties of the collision cross section will be studied more in detail (section 1). As one of the simplest and most typical applications the well-known result for bulk viscosity is shortly recalled (section 2).

1. THE WANG CHANG-UHLENBECK EQUATION

The quantum number which characterizes the internal molecular state is denoted by V, vibrational quantum number. This shall indicate that one has to think of some spherically symmetric vibration of the molecules which gives rise to non-degenerate eigenstates of internal energy E_V. The state of the gas is described by the distribution function f, where

$$f(t\boldsymbol{x}\boldsymbol{p}V)\,\mathrm{d}^3x\;\mathrm{d}^3p$$

is the number of molecules with internal state V, at time t in the pertaining coordinate and momentum intervals. According to a Boltzmann-like Stosszahlansatz one assumes ($\boldsymbol{c}_1 = \boldsymbol{p}_1/m$; m the molecular mass):

$$\frac{\partial f(\boldsymbol{p}_1 V_1)}{\partial t} + \boldsymbol{c}_1 \cdot \frac{\partial f(\boldsymbol{p}_1 V_1)}{\partial \boldsymbol{x}}$$

$$= \sum_{V_2} \sum_{V_1'} \sum_{V_2'} \iiint \mathrm{d}^3 p_2 \, \mathrm{d}^3 p_1' \, \mathrm{d}^3 p_2' \, [w(\boldsymbol{p}_1 V_1 \boldsymbol{p}_2 V_2 \leftarrow \boldsymbol{p}_1' V_1' \boldsymbol{p}_2' V_2') \, f(\boldsymbol{p}_1' V_1')$$

$$\times f(\boldsymbol{p}_2' V_2') - w(\boldsymbol{p}_1' V_1' \boldsymbol{p}_2' V_2' \leftarrow \boldsymbol{p}_1 V_1 \boldsymbol{p}_2 V_2) \, f(\boldsymbol{p}_1 V_1) \, f(\boldsymbol{p}_2 V_2)] \, . \qquad (1.1)$$

This is already the basic form of the Wang Chang-Uhlenbeck (WCU)-equation, written with a general transition probability w per unit time etc. The f's are all taken at the same instant and the same space-point. Properly linearized, eq. (1.1) is a master equation for the one particle distribution function.

The conservation laws for a binary collision and the rotational invariance of the molecules require the following form of the transition probability

$$w(\ldots) = \delta(\boldsymbol{p} - \boldsymbol{p}') \, \delta(E - E') \, \sigma(E \, \theta \, V_1 V_2 V_1' V_2') \, m_{12}^{-2} \, . \qquad (1.2)$$

Here, $\boldsymbol{p}$ means the momentum of a pair

$$\boldsymbol{p} = \boldsymbol{p}_1 + \boldsymbol{p}_2 \, , \qquad (1.3)$$

E denotes the pair's energy in the centre of mass system

$$E = \frac{p_{12}^2}{2m_{12}} + E_{V_1} + E_{V_2} \, , \qquad (1.4)$$

with

$$\boldsymbol{p}_{12} = \tfrac{1}{2}(\boldsymbol{p}_1 - \boldsymbol{p}_2) \, , \qquad m_{12} = \tfrac{1}{2} m \text{ (reduced mass)} \, . \qquad (1.5)$$

The same formulae apply to the primed quantities. The scattering angle

$$\theta = \sphericalangle(\boldsymbol{p}_{12}, \boldsymbol{p}_{12}') \qquad (1.6)$$

is invariant under exchange of primed and unprimed quantities and the same is true for the general (inelastic) collision cross section

$$\sigma(E, \theta, V_1 V_2 V_1' V_2') = \sigma(E, \theta, V_1' V_2' V_1 V_2) \, . \qquad (1.7)$$

This detailed balance relation is a consequence of time reversal invariance of quantum mechanics. We shall come back to this point later.

After inserting (1.2) into (1.1) one can carry out integrations on $\boldsymbol{p}'$ and E', occurring in δ-functions. The first is immediately achieved by use of

$$\mathrm{d}^3 p_1' \, \mathrm{d}^3 p_2' = \mathrm{d}^3 p' \, \mathrm{d}^3 p_{12}' \,, \tag{1.8}$$

a simple consequence of the definitions (1.3 and 5). The second is achieved by observation of

$$m_{12}^{-2} \mathrm{d}^3 p_{12}'^{\,2} = m_{12}^{-2} p_{12}' \, \mathrm{d}p_{12}' \, \mathrm{d}^2 e_{12}' = c_{12}' \, \mathrm{d}E' \, \mathrm{d}^2 e_{12}' \,.$$

Here, c_{12}' means the magnitude of the relative velocity

$$c_{12}' = p_{12}'/m_{12}$$

and $\boldsymbol{e}_{12}'$ means the unit vector of direction of the primed relative momentum (or velocity), so that $\mathrm{d}^2 e_{12}'$ denotes the pertaining solid angle element. Thus, eq. (1.1) takes the elaborate form

$$\frac{\partial f(1)}{\partial t} + \boldsymbol{c}_1 \cdot \frac{\partial f(1)}{\partial \boldsymbol{x}} = \sum_{V_2} \sum_{V_1'} \sum_{V_2'} \iint [f(1')f(2') - f(1)f(2)] \sigma c_{12}' \, \mathrm{d}^2 e_{12}' \, \mathrm{d}^3 p_2 \,. \tag{1.9}$$

The arguments 1, 2 obviously abbreviate $t\,\boldsymbol{x}\,\boldsymbol{p}_1 V_1$ etc. The primed arguments now are dependent variables. They have to be taken from momentum and energy conservation with given values of $\boldsymbol{p}_1 V_1$, $\boldsymbol{p}_2 V_2$, $\boldsymbol{e}_{12}'$ and $V_1' V_2'$:

$$\boldsymbol{p}_{\substack{1 \\ (2)}}' = \tfrac{1}{2}\boldsymbol{p}_1 + \tfrac{1}{2}\boldsymbol{p}_2 \overset{+}{\underset{(-)}{}} p_{12}' \boldsymbol{e}_{12}' \tag{1.10}$$

$$p_{12}' = [p_{12}^2 + 2m_{12}(E_{V_1} + E_{V_2} - E_{V_1'} - E_{V_2'})]^{\frac{1}{2}} \,. \tag{1.11}$$

The WCU-equation (1.9) is very similar to the Boltzmann equation for a monoatomic mixture. The only difference is that σ now also allows for transformations of mixture components into each other through a kind of binary "chemical" reactions (inelastic collisions).

Next, the connection between the cross section and the quantum mechanical scattering amplitude shall be discussed. The asymptotic scattering wave function in the centre of mass system is

$$\psi(\boldsymbol{r}_{12} V_1 V_2) \sim \exp(\mathrm{i}\boldsymbol{r}_{12}\cdot\boldsymbol{p}'_{12}/\hbar)\, \delta_{V_1 V_2,\, V'_1 V'_2} + \frac{1}{r} \exp(\mathrm{i} r p_{12}/\hbar) a\,(E \boldsymbol{e}_{12} V_1 V_2 \boldsymbol{e}'_{12} V'_1 V'_2)\,. \qquad (1.12)$$

The radius vector $\boldsymbol{r}_{12} = r\boldsymbol{e}_{12}$ points from molecule 2 to molecule 1, the incident wave is plane with relative momentum $\boldsymbol{p}'_{12} = p'_{12}\boldsymbol{e}'_{12}$ and internal quantum numbers V'_1, V'_2. Hence, the total energy E has a sharp value, see (1.4) with primed quantities. The momentum p_{12} occurring in the scattered wave has to be taken from (1.11), where primed and unprimed quantities may be interchanged. The scattering amplitude a, of the dimension of a length, characterizes the collisions with respect to intensity and phase, elastic ($p_{12} = p'_{12}$) or inelastic ($p_{12} \neq p'_{12}$). Now let us consider the scattering of a wave packet travelling in direction $\boldsymbol{e}'(= \boldsymbol{e}'_{12})$ with fairly well defined momentum towards the scatterer S, see left side of fig. 1. The internal quantum numbers before scattering are indicated by $V' = (V'_1, V'_2)$. After scattering, right side of fig. 1, the original wave packet, now with a "hole" inside, will be accompanied by several spherical waves corresponding to elastic and inelastic collisions. The "elastic" wave - no change of internal state V' - travels at the same speed as the undisturbed wave packet. Scattering into a state with $E_V \equiv E_{V_1} + E_{V_2} > E_{V'} \equiv E_{V'_1} + E_{V'_2}$ is represented by a wave travelling slowlier, scattering into a state with $E_V < E_{V'}$ is represented by a wave with higher speed. The condition of validity of the WCU-equation is that the waves with $V \neq V'$ are spatially separated before the next collision happens, i.e. before the elapse of time τ_{free} (mean free time, see below). In that case the next collision again starts with a well-defined internal state V (not a superposition of several internal states), such that the phases of the wave function are irrevelant and only the cross section (not the amplitude)

$$\sigma(E\, \theta\, V_1 V_2 V'_1 V'_2) = |a(E\, \boldsymbol{e}_{12} V_1 V_2 \boldsymbol{e}'_{12} V'_1 V'_2)|^2 \qquad (1.13)$$

counts, which indeed only occurs in the WCU-equation. In order to formulate the necessary condition mathematically, we start with the uncertainty relation for the incident wave packet, which has to be fairly stable,

$$\delta x' \delta p' \gg \hbar\,.$$

Here, $\delta x'$ means the length of the incoming packet, see fig. 1, $\delta p'$

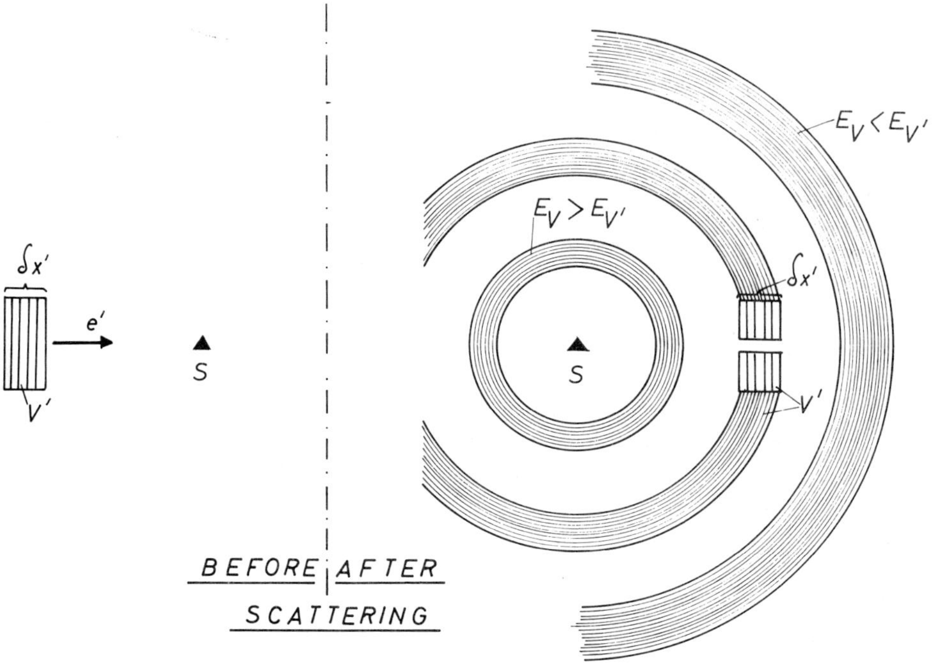

Fig. 1. Elastic and inelastic scattering of a "plane" wave packet with direction of incidence $\boldsymbol{e}'$ and pre-collisional internal quantum number V', by a scatterer S. Time elapse after scattering long enough, such that the spherical wave corresponding to elastic scattering is spatially separated from the waves corresponding to inelastic scattering, i.e. with internal energy $E_V \neq E_{V'}$.

denotes its momentum spread in forward direction and $\hbar$ is Planck's constant. Energy conservation says that there is no change of total energy in scattering

$$0 = \Delta \frac{p^2}{2m_{12}} + \Delta E \simeq c\Delta p + \Delta E = p\Delta c + \Delta E \ .$$

Thus, the difference Δc between velocities of an inelastically scattered wave and of the elastic one may be expressed by the pertaining difference ΔE between internal energies:

$$\Delta c \approx -\frac{\Delta E}{p} \approx -\frac{\Delta E}{p'} \ .$$

Separation of waves within time τ is expressed by requiring

$$\Delta c \mid \tau > \delta x' ,$$

or

$$|\Delta E| \ \tau > p'\delta x' = \delta x'\delta p' \frac{p'}{\delta p'} \gg \hbar .$$

Hence, for the gas to be governed by the WCU-equation, it is necessary that

$$|\Delta E| \ \tau_{\text{free}} \gg \hbar . \tag{1.14}$$

This is the condition of sufficient level spacing exposed earlier (Waldmann [2], p. 491, [3]). For $|\Delta E|$ one has to choose the smallest possible value, i.e. a difference between neighbouring levels. As a matter of course, the levels have to be non-degenerate. Otherwise $\Delta E = 0$ would occur, in contradiction with (1.14). A word should be said why τ_{free} is taken in (1.14). This is the longest time available for separation of waves, hence giving the lowest limit for ΔE. It corresponds to a wave packet which laterally is as small as possible, such that it disappears altogether in one collision and only spherical scattered waves emerge (in fig. 1, right side, the "hole" becomes as large as the original packet in this case). Such a packet obviously travels as far as a mean free path. The scattered waves then are deliberately broken into similar pieces and the game is repeated. All this is pretty qualitative. A more quantitative treatment is found in the papers by Waldmann [2] and Hess [4].

What about the validity of (1.14) in practical cases? The frequencies of oscillation of the H_2- or N_2-molecules are about

$$\omega \approx 10^{15} \text{ s}^{-1} .$$

As $|\Delta E|/\hbar = \omega$ and with $\tau_{\text{free}} \approx 10^{-10}$ s for 273°K, 1 at, one has $\omega\tau_{\text{free}} \approx 10^5$. So (1.14) is amply fulfilled. At lower temperature and pressure it is even better. The condition (1.14) would also be fulfilled for rotational levels if these were non-degenerate. The important rotational frequency at temperature T is

$$\omega_{\text{rot}} \approx \sqrt{\frac{kT}{\theta}}$$

in the classical equipartition regime (θ = molecular moment of inertia, k Boltzmann's constant). For N_2 at 273°K one has $\omega_{\text{rot}} \approx 10^{13} \text{ s}^{-1}$ by order of magnitude, hence still $\omega_{\text{rot}}\tau_{\text{free}} \approx 10^3$.

But the inherent degeneracy of levels in this case prohibits indiscriminate application of (1.14 and 9). Despite of this, one can draw a conclusion from this estimate for rotational states. After having dealt with degeneracy in an appropriate way, by using a distribution matrix description with respect to magnetic (degeneracy) quantum number as it is done in part II for particles with spin, then it is no more necessary to treat the energetic quantum number, which numbers the different rotational levels, in the same unwieldy way. With respect to this, one can restrict oneself to the diagonal elements of the pertaining matrix as it is done in the WCU-equation from the very beginning.

Finally, the detailed balance relation (1.7) is discussed. This relation follows immediately from (1.13) and from the symmetry property of the scattering amplitude in connection with isotropy of space

$$a(E\, \boldsymbol{e}_{12}\, V_1\, V_2\, \boldsymbol{e}'_{12}\, V'_1 V'_2) = a(E - \boldsymbol{e}'_{12}\, V'_1 V'_2 - \boldsymbol{e}_{12}\, V_1\, V_2) \,.$$

This symmetry again results from time reversal invariance of quantum mechanics (symmetry of the scattering matrix on the energy shell). The cross section σ used in (1.9), emerging from the representation (1.12) of the asymptotic wave function and having the simple symmetry (1.7), seems to be the most appealing one, from the theoretical point of view. It should be stressed that, in connection with this σ, c'_{12} occurs in (1.9). Wang Chang and Uhlenbeck instead write $\sigma_{\mathrm{WCU}}(V_1 V_2 V'_1 V'_2) c_{12}$. Their cross section therefore has the detailed balance symmetry (Waldmann [2], p. 492)

$$\sigma_{\mathrm{WCU}}(V_1\, V_2\, V'_1 V'_2)\, \frac{c_{12}}{c'_{12}} = \sigma = \sigma_{\mathrm{WCU}}(V'_1\, V'_2\, V_1\, V_2)\, \frac{c'_{12}}{c_{12}} \,,$$

or

$$\sigma_{\mathrm{WCU}}(V_1\, V_2\, V'_1\, V'_2)\, c^2_{12} = \sigma_{\mathrm{WCU}}(V'_1\, V'_2\, V_1\, V_2)\, c'^2_{12} \,.$$

That c'_{12} occurs in (1.9) is quite natural. Let us look at a loss collision first. The "channel velocity" after such a collision is c'_{12} and, in (1.9), is multiplied with σ which after (1.12 and 1.13) indeed is a measure for the scattered particle density. On the other hand, the rate of gain collisions starting with particles in intervals $\mathrm{d}^3 p'_1\, \mathrm{d}^3 p'_2$ and leading to the interval $\mathrm{d}^2 e_{12}$, primarily is

$$f(1')\,f(2')\,d^3p_1'\,d^3p_2'\,\sigma\,c_{12}\,d^2e_{12}\,.$$

Again, one has to use the "channel velocity" after collision, i.e. c_{12} in this case. But for a binary collision in which total momentum is conserved, one concludes from the identity

$$d^3p_{12}'\,d^3p_{12} = d^3p_{12}\,d^3p_{12}'\,,$$

by multiplying both sides with $d^3p' = d^3p$ respectively and using (1.8), that

$$d^3p_1'\,d^3p_2'\,d^3p_{12} = d^3p_1\,d^3p_2\,d^3p_{12}'\,. \tag{1.15}$$

By introducing the unit vectors $\boldsymbol{e}_{12}$, $\boldsymbol{e}_{12}'$ and the pertaining solid angle elements one has

$$d^3p_{12} = p_{12}^2\,dp_{12}\,d^2e_{12} = \tfrac{1}{2}p_{12}\,dp_{12}^2\,d^2e_{12}$$

and (1.16)

$$d^3p_{12}' = \tfrac{1}{2}p_{12}'\,dp_{12}'^2\,d^2e_{12}'\,.$$

Now, due to energy conservation the intervals of kinetic energy, for fixed $V_1V_2V_1'V_2'$, are equal after and before collision:

$$d\,p_{12}^2 = d\,p_{12}'^2\,.$$

Thus, after (1.16), eq. (1.15) can be rewritten as

$$d^3p_1'\,d^3p_2'\,c_{12}\,d^2e_{12} = d^3p_1\,d^3p_2\,c_{12}'\,d^2e_{12}'\,. \tag{1.17}$$

With this kinematic identity (Waldmann [2], p. 336), the rate of gain collisions also is

$$f(1')\,f(2')\,d^3p_1\,d^3p_2\,\sigma\,c_{12}'\,d^2e_{12}'\,,$$

which in fact underlies eq. (1.9).

2. THE BULK VISCOSITY

The evaluation of the WCU-equation along the lines of the Chapman-Enskog method is exposed in the article by Wang Chang, Uhlenbeck and de Boer [5]. For illustration, we quote just one result, the formula for the bulk viscosity κ, defined in the usual way

$$\mathbf{p}' = -\kappa \nabla \cdot \boldsymbol{v}\, \mathbf{U} - 2\eta \overset{\circ}{\overline{\nabla \cdot \boldsymbol{v}}},$$

($\mathbf{p}'$ frictional pressure tensor, $\boldsymbol{v}$ gas velocity, $\mathbf{U}$ unit tensor, η shear viscosity). One finds

$$\kappa = \left(\frac{c_{\text{int}}}{c_v}\right)^2 \frac{kT}{2(kT/\pi m)^{\frac{1}{2}} \bar{\sigma}_\kappa}, \tag{2.1}$$

with

$$\bar{\sigma}_\kappa = \left[\sum_V \exp\left(-\frac{E_V}{kT}\right)\right]^2 \sum_{V_1} \ldots \sum_{V_2'} 2\pi \int_{-\infty}^{\infty} \frac{\mathrm{d}E}{kT} \int_0^{\pi} \sin\theta \mathrm{d}\theta$$

$$\times \exp\left(-\frac{E}{kT}\right)\left(\frac{\Delta E}{kT}\right)^2 \frac{(E - E_{V_1} - E_{V_2})^{\frac{1}{2}}(E - E_{V_1'} - E_{V_2'})^{\frac{1}{2}}}{kT}\,\sigma$$

and

$$\Delta E = E_{V_1} + E_{V_2} - E_{V_1'} - E_{V_2'}\,. \tag{2.2}$$

The c's in (2.1) denote the internal and total (for constant volume) specific heats of the gas. The effective cross section $\bar{\sigma}_\kappa$ is based on the σ used in (1.9). For convenience we think of σ as defined for all values of E between $\pm\infty$. Then, as the relative kinetic energies $p_{12}^2/2m_{12}$ and $p_{12}'^2/2m_{12}$ have to be positive, one has to take

$$\sigma = 0 \quad \text{for } E < \begin{cases} E_{V_1} + E_{V_2} \\ E_{V_1'} + E_{V_2'} \end{cases}.$$

The integrand in (2.2) is written in such a way as to show its perfect symmetry in primed and unprimed variables.

For monoatomic gases the bulk viscosity is zero. But for polyatomic gases it usually exceeds the shear viscosity and gives the main contribution to sound absorption at low frequency.

Part II. GAS COMPOSED OF MOLECULES WITH DEGENERATE STATES OF INTERNAL ENERGY, ESPECIALLY OF PARTICLES WITH SPIN

In order to deal with this case one has to go deeper into the foundations of statistical quantum mechanics. In it, a molecule is primarily described by a Hermitian positive density matrix (or operator) $\rho(t\boldsymbol{x}I,\ \boldsymbol{x}'I')$, with trace 1, where $\boldsymbol{x}$ is the space coordinate and I the internal quantum number. The latter stands for a triple V, L, M, vibrational, rotational and magnetic, of quantum numbers. The one particle density matrix is also appropriate to describe a (dilute) gas of independent particles. In this case we multiply ρ by the total number of particles and then write formally $f(t\boldsymbol{x}\boldsymbol{p})_{II'}$, the one particle distribution matrix (or operator). It is Hermitian again. Originally, $\boldsymbol{x}$ and $\boldsymbol{p}$ are non-commutative operators, but in practical cases the translational motion is essentially unquantized. So, $\boldsymbol{x}$ and $\boldsymbol{p}$ will be treated as classical variables henceforth.

In section 1, from the triple number I only the part V was assumed to exist. It was shown that with sufficient level spacing only the number density of particles in different internal states (intensities of waves!) is important. This means that only the diagonal elements

$$f(t\boldsymbol{x}\boldsymbol{p})_{VV} \equiv f(t\boldsymbol{x}\boldsymbol{p}V)$$

occur in the kinetic equation, i.e. the WCU-equation.

With degenerate internal levels whose energy depends on V and L but not on M, one will expect that again the pertaining diagonal elements only will play a role:

$$f(t\boldsymbol{x}\boldsymbol{p})_{VLM\ VLM'} \equiv f(t\boldsymbol{x}\boldsymbol{p}\,VL)_{MM'}\ .$$

Scattered waves with equal V and L but different M's will not separate in free motion, their phase relations remain important and therefore the matrix character of f with respect to magnetic quantum numbers (orientational degeneracy) has to be preserved.

The simplest case of a degenerate level occurs with particles with spin. Such a gas is described by

$$f(t\boldsymbol{x}\boldsymbol{p})_{MM'}\ , \qquad \text{or } f(t\boldsymbol{x}\boldsymbol{p}\boldsymbol{s})\ .$$

The second form, with the spin operator $\boldsymbol{s}$ in the argument, is the most convenient one for the general theory. Any $(2S+1)\times(2S+1)$ matrix $f_{MM'}$, where S is the spin value ($\frac{1}{2}$ for the electron), can be expressed by a polynomial $f(s)$ in the three spin matrices s_x, s_y, s_z. The highest necessary degree in the polynomial is $2S$.

The pertaining kinetic equation has to be an equation for this matrix- or operator-f. The necessary preservation of phase relations will imply the occurrence of the scattering amplitude rather than of the cross section in the kinetic equation. A qualitative derivation of the latter is given in section 3 for particles with spin. The generalisation to real molecules with different rotational levels, sufficiently spaced, is then obvious (Waldmann [2], p. 491, Snider [6]) and may be omitted.

3. KINETIC EQUATION FOR A DILUTE GAS OF PARTICLES WITH SPIN

Let us look at the collision of two particles with spin, in the centre of mass system. First, the incoming wave shall be taken as spherical with relative momentum $p_{12} = \hbar k$. The relative position vector is called $\boldsymbol{r}_{12} = \boldsymbol{r}_1 - \boldsymbol{r}_2 = r\boldsymbol{e}_{12}$, where $\boldsymbol{e}_{12}$ is the unit vector of direction. The ingoing spinor amplitude is called $\chi_-(-\boldsymbol{e}_{12})$. It depends, besides $\boldsymbol{e}_{12}$, tacitly on the magnetic quantum numbers M_1, M_2. The outgoing amplitude is called $\chi_+(\boldsymbol{e}_{12})$. The asymptotic form of this wave function then is

$$\psi \approx \frac{1}{r}\exp(-ikr)\,\chi_-(-\boldsymbol{e}_{12}) - \frac{1}{r}\exp(ikr)\,\chi_+(\boldsymbol{e}_{12}) \,. \tag{3.1}$$

Without scattering one would have

$$\chi_+(\boldsymbol{e}_{12}) = \chi_-(\boldsymbol{e}_{12}) \,.$$

With scattering, χ_+ still is a linear functional of χ_-:

$$\chi_+(\boldsymbol{e}_{12}) = \int S(\boldsymbol{e}_{12}\boldsymbol{e}''_{12}\,s_1 s_2)\,\chi_-(\boldsymbol{e}''_{12})\,\mathrm{d}^2 e''_{12} \,, \tag{3.2}$$

or briefly

$$\chi_+ = S\chi_- \,. \tag{3.3}$$

The S-matrix is unitary

$$S^\dagger S = 1 \,; \qquad SS^\dagger = 1 \,. \tag{3.4}$$

The first (orthonormality) property is in detail

$$\int S^\dagger(\boldsymbol{e}_{12}\, \boldsymbol{e}''_{12}\, s_1\, s_2)\, S(\boldsymbol{e}''_{12}\, \boldsymbol{e}'_{12}\, s_1\, s_2)\, \mathrm{d}^2 e''_{12} = \delta(\boldsymbol{e}_{12} - \boldsymbol{e}'_{12}) .$$

The second (completeness) property has similar detailed meaning. Now, to obtain the analogue of the scattering amplitude from (1.12), we specialize for scattering of a plane wave with momentum $\boldsymbol{p}'_{12} = \hbar k \boldsymbol{e}'_{12}$ and spinor amplitude χ, tacitly dependent on M_1, M_2. The ingoing part of this wave is asymptotically

$$\chi_-(-\boldsymbol{e}_{12}) = \delta(-\boldsymbol{e}_{12} - \boldsymbol{e}'_{12})\chi .$$

The full wave, after (3.1 and 2), is in this case

$$\psi \approx [\frac{1}{r} \exp(-ikr)\, \delta(-\boldsymbol{e}_{12} - \boldsymbol{e}'_{12}) - \frac{1}{r} \exp(ikr)\, S(\boldsymbol{e}_{12}\, \boldsymbol{e}'_{12}\, s_1\, s_2)]\, \chi . \quad (3.5)$$

Without scattering one would have ($S = 1$!)

$$\psi = \psi_0 \approx [\frac{1}{r} \exp(-ikr)\, \delta(-\boldsymbol{e}_{12} - \boldsymbol{e}'_{12}) - \frac{1}{r} \exp(ikr)\, \delta(\boldsymbol{e}_{12} - \boldsymbol{e}'_{12})]\, \chi$$

$$= \left[\frac{k}{2\pi \mathrm{i}} \exp(\mathrm{i}\, \boldsymbol{r}_{12} \cdot \boldsymbol{p}'_{12}/\hbar)\right]_{\text{asympt.}} \chi . \quad (3.6)$$

The latter equality is the asymptotic form of Rayleigh's decomposition of the plane wave into spherical waves. By use of (3.6) the wave (3.5) can be written as

$$\psi \approx \frac{k}{2\pi \mathrm{i}} [\exp(\mathrm{i}\, \boldsymbol{r}_{12} \cdot \boldsymbol{p}'_{12}/\hbar) + \frac{1}{r} \exp(ikr)\, a(\boldsymbol{e}_{12}\, \boldsymbol{e}'_{12}\, s_1\, s_2)]\, \chi , \quad (3.7)$$

where the plane wave scattering amplitude is

$$a(\boldsymbol{e}_{12}\, \boldsymbol{e}'_{12}\, s_1\, s_2) = \frac{2\pi \mathrm{i}}{k} [\delta(\boldsymbol{e}_{12} - \boldsymbol{e}'_{12}) - S(\boldsymbol{e}_{12}\, \boldsymbol{e}'_{12}\, s_1\, s_2)] . \quad (3.8)$$

Eq. (3.7) may be compared with (1.12). The only difference is that χ now means an arbitrary spinor whereas in (1.12) a specified incoming internal state was taken. From unitarity (3.4) follows

$$a^\dagger a = a a^\dagger = \frac{2\pi}{k} \frac{a - a^\dagger}{\mathrm{i}} . \quad (3.9)$$

This is known as the optical theorem; the same brief notation as in (3.4) is used here.

Next, the question is asked what happens in scattering with an ingoing statistical state characterized by a two-particle Hermitian distribution matrix f_-. It may be represented by pure states in the way

$$f_- = \sum_i \chi_-^{(i)} \chi_-^{(i)\dagger} .$$

What comes out after (3.3) is

$$f_+ = \sum_i \chi_+^{(i)} \chi_+^{(i)\dagger} = \sum_i S\chi_-^{(i)} \chi_-^{(i)\dagger} S^\dagger = S f_- S^\dagger .$$

The change in scattering therefore is

$$f_+ - f_- = S f_- S^\dagger - f_- \propto a f_- a^\dagger - \frac{2\pi}{ik}(a f_- - f_- a^\dagger)$$

$$= a f_- a^\dagger - \frac{\hbar}{i m_{12} c_{12}'}(a f_- - f_- a^\dagger) . \quad (3.10)$$

For m_{12} see (1.5). That much about the underlying quantum mechanics.

The changes (3.10) in the two particle distribution now have to be summed per unit time, as always in a more or less intuitive way. Making use of a Stosszahlansatz, one obtains the following kinetic equation for a spin gas (Waldmann [7, 8], Snider [6]) in which also the well-known precession term in a magnetic field is incorporated:

$$\frac{\partial f(1)}{\partial t} + \boldsymbol{c}_1 \cdot \frac{\partial f(1)}{\partial \boldsymbol{x}} - \boldsymbol{\omega}_H \cdot [\boldsymbol{s}_1, f(1)]$$

$$= \mathrm{tr}_2 \int \{ \int a(\boldsymbol{e}_{12}\, \boldsymbol{e}_{12}'\, \boldsymbol{s}_1\, \boldsymbol{s}_2) f(1') f(2') a^\dagger(\boldsymbol{e}_{12}'\, \boldsymbol{e}_{12}\, \boldsymbol{s}_1\, \boldsymbol{s}_2)\, c_{12}'\, \mathrm{d}^2 e_{12}'$$

$$- \frac{\hbar}{i m_{12}} [a(\boldsymbol{e}_{12}\, \boldsymbol{e}_{12}\, \boldsymbol{s}_1\, \boldsymbol{s}_2) f(1) f(2) - f(1) f(2) a^\dagger(\boldsymbol{e}_{12}\, \boldsymbol{e}_{12}\, \boldsymbol{s}_1\, \boldsymbol{s}_2)] \}\, \mathrm{d}^3 p_2 . \quad (3.11)$$

Here, $f(1)$ is an abbreviation for $f(t\boldsymbol{x}\boldsymbol{p}_1\, \boldsymbol{s}_1)$, etc. The primes have the same meaning as in (1.9). The symbol tr_2 means the trace on the indices of spin $\boldsymbol{s}_2$. For particles without spin the kinetic equation (3.11) is identical with the Boltzmann equation, as it should be.

This follows from the optical theorem (3.9).

An explanation has still to be given of the magnetic term. The precession frequency vector ω_H is

$$\omega_H = -\frac{\mu}{S\hbar} \boldsymbol{H} , \tag{3.12}$$

where μ is the magnetic moment of a particle, S its spin value ($\frac{1}{2}$ for electrons) and $\boldsymbol{H}$ the magnetic field. The commutator bracket

$$[s_1, f(1)] \equiv \frac{1}{i}(s_1 f(1) - f(1)\, s_1) \tag{3.13}$$

is Hermitian, s and f being so. There is no way to introduce such a commutator into a kinetic equation of type (1.9), which knows of diagonal elements only of the distribution (density) matrix.

The considerations of this section merely give the skeleton or idea of a derivation. For more details and preciseness the original papers should be consulted.

But eq. (3.11), with magnetic field, requires some further remarks. The scattering amplitude was taken to be field-independent. This implies conservation of kinetic energy in spin-flip collisions, when, strictly speaking, the sum of kinetic and magnetic energy is conserved. An amendment of the collision term in this respect was proposed by Hess [4]. His amended collision term automatically leads to the Bloch equation of magnetic resonance for which magnetic energy is decisive. - The Senftleben effect, to be discussed in the next section, is practically not influenced by this correction. Nevertheless, this case too needs closer inspection. The Senftleben effect is saturated (see section 4) if the Zeeman splitting, measured by ω_H, is large compared with the reciprocal time of free flight

$$\omega_H \gg \omega_{\text{free}} .$$

This is just the condition (1.14) of validity for a WCU-equation (separation of wave packets during time of free flight). So, if eq. (3.11) is correct, it must become a WCU-equation in the high field limit. This in fact happens. If $|\omega_H|$ is large, the magnetic term on the left side of (3.11) will dominate and one has the approximate solution

$$f(t) \approx \exp(-i\omega_H \cdot s\, t)\, f(0) \exp(i\,\omega_H \cdot s\; t) .$$

With the magnetic field $\boldsymbol{H}$ as the quantization axis and M as the pertaining quantum number, this is explicitly

$$f(t)_{MM'} \approx \exp(i\omega_H(M-M')t)\, f(0)_{MM'} ,$$

with $\omega_H = \mu H/S\hbar$. The diagonal elements $M = M'$ of f are time-independent in this approximation, the non-diagonal elements are rapidly oscillating, as $|\omega_H|$ is large. Under these circumstances, returning to the full eq. (3.11) and writing it down for the diagonal element

$$f(t\,\boldsymbol{x}\,\boldsymbol{p}_1)_{M_1M_1} \equiv \bar{f}(t\,\boldsymbol{x}\,\boldsymbol{p}_1\,M_1) = \bar{f}(1) ,$$

one may neglect the rapidly oscillating elements in the collision term on the right side. This yields

$$\left(\frac{\partial}{\partial t} + \boldsymbol{c}_1 \cdot \frac{\partial}{\partial \boldsymbol{x}}\right) \bar{f}(1)$$

$$= \sum_{M_2} \sum_{M_1'} \sum_{M_2'} \iint [\bar{f}(1')\, \bar{f}(2') - \bar{f}(1)\, \bar{f}(2)]\, \sigma\, c_{12}'\, \mathrm{d}^2 e_{12}'\, \mathrm{d}^3 p_2 ,$$

with

$$\sigma = |a_{M_1M_2M_1'M_2'}|^2 .$$

Again, the optical theorem has been used. Hence, for high enough field, eq. (3.11) indeed is a WCU-equation. As such, it automatically accounts for the separation of waves packets which takes place in this case. So, eq. (3.11) covers both, the zero field case anyway and the high field case too. It seems reasonable to expect that eq. (3.11) also describes the intermediate field case correctly. Comparison of the theoretical results for the Senftleben effect, derived from (3.11), with experiment justify this expectation.

4. TRANSPORT PHENOMENA IN A GAS OF PARTICLES WITH SPIN

The kinetic equation (3.11), which fulfills an H-theorem (Waldmann [2], p. 484), uniquely determines the distribution f for $t > 0$ if it is known at $t = 0$ (initial value problem). From f, all macroscopic quantities (mean values) can be inferred. The number density of particles e.g., is

$$n(t\,\boldsymbol{x}) = \mathrm{tr} \int f(t\,\boldsymbol{x}\,\boldsymbol{p}\,\boldsymbol{s})\, \mathrm{d}^3 p . \qquad (4.1)$$

The instantaneous local mean value of any quantity per particle $g(t\,\boldsymbol{x}\,\boldsymbol{p}\,s)$ is

$$\langle g\rangle = \mathrm{tr} \int g \frac{f}{n} \mathrm{d}^3 p \,. \qquad (4.2)$$

The time-space behaviour of such quantities, e.g. of mean momentum, energy or spin and of the corresponding fluxes, is what one is mainly interested in.

The physical contents of (3.11) are richer than those of the classical Boltzmann equation. In order to obtain some insight two methods offer themselves. 1) Expansion of the distribution operator with respect to a complete orthonormalized set of irreducible tensors in $\boldsymbol{p}$ and s about a standard equilibrium distribution f_0; the kinetic equation then is replaced by a set of "transport-relaxation equations" for the t, $\boldsymbol{x}$-dependent coefficients of the expansion. 2) Construction of "normal solutions" in the vein of the familiar Chapman-Enskog method, i.e. expansion about a local equilibrium distribution with asymptotic state of microscopic relaxation. In the following the first method is indicated. The second one has been treated by Mc Court and Snider [9,10]. In applications to transport phenomena both methods are finally equivalent.

The distribution f is written as

$$f(t\,\boldsymbol{x}\,\boldsymbol{p}\,s) = f_0(\boldsymbol{p}) \left(1 + \Phi\,(t\,\boldsymbol{x}\,\boldsymbol{p}\,s)\right) ,$$

where f_0 is a Maxwellian distribution and Φ is treated as "small", so that the collision term of eq. (3.11) becomes linear in Φ, yielding

$$\frac{\partial \Phi}{\partial t} + \boldsymbol{c} \cdot \frac{\partial \Phi}{\partial \boldsymbol{x}} - \boldsymbol{\omega}_H \cdot [\,s\,,\Phi] = -\omega(\Phi) \,. \qquad (4.3)$$

The linearized collision operator $\omega(\Phi)$ has several general properties which have been explored previously (Waldmann [11]). It is positive-semidefinite, it has symmetries because of parity conservation and of time reversal invariance built into the scattering amplitude a of (3.11) and it is isotropic in the sense that from a $\boldsymbol{p}$, s-dependent irreducible tensor it produces one of the same genre. From (3.11), $\omega(\Phi)$ can be written down explicitly. But this is omitted here.

Now, let us choose a complete set of functions $\Phi_k(\boldsymbol{p}\,s)$, orthonormalized such that

$$\langle \Phi_i \Phi_k \rangle_0 \equiv \mathrm{tr} \int \Phi_i \Phi_k \frac{f_0}{n_0} \mathrm{d}^3 p = \delta_{ik} \,, \qquad (4.4)$$

and let us put

$$\Phi = \sum_k a_k(t\,\boldsymbol{x})\,\Phi_k(\boldsymbol{p}\,\boldsymbol{s})\,. \tag{4.5}$$

The expansion coefficients a_i have a simple physical meaning. They are essentially the instantaneous local mean values of the pertaining functions Φ_i, as follows from (4.2 and 4):

$$n\langle\Phi_i\rangle = n_0\,(\langle\Phi_i\rangle_0 + a_i)\,. \tag{4.6}$$

Inserting into (4.3) and applying the orthonormalisation (4.4) gives the "transport-relaxation equations" for the expansion coefficients a:

$$\frac{\partial a_i}{\partial t} + \sum_k \boldsymbol{c}_{ik}\cdot\frac{\partial a_k}{\partial \boldsymbol{x}} - \omega_H\cdot\sum_k \boldsymbol{s}_{ik}\,a_k = -\sum_k \omega_{ik}a_k\,,$$

with (4.7)

$$\boldsymbol{c}_{ik} = \langle\Phi_i\,\boldsymbol{c}\,\Phi_k\rangle_0\,, \qquad \boldsymbol{s}_{ik} = \langle\Phi_i[\,\boldsymbol{s},\Phi_k]\rangle_0\,, \qquad \omega_{ik} = \langle\Phi_i\,\omega(\Phi_k)\rangle_0\,.$$

The coefficients in the first two sums are unproblematic. Those in the last sum, the relaxation matrix ω_{ik}, comprise all the interesting properties of the collisions. Of course completeness of the expansion system is now readily abandoned and for tractability, (4.7) is truncated down to a few equations which hopefully will be the most important ones.

In doing that, let us make a choice of expansion functions Φ_k. As the collision operator is isotropic, it is appropriate to use a sequence of scalars ($L=0$), vectors ($L=1$), irreducible second rank tensors ($L=2$), etc. Again, these should be chosen such as to have well defined parity, i.e. the parity operation $\boldsymbol{p}, \boldsymbol{s} \to -\boldsymbol{p}, \boldsymbol{s}$, shall multiply them by $P=+1$ or -1. Tensors with $(-1)^L P = 1$ are henceforth denoted by Φ's, those with $(-1)^L P = -1$ by ψ's (pseudo). Up to second power these tensors are (we now use the velocity $\boldsymbol{c}$ instead of momentum $\boldsymbol{p}$ and omit the normalisation factors for brevity):

scalars $$\Phi^{(1)} = 1\,, \quad \Phi^{(2)} \propto \frac{mc^2}{2kT_0} - \tfrac{3}{2}\,, \tag{4.8}$$

pseudo scalar $$\psi^{(1)} \propto \boldsymbol{c}\cdot\boldsymbol{s}\,, \tag{4.9}$$

vectors	$\boldsymbol{\Phi}^{(1)} \propto \boldsymbol{c}\ , \quad \boldsymbol{\Phi}^{(2)} \propto \boldsymbol{c}\times\boldsymbol{s}\ ,$	(4.10)
pseudo vector	$\boldsymbol{\Psi}^{(1)} \propto \boldsymbol{s}\ ,$	(4.11)
2nd rank tensors	$\overset{\leftrightarrow}{\Phi}^{(1)} \propto \overline{\boldsymbol{c}\,\boldsymbol{c}}\ , \quad \overset{\leftrightarrow}{\Phi}^{(2)} \propto \overline{\boldsymbol{s}\,\boldsymbol{s}}\ ,$	(4.12)
2nd rank pseudo tensor	$\overset{\leftrightarrow}{\Psi}^{(1)} \propto \overline{\boldsymbol{c}\,\boldsymbol{s}}\ .$	(4.13)

The sign $\overline{\quad}$ means irreducibility, e.g.

$$\overline{\boldsymbol{c}\,\boldsymbol{s}} = \tfrac{1}{2}(\boldsymbol{c}\,\boldsymbol{s} + \boldsymbol{s}\,\boldsymbol{c}) - \tfrac{1}{3}\boldsymbol{c}\cdot\boldsymbol{s}\,\mathsf{U}\ ,$$

where U is the unit tensor of three dimensional space.

The pertaining coefficients in the expansion (4.5) are of the same tensorial behaviour. The scalars are denoted by $a^{(1)}$, $a^{(2)}$, $b^{(1)}$ respectively, the vectors by $\boldsymbol{a}^{(1)}$, $\boldsymbol{a}^{(2)}$, $\boldsymbol{b}^{(1)}$, the tensors by $\mathsf{a}^{(1)}$, $\mathsf{a}^{(2)}$, $\mathsf{b}^{(1)}$. After (4.6) their physical meaning is

scalars	$a^{(1)} \propto n - n_0$ number density, $\quad a^{(2)} \propto T - T_0$ temperature
pseudo scalar	$b^{(1)} \propto \langle \boldsymbol{c}\cdot\boldsymbol{s}\rangle$ mean "helicity"
vectors	$\boldsymbol{a}^{(1)} \propto \langle \boldsymbol{c}\rangle$ gas velocity, $\quad \boldsymbol{a}^{(2)} \propto \langle \boldsymbol{c}\times\boldsymbol{s}\rangle$ "azimuthal polarization"
pseudo vector	$\boldsymbol{b}^{(1)} \propto \langle \boldsymbol{s}\rangle$ mean spin ,

etc. Those mean values, which have both $\boldsymbol{c}$ and $\boldsymbol{s}$ between the brackets, measure the velocity-spin correlation contained in the general distribution function $f(\boldsymbol{c}, \boldsymbol{s})$. The correlation described by the vector $\boldsymbol{a}^{(2)}$ is naively pictured in fig. 2.

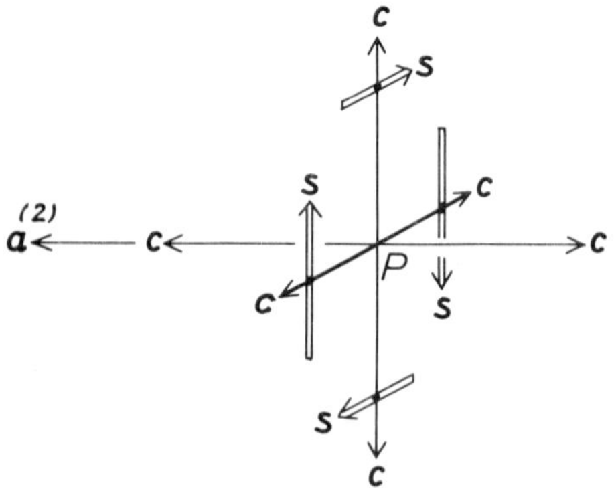

Fig. 2. Molecular velocity-spin correlation at point P with azimuthal polarization characterized by polar vector $\boldsymbol{a}^{(2)}$. Six orthogonal directions of velocities $\boldsymbol{c}$ and accompanying spins $\boldsymbol{s}$ indicated.

The above set of expansion tensors up to second power is just sufficient to describe, in the lowest significant approximation, the diffusion of a Lorentzian gas of particles with spin in a background of spinless scatterers fixed at random locations. Again, this Lorentzian gas is also governed by a kinetic equation of type (4.3) or the equivalent set (4.7). This case is mathematically somewhat simpler than the pure gas which, in the lowest significant approximation, requires the use of expansion tensors up to third power. For this reason the following methodical considerations are mainly concerned with the diffusion of the Lorentzian gas (Waldmann [11], Waldmann and Kupatt [12]*). For the pure gas a few results only will be mentioned later.

The first two equations of (4.7), with the abbreviation

$$c_0 = \sqrt{3kT_0/m} \ ,$$

then are (notation from Hess and Waldmann [13])

$$\frac{\partial a^{(1)}}{\partial t} + \frac{1}{\sqrt{3}} c_0 \nabla \cdot \boldsymbol{a}^{(1)} = 0 \ ; \qquad \frac{\partial a^{(2)}}{\partial t} + \sqrt{\tfrac{2}{3}}\, c_0 \nabla \cdot \boldsymbol{a}^{(1)} = 0 \ . \tag{4.8a}$$

These are, in the present approximation, the conservation equations for particle number and energy. Next comes the transport-relaxation equation for the pseudo scalar. Denoting the unit vector in the direction of the magnetic field by $\boldsymbol{h}$ and putting

$$\boldsymbol{\omega}_H = -\omega_H \boldsymbol{h} \ , \qquad \omega_H = \mu H/\hbar S \ ,$$

one finds

$$\frac{\partial b^{(1)}}{\partial t} + \tfrac{1}{3} c_0 \nabla \cdot \boldsymbol{b}^{(1)} - \sqrt{\tfrac{2}{3}}\, \omega_H \boldsymbol{h} \cdot \boldsymbol{a}^{(2)} = -\omega_{-0}^{(11)} b^{(1)} \ . \tag{4.9a}$$

Hence, a mean helicity, present alone, relaxes with the characteristic time $1/\omega_{-0}^{(11)}$ which in principle can be calculated from the collision operator.

The equations for the two polar vectors turn out to be

* In this work the second rank tensors had been omitted. This is unjustified. I thank Dr. Hess for pointing this out to me.

$$\frac{\partial \boldsymbol{a}^{(1)}}{\partial t} + c_0 \nabla \left(\frac{1}{\sqrt{3}} a^{(1)} + \frac{\sqrt{2}}{3} a^{(2)} \right) = - \sum_1^2 \omega_{-1}^{(1k)} \boldsymbol{a}^{(k)} ,$$

$$\frac{\partial \boldsymbol{a}^{(2)}}{\partial t} + \frac{1}{\sqrt{6}} c_0 \nabla \times \boldsymbol{b}^{(1)} + \omega_H \left(\sqrt{\tfrac{2}{3}}\, \boldsymbol{h} b^{(1)} + \tfrac{1}{2} \boldsymbol{h} \times \boldsymbol{a}^{(2)} - \frac{1}{\sqrt{2}} \boldsymbol{h} \cdot \mathsf{b}^{(1)} \right) \qquad (4.10a)$$

$$= - \sum_1^2 \omega_{-1}^{(2k)} \boldsymbol{a}^{(k)} .$$

For the pseudo vector one finds

$$\frac{\partial \boldsymbol{b}^{(1)}}{\partial t} + c_0 \left(\tfrac{1}{3} \nabla b^{(1)} - \frac{1}{\sqrt{6}} \nabla \times \boldsymbol{a}^{(2)} + \frac{1}{\sqrt{3}} \nabla \cdot \mathsf{b}^{(1)} \right) + \omega_H \boldsymbol{h} \times \boldsymbol{b}^{(1)}$$

$$= - \omega_{+1}^{(11)} \boldsymbol{b}^{(1)} . \qquad (4.11a)$$

The latter equation, in a spatially homogeneous situation, describes the motion of the mean spin. It precesses with the angular velocity ω_H around the magnetic field and relaxes according to the positive constant $\omega_{+1}^{(11)}$. The second eq. (4.10a) says that the azimuthal polarization $\boldsymbol{a}^{(2)}$ precesses with $\frac{1}{2}\omega_H$. This is physically plausible from fig. 2. Let $\boldsymbol{a}^{(2)}$ be perpendicular to the magnetic field. The preferential spins of the particles flying in $\boldsymbol{H} \times \boldsymbol{a}^{(2)}$-direction then are parallel with $\boldsymbol{H}$ and don't precess. Only the spins of particles flying in $\boldsymbol{H}$-direction do precess. So, an average precession with $\frac{1}{2}\omega_H$ is plausible.

The equations for the irreducible second rank tensors are

$$\frac{\partial \mathsf{a}^{(1)}}{\partial t} + \sqrt{\tfrac{2}{3}}\, c_0 \overline{\nabla \boldsymbol{a}^{(1)}} = - \sum_1^2 \omega_{+2}^{(1k)} \mathsf{a}^{(k)} ,$$

$$\frac{\partial \mathsf{a}^{(2)}}{\partial t} + 2\omega_H \overline{\boldsymbol{h} \times \mathsf{a}^{(2)}} = - \sum_1^2 \omega_{+2}^{(2k)} \mathsf{a}^{(k)} . \qquad (4.12a)$$

The generalized cross product is defined by ($\mu, \nu, \lambda, \kappa = 1, 2, 3$ Cartesian components; summation rule)

$$(\boldsymbol{h} \times \mathsf{a})_{\mu\nu} = \epsilon_{\mu\lambda\kappa} h_\lambda a_{\kappa\nu} .$$

Finally the pseudo tensor equation is

$$\frac{\partial \mathbf{b}^{(1)}}{\partial t} + \frac{1}{\sqrt{3}} c_0 \overline{\nabla \boldsymbol{b}^{(1)}} + \omega_H \left(\frac{1}{\sqrt{2}} \overline{\boldsymbol{h}\, \boldsymbol{a}^{(2)}} + \overline{\boldsymbol{h} \times \mathbf{b}^{(1)}} \right) = - \omega_{-2}^{(11)} \mathbf{b}^{(1)} . \tag{4.13a}$$

Now, stationary diffusion in an infinite gas due to a concentration gradient shall be analyzed. In this situation, all time and space derivatives in eqs. (4.8a-4.13a) are zero, with the exception of the gradient of the number density. Then one has a linear algebraic system with $\nabla a^{(1)}$ as the inhomogeneity.

Eqs. (4.11a and 4.12a) become homogeneous and uncoupled from the rest; so $\boldsymbol{b}^{(1)}$, $\mathbf{a}^{(1)}$ and $\mathbf{a}^{(2)}$ vanish.

From (4.9a) the pseudo scalar $b^{(1)}$ occurring in (4.10a) can be replaced by $\boldsymbol{h} \cdot \boldsymbol{a}^{(2)}$:

$$b^{(1)} = \sqrt{\tfrac{2}{3}}\, \varphi_{-0}\, \boldsymbol{h} \cdot \boldsymbol{a}^{(2)} ,$$

with

$$\varphi_{-0} = \omega_H / \omega_{-0}^{(11)} , \tag{4.14}$$

which is the angle of spin precession during the relaxation time $1/\omega_{-0}^{(11)}$.

The second eq. (4.10a) contains the polar vector

$$\tilde{\boldsymbol{a}} = \boldsymbol{h} \cdot \mathbf{b}^{(1)} .$$

It can be expressed by $\boldsymbol{a}^{(2)}$ with the aid of eq. (4.13a) by dotting it with $\boldsymbol{h}$:

$$\omega_{-2}^{(11)} \tilde{\boldsymbol{a}} + \tfrac{1}{2} \omega_H\, \boldsymbol{h} \times \tilde{\boldsymbol{a}} = - \frac{1}{\sqrt{2}} \omega_H\, \boldsymbol{h} \cdot \overline{\boldsymbol{h}\, \boldsymbol{a}^{(2)}} . \tag{4.15}$$

To solve an equation like this for $\tilde{\boldsymbol{a}}$, an effective method has been invented by Hess (Hess and Waldmann, to be published). One introduces the three second rank projectors ($\mathbf{U}$ = unit tensor)

$$\mathbf{P}_{\pm} = \tfrac{1}{2}(\mathbf{U} \pm \mathrm{i}\, \boldsymbol{h} \times - \boldsymbol{h}\boldsymbol{h}) , \qquad \mathbf{P}_h = \boldsymbol{h}\, \boldsymbol{h} , \tag{4.16}$$

with the properties

$$\mathbf{P}_+ + \mathbf{P}_- + \mathbf{P}_h = \mathbf{U}; \qquad \mathbf{P}_+ \mathbf{P}_+ = \mathbf{P}_+ \text{ etc.}; \qquad \mathbf{P}_+ \mathbf{P}_- = 0 \text{ etc.}$$

Then the vector equation (the α's are complex numbers)

$$(\alpha_+ \mathsf{P}_+ + \alpha_- \mathsf{P}_- + \alpha_h \mathsf{P}_h) \cdot \tilde{\boldsymbol{a}} = \boldsymbol{a}$$

obviously has the solution

$$\tilde{\boldsymbol{a}} = \left(\frac{1}{\alpha_+} \mathsf{P}_+ + \frac{1}{\alpha_-} \mathsf{P}_- + \frac{1}{\alpha_h} \mathsf{P}_h\right) \cdot \boldsymbol{a} \ .$$

Hence, eq. (4.15) and afterwards eqs. (4.10a) only have to be cast into a corresponding form, in order to be readily solved.

With the precession angle

$$\varphi_{-2} = \omega_H / 2\omega_{-2}^{(11)} \ , \tag{4.17}$$

eq. (4.15) takes the form

$$[(1 - i\varphi_{-2}) \mathsf{P}_+ + (1 + i\varphi_{-2}) \mathsf{P}_- + \mathsf{P}_h] \cdot \tilde{\boldsymbol{a}} = -\frac{1}{\sqrt{2}} \varphi_{-2} (\mathsf{P}_+ + \mathsf{P}_- + \tfrac{1}{3} \mathsf{P}_h) \cdot \boldsymbol{a}^{(2)} \ .$$

This gives

$$\tilde{\boldsymbol{a}} = -\frac{1}{\sqrt{2}} \varphi_{-2} \left(\frac{1}{1 - i\varphi_{-2}} \mathsf{P}_+ + \frac{1}{1 + i\varphi_{-2}} \mathsf{P}_- + \tfrac{1}{3} \mathsf{P}_h\right) \cdot \boldsymbol{a}^{(2)} \ .$$

Inserting this and the $b^{(1)}$-value, taken from (4.9a), into the second eq. (4.10a) yields

$$\sum \omega_{-1}^{(2k)} \boldsymbol{a}^{(k)} = -\tfrac{1}{2} \omega_H \left[\frac{1}{\varphi_{-2} + i} \mathsf{P}_+ + \frac{1}{\varphi_{-2} - i} \mathsf{P}_- + \tfrac{4}{3} (\varphi_{-0} + \varphi_{-2}) \mathsf{P}_h\right] \cdot \boldsymbol{a}^{(2)} \ . \tag{4.18}$$

The first eq. (4.10a) simply is

$$\sum \omega_{-1}^{(1k)} \boldsymbol{a}^{(k)} = -\frac{1}{\sqrt{3}} c_0 \nabla a^{(1)} \ . \tag{4.19}$$

These are two linear equations for the vectors $\boldsymbol{a}^{(1)}$, essentially the particle diffusion flux, and $\boldsymbol{a}^{(2)}$, the "azimuthal polarization".

Introducing the free path matrix l_{-1} connected with the relaxation matrix ω_{-1} by inversion

$$l_{-1} = c_0 (\omega_{-1})^{-1}$$

and the precession angle

$$\varphi_{-1} = l_{-1}^{(22)} \omega_H / 2c_o , \tag{4.20}$$

one obtains from (4.18) and (4.19) for the azimuthal polarization

$$\boldsymbol{a}^{(2)} = -\left[\frac{\varphi_{-2}+i}{\varphi_{-1}+\varphi_{-2}+i} \mathsf{P}_+ + \frac{\varphi_{-2}-i}{\varphi_{-1}+\varphi_{-2}-i} \mathsf{P}_- + \frac{1}{1+\frac{4}{3}\varphi_{-1}(\varphi_{-o}+\varphi_{-2})} \mathsf{P}_h\right]$$

$$\cdot \frac{1}{\sqrt{3}} l_{-1}^{(21)} \nabla a^{(1)} . \tag{4.21}$$

The diffusion flux then can be taken from (4.19). Replacing the projectors according to their original meaning (4.16) leads to the diffusion tensor. It depends on the magnetic field through the precession angles φ.

The results simplify if the scattering amplitude contained in the collision operator ω of (4.3) depends only weakly on the spin (small "non-sphericity"). Then one has

$$\omega_{-o}^{(11)} \approx \omega_{-1}^{(22)} \approx \omega_{-2}^{(11)} \approx \omega_{\text{free}} , \tag{4.22}$$

hence after (4.14), (4.20) and (4.17)

$$\varphi_{-o} \approx 2\varphi_{-1} \approx 2\varphi_{-2} \approx \varphi , \qquad \varphi \approx \omega_H/\omega_{\text{free}} . \tag{4.23}$$

Furthermore, let us replace in (4.21) $a^{(1)}$ and $\boldsymbol{a}^{(2)}$ by their corresponding mean values

$$n = n_o(1 + a^{(1)}) , \qquad \langle \boldsymbol{c} \times \boldsymbol{s} \rangle = \tfrac{1}{3} c_o \sqrt{2S(S+1)}\, \boldsymbol{a}^{(2)} ,$$

where S is the spin value ($\frac{1}{2}$ for the electron). The projectors too are replaced by their original meaning. Eq. (4.21) then is

$$n_o \langle \boldsymbol{c} \times \boldsymbol{s} \rangle = D_{\text{iso}} \frac{\omega_{-1}^{(21)}}{\omega_{-1}^{(22)}} \sqrt{\tfrac{2}{3} S(S+1)} \left(\frac{1+\frac{1}{2}\varphi^2}{1+\varphi^2} - \frac{\frac{1}{2}\varphi}{1+\varphi^2} \boldsymbol{h} \times - \frac{\frac{1}{2}\varphi^2}{1+\varphi^2} \boldsymbol{h}\boldsymbol{h} \cdot \right) \nabla n , \tag{4.24}$$

with a theoretical diffusion coefficient (cf. 4.26)

$$D_{\text{iso}} = \tfrac{1}{3} c_o^2 / \omega_{-1}^{(11)} .$$

The particle flux

$$j = n_0 \langle c \rangle = \frac{1}{\sqrt{3}} n_0 c_0 \, a^{(1)}$$

is then inferred from (4.19). The result is

$$j = -D \left[1 - \tfrac{1}{2} A_{-1}^{(12)} \left(\frac{\varphi^2}{1+\varphi^2} + \frac{\varphi}{1+\varphi^2} \, h \times + \frac{\varphi^2}{1+\varphi^2} \, h h \cdot \right) \right] \nabla n \, , \qquad (4.25)$$

with

$$A_{-1}^{(12)} = \omega_{-1}^{(12)} \omega_{-1}^{(21)} / \omega_{-1}^{(11)} \omega_{-1}^{(22)} < 0 \, ; \qquad |A| \ll 1 \, .$$

The number A is negative because of the Onsager-Casimir relation

$$\omega_{-1}^{(21)} = - \omega_{-1}^{(12)} \, ,$$

due to the fact that the velocity $c \propto \Phi^{(1)}$ is odd, whereas $c \times s \propto \Phi^{(2)}$ is even under time reversal. The scalar diffusion coefficient is given by

$$D = D_{\text{iso}} \, (1 + A_{-1}^{(12)}) = \tfrac{1}{3} c_0 l_{-1}^{(11)} \, . \qquad (4.26)$$

These results shall be discussed now.

Without magnetic field, $\varphi = 0$, the diffusion coefficient is a scalar, the particle flux has the direction of the concentration gradient:

$$D_{H=0} = D \, . \qquad (4.27)$$

The experimental D is different from the theoretical quantity D_{iso}. One would obtain the latter as the diffusion coefficient by using a spin-independent (-isotropic) distribution function from the beginning, retaining only first power quantities, no $c \times s$, in the expansion (4.5). The difference is due to the spin-dependent part of the scattering amplitude which shows up in a non-vanishing $\omega_{-1}^{(12)}$ or $A_{-1}^{(12)}$ and in the velocity-spin correlation described by (4.24) for $\varphi = 0$. This is not directly detectable. But with a magnetic field, the velocity-spin correlation is changed after (4.24) according to the precession angle φ. At the same time the diffusion coefficient becomes a field-dependent tensor. Eq. (4.25) shows that it has an isotropic, but field-dependent part - first term in the parenthesis -, an antisymmetric part - second term - and a symmetric part - third term. Hence the field-dependence offers the possibility to determine $A_{-1}^{(12)}$. With the field parallel to the concentration gradient and high enough ($|\varphi| \gg 1$; saturation), one has after (4.25) and (4.26)

$$D_{H=\infty,\text{parallel}} = D(1 - A_{-1}^{(12)}) = D_{\text{iso}} > D . \tag{4.28}$$

Astonishingly, no (Senftleben) effect with diffusion has been found in experiment up to now, despite of an effort by Senftleben and Schult [13a] - which however is inconclusive - and by Vugts, Tip and Los (to be published). So, the above considerations, although showing the essential theoretical points with the simplest mathematical apparatus, rest somewhat academic for the time being.

The generalisation to transport of heat and momentum in pure or mixture gases is straightforward. It is too cumbersome to be displayed here in detail. The principles remain the same. The results of this generalisation cover a large amount of experimental material - Senftleben-Beenakker effect -, at least phenomenologically. The proper Senftleben effect has been observed with heat conductivity and viscosity of pure paramagnetic gases, O_2, NO (Senftleben [14]; Engelhardt and Sack [15]). It has been shown experimentally that these transport coefficients indeed become tensors in a magnetic field, the decisive parameters being a few precession angles $\varphi \propto H/p$, as met before. With high enough field a further increase no more matters (saturation). In all observed cases, including the diamagnetic gases to be mentioned immediately, the saturation values of the transport coefficients were smaller, by 1% or so, than the zero-field values. The realm of the phenomenon has been widened considerably by Beenakker and co-workers (Beenakker, Scoles, Knaap and Jonkman [16]). Their basic idea was that with diamagnetic gases, N_2, H_2, CH_4 etc., a saturation change in viscosity or heat conductivity will occur of comparable size as with O_2 because of comparable non-sphericity. The only difference will be that higher H/p-values have to be applied to obtain saturation, due to the smallness of the diamagnetic moment linked with the rotation of the molecules. (The requirement for saturation generally is $\varphi \approx \mu H/J\omega_{\text{free}} \gg 1$, where μ stands for the magnetic moment, J for a molecular angular momentum, $\omega_{\text{free}} \propto p$ for some collisional relaxation constant.) So, the Senftleben-Beenakker effect is common to all kinds of polyatomic gases and gives valuable information about the non-spherical part of binary molecular interaction (Korving, Thesis [17]).

Theoretically the Senftleben effect was first treated by Zernike and Van Lier [18] after an idea of Gorter [19] by a mean free path method. A kinetic theory starting from a Boltzmann equation was first developed by Kagan and Maksimov [20]. In their paper a somewhat artificial classical model was used, with a dependence of the collisional cross section on the orientation of molecular angular momenta but without change of this orientation during collision. A generalized Chapman-Enskog method was applied. Next, the quantum case was treated by Waldmann and Kupatt [12], for the example of the diffusion of particles with spin in a magnetic field, essentially the theory described above. Thermal conductivity of a gas with rotating molecules was treated in two comprehensive papers by McCourt and Snider [9,10], without and with magnetic field. A classical theory of the magnetic field influence on heat conductivity and viscosity of diamagnetic gases especially was given by Knaap and Beenakker [21].

The description of heat conduction at least requires vectors of third power in $\boldsymbol{c}$, $\boldsymbol{s}$ in an expansion like (4.7). So, the list (4.8)-(4.13) has to be enlarged. The most important new quantities are the vector

$$\boldsymbol{\Phi}^{(3)} \propto \left(\frac{mc^2}{2kT_0} - \tfrac{5}{2}\right)\boldsymbol{c}\,,$$

whose mean value is the heat flux, and the "Kagan vector"

$$\boldsymbol{\Phi}^{(4)} \propto \boldsymbol{c}\cdot\overline{\boldsymbol{s}\,\boldsymbol{s}}\,,$$

which implies a velocity-spin correlation of a type different from the "azimuthal polarization" discussed before. The latter plays no role with heat conduction because $\omega^{(32)}_{-1}$ vanishes for collisions between identical spin particles whatever the spin-dependence of the scattering amplitude may be. This general observation is due to Hess [22]. The result for the heat flux in terms of the temperature gradient then is

$$\boldsymbol{q} = -\lambda\left\{1 - \tfrac{3}{10}A^{(34)}_{-1}\left[\frac{\varphi^2}{1+\varphi^2} + \frac{8\varphi^2}{1+4\varphi^2} + \left(\frac{\varphi}{1+\varphi^2} + \frac{4\varphi}{1+4\varphi^2}\right)\boldsymbol{h}\,\times\right.\right.$$
$$\left.\left. + \left(\frac{\varphi^2}{1+\varphi^2} - \frac{8\varphi^2}{1+4\varphi^2}\right)\boldsymbol{h}\boldsymbol{h}\cdot\right]\right\}\nabla T\,, \qquad (4.29)$$

with

$$\varphi = \omega_H/\omega_{-1}^{(44)}\,, \qquad A_{-1}^{(34)} = \omega_{-1}^{(34)}\omega_{-1}^{(43)}/\omega_{-1}^{(33)}\omega_{-1}^{(44)} > 0\,, \quad |A| \ll 1\,.$$

Again, small non-sphericity was assumed. The structure of this formula is similar to eq. (4.25), with the important difference that the number A now is positive because of the Onsager relation

$$\omega_{-1}^{(34)} = \omega_{-1}^{(43)}\,.$$

This is due to the fact that the vectors $\boldsymbol{\Phi}^{(3)}$ and $\boldsymbol{\Phi}^{(4)}$ are both even under time reversal. This sign of A is necessary to explain the experiments which exhibit a decrease in heat conductivity with increasing field. Again one has, in analogy with (4.26),

$$\lambda = \lambda_{\text{iso}}\left(1 + A_{-1}^{(34)}\right), \qquad \lambda_{\text{iso}} = \tfrac{5}{2}\,kp_0/m\omega_{-1}^{(33)}\,. \tag{4.30}$$

The analogon of eq. (4.28) is

$$\lambda_{H=\infty,\,\text{parallel}} = \lambda\left(1 - \tfrac{3}{5}A_{-1}^{(34)}\right) < \lambda\,. \tag{4.31}$$

So, the heat conductivity measurements with field give the number $A_{-1}^{(34)}$ which again is a measure for the non-sphericity of particles. Formula (4.29) has been derived by Knaap and Beenakker [21] for the classical Kagan model and it is contained in the quantum theoretical treatment by McCourt and Snider [10]. In the paper by Kagan and Maksimov [20], which opened the new theoretical work on the Senftleben effect, the antisymmetric part of the heat conductivity tensor (see below) was missing. The above form is especially for a pure gas of particles with spin which were treated by Hess [22].

Of course, introducing the Kagan vector $\boldsymbol{\Phi}^{(4)}$ into diffusion theory would give additional terms in (4.25) with a coefficient $A_{-1}^{(14)} > 0$. Due to such terms the diffusion coefficient should decrease with increasing field. Maybe that both contributions $A_{-1}^{(12)}$ and $A_{-1}^{(14)}$ happen to cancel within the accuracy of the available experiments which are conclusive at all.

The tensors occurring in (4.25) and (4.29) also contain an antisymmetric part, giving transverse fluxes. The transverse diffusion flux first appeared, in kinetic theory, in the paper by Waldmann and Kupatt [12]. The transverse heat flux - Righi-Leduc effect with polyatomic gases - and its counterpart in the pressure tensor was found experimentally by Korving, Hulsman, Knaap and Beenakker [23]. It offers the possibility to determine the sign of φ and herewith of the magnetic moment.

In conclusion the magneto-thermal momentum transfer discovered experimentally by Scott, Sturner and Williamson [24] shall be mentioned. These authors observed that polyatomic gases, O_2, N_2, CH_4 etc., exert a torque on a heated cylinder if a homogeneous magnetic field parallel to the axis is applied. With nitrogen at 0.10 torr the maximal torque, of about $1{,}6 \cdot 10^{-2}$ erg, was obtained for 140 Oe. The temperature difference between cylinder (1 cm radius, 20 cm length) and outer wall (coaxial glass cylinder of 3,8 cm radius) was 30°K.

The effect is largest in the transition region. It disappears in the Knudsen ($p \to 0$) and in the hydrodynamical ($p \to \infty$) regimes. In the almost-hydrodynamical regime (mean free path small, but not negligible on length scale of apparatus) two contributions offer themselves for explanation. 1) At the surface a slip velocity might exist proportional to the tangential component of some of the polar vectors, $\boldsymbol{a}^{(2)}$ or moreover $\boldsymbol{a}^{(4)}$, which are linked with the temperature gradient. Such a tangential component may indeed exist for the above cylinder arrangement, if a magnetic field is present. In consequence of this magneto-thermal slip at the surfaces, the gas should move circularly. This in fact gives a torque with the right pressure and field dependence. But a numerical slip coefficient is left open (Waldmann [25]). 2) The friction pressure tensor, hydrodynamically proportional to the velocity gradient, in the next approximation contains contributions from the second order space derivatives of temperature (thermal stresses). The pertaining fourth rank "thermal viscosity" tensor is field dependent, as the ordinary viscosity tensor is. It leads to a torque of the same type as the magneto thermal slip assumption did. But now, no unknown coefficients enter. It is only necessary to know the coefficients governing the Senftleben effect. The calculated torque has the right sign and order of magnitude (Levi and Beenakker [26]). - How large the contribution of a magneto-thermal slip possibly is, could be tested from a pressure experiment in which the temperature is linear in space coordinates so that no thermal stresses exist.

REFERENCES

[1] C. S. Wang Chang and G. E. Uhlenbeck, Eng. Res. Inst. Univ. Michigan Report CM 681 (1951).

[2] L. Waldmann, *Transporterscheinungen in Gasen von mittlerem Druck*, *Handbuch der Physik*, ed. S. Flügge, vol. 12, Springer, Berlin (1958).

[3] L. Waldmann in *Statistical Mechanics of Equilibrium and Non-Equilibrium*, ed. J. Meixner, p. 177, North-Holland Publ. Co., Amsterdam (1965).

[4] S. Hess, Z. Naturforsch. 22a (1967) 1871.

[5] C.S.Wang Chang, G.E.Uhlenbeck and J. De Boer in *Studies in Statistical Mechanics*, ed. J. De Boer and G.E.Uhlenbeck, vol. II, North-Holland Publ. Co., Amsterdam (1964).
[6] R.F.Snider, J. Chem. Phys. 32 (1960) 1051.
[7] L.Waldmann, Z. Naturforsch. 12a (1957) 660.
[8] L.Waldmann, Z. Naturforsch. 13a (1958) 609.
[9] F.R.McCourt and R.F.Snider, J. Chem. Phys. 41 (1964) 3185.
[10] F.R.McCourt and R.F.Snider, J. Chem. Phys. 46 (1967) 2387.
[11] L.Waldmann, Z. Naturforsch. 15a (1960) 19.
[12] L.Waldmann and H.D.Kupatt, Z. Naturforsch. 18a (1963) 86.
[13] S.Hess and L.Waldmann, Z. Naturforsch. 21a (1966) 1529.
[13a] H.Senftleben and H.Schult, Ann. d. Physik 7 (1950) 103.
[14] H.Senftleben, Z. Physik 31 (1930) 822.
[15] H.Engelhardt and H.Sack, Physikal. Z. 33 (1932) 724.
[16] J.J.M.Beenakker, G.Scoles, H.F.P.Knaap and R.M.Jonkman, Phys. Letters 2 (1962) 5.
[17] J.Korving, *Thesis*, University of Leiden (1967).
[18] F.Zernike and C. Van Lier, Physica 6 (1939) 961.
[19] C.J.Gorter, Naturwissenschaften 26 (1938) 140.
[20] Y.Kagan and L.Maksimov, Soviet Phys. JETP 14 (1962) 604.
[21] H.F.P.Knaap and J.J.M.Beenakker, Physica 33 (1967) 643.
[22] S.Hess, *Diplomarbeit*, Institut für Theoretische Physik der Universität Erlangen-Nürnberg (1964).
[23] J.Korving, H.Hulsman, H.F.P.Knaap and J.J.M.Beenakker, Phys. Letters 21 (1966) 5.
[24] G.G.Scott, H.W.Sturner and R.M.Williamson, Phys. Rev. 158 (1967) 117.
[25] L.Waldmann, Z. Naturforsch. 22a (1967) 1678.
[26] A.C.Levi and J.J.M.Beenakker, Phys. Letters 25A (1967) 350.

STATISTICAL MECHANICS OF CLASSICAL PLASMAS

N. G. VAN KAMPEN
Institute for Theoretical Physics, University of Utrecht, The Netherlands

1. INTRODUCTION

1.1. *Characterization of a plasma*

A plasma is a fluid in which a sufficient number of charged particles (ions and free electrons) are present to affect the macroscopic properties of the fluid materially. Any gas will be a plasma when it is sufficiently hot and/or sufficiently dilute, but often the ionization is maintained by external agencies, like in ordinary gas discharges. The higher the degree of ionization the more pronounced are the plasma properties. Most theories therefore deal with a fully ionized plasma, so that no neutral particles need be taken into account. Accordingly, the word "particle" will hereafter mean "charged particle" unless stated otherwise.

The statistical mechanics of a plasma *in equilibrium* is in reasonably good shape, although a few difficulties remain. The treatment of *motions and non-equilibrium processes* in plasmas turns out to be much harder than for ordinary gases. There are two reasons why the kinetic theory of gases, as developed by the other speakers, does not apply to plasmas.

(i) The interaction between two charged particles decreases typically as $1/r$ with increasing distance r between both particles, and represents therefore a *long range interaction force* *. In the derivation of the Boltzmann equation for dilute gases one makes essential use of the assumption that the duration of a collision is much shorter than the average time of free flight between two collisions. In plasmas, however, each particle always interacts with *all* other particles, owing to the long range of the Coulomb force. To show this, select one particle and consider the force it undergoes from

* The theory of the Van der Waals gas is also based on a long range interaction [1]. However, this long range attractive potential decreases faster than $1/r$, and in fact it must fall off so rapidly that the potential can be integrated over all space, as the integral represents the a of Van der Waals. The essential features of a plasma are due to the fact that this integral does not exist for a Coulomb potential.

all particles lying at a distance between r and $r + \Delta r$ from it in a certain spherical angle $\Delta\Omega$. The number of such particles is $n_0 r^2 \Delta\Omega \Delta r$, where n_0 is the particle density. They exert a Coulomb force $n_0 e^2 \Delta\Omega \Delta r$, when e is the charge per particle. This force is independent of r, which shows that all particles (up to the boundary of the plasma) act on the one selected particle. Hence the derivation of the Boltzmann equation does not apply to a plasma.

It is true that the effective interaction has a shorter range, owing to the "Debye screening", but this is a statistical effect, *which should result from the theory* and cannot be used a priori. Moreover, this Debye radius is still much longer than the average distance between particles, so that it could not be used to justify the derivation of Boltzmann's equation anyway.

The difficulty due to the long range cannot be circumvented by studying very dilute plasmas. The reason is that the interaction contains no parameter of the dimension of length, so that it is impossible to make the mean distance between particles, $d \approx n_0^{-\frac{1}{3}}$, large compared to the range. The only other length parameter in the (classical) plasma is e^2/kT, so that "dilute" can only mean

$$n_0^{-\frac{1}{3}} \gg \frac{e^2}{kT} , \quad \text{or} \quad \frac{e^2}{d} \ll kT . \tag{1}$$

This states that the mean potential energy of any two particles is small compared to their mean kinetic energy. Statistical theory of plasmas is essentially based on this inequality; we shall refer to such plasmas as *hot dilute plasmas*.

(ii) The second reason why a plasma is not covered by the kinetic theory of gases is that the interaction cannot be described by just a potential between any two particles. Rather, the interaction is transmitted by the *electromagnetic field*, which has its own degrees of freedom and its own equations of motion, viz., Maxwell's equations. These additional degrees of freedom give rise to a wealth of phenomena, in particular waves and instabilities, which do not exist in ordinary gases. Only in the case of slow motions is it possible to replace these field equations by an instantaneous interacting force. The kinetic theory of plasmas is mainly developed for this case (section 3).

We shall only study classical plasmas. The condition that quantum effects are negligible is that the mean de Broglie wave length is much less than the mean distance between particles:

$$\frac{h}{\sqrt{mkT}} \ll d , \quad \text{or} \quad kT \gg \frac{h^2}{md^2} . \tag{2}$$

If (1) holds it is sufficient for (2) that $d \geqslant h^2/me^2$, that is, the interparticle distance is not less than the Bohr radius. For most plasmas this is true, but the electrons in metals obey neither (1) nor (2).

1.2. *Survey of theoretical approaches*

Various approaches are used to describe the motions and non-equilibrium processes, in analogy with the various levels of description in gas theory.

I. The hydrodynamical description [2]

The plasma is regarded as a compressible, or sometimes incompressible fluid, with an electric current density $\boldsymbol{J}$, subject to Ohm's law. The current gives rise to a magnetic field $\boldsymbol{B}$, which in turn exerts a force on the current. This is called magnetohydrodynamics (MHD). The MHD equations are the following:

$$\frac{\partial \rho}{\partial t} + \operatorname{div} \rho \boldsymbol{V} = 0 . \tag{3}$$

(Conservation of mass; ρ is mass density, $\boldsymbol{V}$ is flow velocity.)

$$\rho \frac{\mathrm{D}\boldsymbol{V}}{\mathrm{D}t} = - \operatorname{grad} p + \frac{1}{c} \boldsymbol{J} \times \boldsymbol{B} . \tag{4}$$

(Conservation of momentum; p is ordinary pressure, $\boldsymbol{J} \times \boldsymbol{B}/c$ is the Lorentz force.)

$$p = f(\rho) . \tag{5}$$

(A rather simplistic equation of state, which for steady states is usually taken to be the isothermal equation, and for oscillations the adiabatic equation.)

$$\boldsymbol{J} = \sigma (\boldsymbol{E} + \frac{1}{c} \boldsymbol{V} \times \boldsymbol{B}) . \tag{6}$$

(Ohm's law; σ is the conductivity, $\boldsymbol{E} + \boldsymbol{V} \times \boldsymbol{B}/c$ is the electric field experienced by an observer moving with the fluid.)

$$\operatorname{curl} \boldsymbol{B} = \frac{4\pi}{c} \boldsymbol{J} . \tag{7}$$

(Biot and Savart; Maxwell's term $\partial \boldsymbol{D}/\partial ct$ is neglected, because the whole description applies only to slow phenomena; moreover the magnetic permeability is taken unity, which is a good approximation in most cases.)

$$\text{curl}\, \boldsymbol{E} = -\frac{1}{c}\frac{\partial \boldsymbol{B}}{\partial t}\,, \quad \text{div}\, \boldsymbol{B} = 0\,. \tag{8}$$

(Faraday's induction law.)

This MHD description is the counterpart of hydrodynamics in ordinary gas theory. It is based on the following essential assumptions.

(i) *The hydrodynamic assumption* that the collisions of the ions (and possibly neutral molecules) are so frequent that a local equilibrium is established. This makes it possible to describe the fluid at a given temperature completely by ρ and $\boldsymbol{V}$, and to have a scalar pressure p determined by ρ alone. This assumption is reasonable for slow motions in cold, dense plasmas, and incorrect for fast motions in hot, dilute plasmas.

(ii) *The electrical conductivity assumption* that the electrons collide so frequently with the fluid particles that a drift is established proportional to the instantaneous force on the electrons, so that Ohm's law holds.

(iii) *The quasi-neutrality assumption* that the charge density τ is small enough to omit the term $\tau\boldsymbol{E}$ on the right of (4). A dimensional analysis shows that the ratio of this term to $\boldsymbol{J}\times\boldsymbol{B}/c$ is of the order $(\sigma L/c)^2$, where L is a typical length, characterizing the distances over which the macroscopic state of the plasma changes appreciably. Thus quasi-neutrality holds in case of very good conductivity, and in the case of very large plasmas, like the earth's core *. It does not apply, however, to oscillations with short wave lengths (L is small) or to motions in which $\boldsymbol{J}\times\boldsymbol{B}$ happens to vanish.

In addition some less essential assumptions have been incorporated in the equations (3)-(8), such as the absence of viscosity, the neglect of the Maxwell term in (7), and the use of a relation between p and ρ rather than a more elaborate description involving the temperature and an equation for heat transport. These complications may be added when needed **.

II. Two component theory [3]

Plasma waves are defined as waves in which the positive and negative charge carriers in the plasma oscillate with respect to

* For the same reason it was permissible to write $\boldsymbol{J}$ on the left-hand side of (6) rather than the current density $\boldsymbol{J}-\tau\boldsymbol{V}$ measured by the co-moving observer.

** For example, in experiments with liquid metal flow in an external magnetic field the viscosity term has to be added.

each other. Such waves cannot be described by MHD, since Ohm's law uniquely relates the current to the field *. The reason why plasma waves can exist is that for rapid motions this is no longer true; that is, assumption (ii) above breaks down. Therefore, in order to describe plasma waves the plasma is regarded as consisting of two interpenetrating components, the ion gas and the electron gas. Each obeys its own magnetohydrodynamic equations, but of course they are imbedded in the same electromagnetic field. In addition there may be some interaction through collisions, which is taken into account by inserting a term describing friction between both fluids, i.e. electrical resistivity.

The result is that Ohm's law is replaced with the *generalized Ohm's law*

$$\frac{\partial \boldsymbol{J}}{\partial t} = \frac{e}{m_e} \nabla p_e + \frac{\rho Z e^2}{m_e m_i} \left(\boldsymbol{E} + \frac{1}{c} \boldsymbol{V} \times \boldsymbol{B} - \frac{1}{\sigma} \boldsymbol{J} \right) - \frac{e}{m_e c} \boldsymbol{J} \times \boldsymbol{B} \, .$$

Here m_i and m_e are the masses of the ions and the electrons, Ze and $-e$ their charges, and p_e the pressure of the electron gas. The main innovation is the new term $\partial \boldsymbol{J}/\partial t$, which exhibits the fact that, in contrast to the MHD description, the electron gas is now treated as a separate physical system with its own degrees of freedom.

The two-component theory is based on the essential assumption that the particles of each component collide so frequently *among themselves* that each component may be described like a fluid by hydrodynamical equations. On the other hand, collisions involving particles of *different* components must be so rare that each component can move largely unhampered by the other, apart from the interaction through the electromagnetic field. The physical reality of the picture seems to me rather questionable, although it is true that two-component theory often arrives at qualitatively correct results.

III. The collisionless or Vlasov plasma [4]

This approach begins by neglecting collisions altogether and treats the plasma particles as moving freely in the over-all electromagnetic field created by themselves, and possibly external agencies. This is a more consistent model than the two-component theory, and for rapid oscillations in a hot dilute plasma also more realistic. The effect of collisions is later added as a correction

* The waves found in MHD are the ordinary sound waves and the transverse Alfvén waves, and mixtures between both. Conventionally they are not included in the term "plasma wave".

term. This term is indispensable for the description of irreversible effects like electrical and thermal resistivity.

Owing to the absence or rarity of the collisions no local equilibrium is established and it is not possible to use the hydrodynamic description in terms of local density, velocity and pressure. A more detailed description is required in terms of the distribution functions $f_\nu(\boldsymbol{r}, \boldsymbol{v}, t)$ for the particles of the several components ν. They obey equations of Boltzmann type

$$\frac{\partial f_\nu}{\partial t} + \boldsymbol{v} \cdot \frac{\partial f_\nu}{\partial \boldsymbol{r}} + \frac{e_\nu}{m_\nu} \left\{ \boldsymbol{E} + \frac{1}{c} \boldsymbol{v} \times \boldsymbol{B} \right\} \frac{\partial f_\nu}{\partial \boldsymbol{v}} = \left[\frac{\partial f_\nu}{\partial t} \right]_{\text{collisions}} . \tag{9}$$

The electromagnetic field $\boldsymbol{E}, \boldsymbol{B}$ is determined by

$$\operatorname{div} \boldsymbol{E} = 4\pi \sum_\nu e_\nu \int f_\nu (\boldsymbol{r}, \boldsymbol{v}, t) \, \mathrm{d}^3 \boldsymbol{v} \, ,$$

$$\operatorname{curl} \boldsymbol{B} - \frac{1}{c} \frac{\partial \boldsymbol{E}}{\partial t} = \frac{4\pi}{c} \sum_\nu e_\nu \int \boldsymbol{v} \, f_\nu (\boldsymbol{r}, \boldsymbol{v}, t) \, \mathrm{d}^3 \boldsymbol{v} \, ,$$

$$\operatorname{curl} \boldsymbol{E} = - \frac{1}{c} \frac{\partial \boldsymbol{B}}{\partial t} , \qquad \operatorname{div} \boldsymbol{B} = 0 \, .$$

In the Vlasov approach one replaces the right-hand side of (9) with zero. The remaining set of equations is still non-linear, and only a few exact solutions are known [5]. However, if one starts from a known steady state solution, and studies deviations from it which are small enough to be treated in linear approximation, longitudinal [6] and transverse [7] plasma waves can be derived.

In addition an approximation for the collision term on the right of (9) has been found by Landau [8] by substituting the Coulomb interaction in Boltzmann's collision term. Of course this is not justified since the derivation of Boltzmann's collision term is based on the assumption that the interaction has a short range. Alternative derivations, however, have led to the same result [9,10,11], see section 3.

IV. Kinetic or statistical mechanical approach

Here one starts from the equations of motion for all particles, and replaces the actual physical plasma by an ensemble. The ensemble obeys the Liouville equation, which may be written in the form of the BBGKY hierarchy. In principle this contains all previous more intuitive approaches, and therefore a means to justify them and compute higher corrections to them. In practice, however, this is a very arduous road and only limited results have been obtained (section 3). Yet the equilibrium case can be treated satisfactorily on the basis of standard statistical mechanics, as will be shown in the next chapter.

2. EQUILIBRIUM STATE OF A PLASMA

2.1. *The partition function*

The N_ν particles of component ν will be distinguished by the index σ, their positions and momenta are $\boldsymbol{q}_{\nu\sigma}$ and $\boldsymbol{p}_{\nu\sigma}$. The longitudinal part of the electromagnetic field consists of the Coulomb interaction between these particles and possibly a potential φ_{ex} due to external sources. The transverse part will be described by a divergence-free vector potential $\boldsymbol{A}^T$ and the corresponding transverse electric field $\boldsymbol{E}^T = -\dot{\boldsymbol{A}}^T/c$. The Hamilton function describing the total system is

$$\mathcal{H} = \mathcal{H}_{\text{kin}} + \mathcal{H}_{\text{pot}} + \mathcal{H}_{\text{rad}} ,$$

$$\mathcal{H}_{\text{kin}} = \sum_\nu \frac{1}{2m_\nu} \sum_\sigma \{\boldsymbol{p}_{\nu\sigma} - e_\nu \boldsymbol{A}^T(\boldsymbol{q}_{\nu\sigma})\}^2 ,$$

$$\mathcal{H}_{\text{pot}} = \tfrac{1}{2} \sum_{\nu\nu'} e_\nu e_{\nu'} \sum_{\substack{\sigma,\sigma' \\ \sigma\neq\sigma'}} \frac{1}{|\boldsymbol{q}_{\nu\sigma} - \boldsymbol{q}_{\nu\sigma'}|} + \sum_\nu e_\nu \sum_\sigma \varphi_{\text{ex}}(\boldsymbol{q}_{\nu\sigma}) ,$$

$$\mathcal{H}_{\text{rad}} = \frac{1}{8\pi} \int \{(\boldsymbol{E}^T)^2 + \text{curl}\, \boldsymbol{A}^T)^2\} \, \mathrm{d}\boldsymbol{r} .$$

To each point $\boldsymbol{r}$ in space corresponds a canonical variable $\boldsymbol{A}^T(\boldsymbol{r})$ and its conjugate momentum $-\boldsymbol{E}^T(\boldsymbol{r})/4\pi c$.

In order to find the thermodynamic properties of the equilibrium state one has to evaluate the partition function

$$\mathrm{e}^{-\beta F} = \frac{1}{\prod_\nu N_\nu!} \int \mathrm{e}^{-\beta\mathcal{H}} \prod_{\nu\sigma} \mathrm{d}\boldsymbol{p}_{\nu\sigma}\, \mathrm{d}\boldsymbol{q}_{\nu\sigma} \prod_{\boldsymbol{r}} \mathrm{d}\boldsymbol{A}^T(\boldsymbol{r}) \frac{\mathrm{d}\boldsymbol{E}^T(\boldsymbol{r})}{4\pi c} , \tag{10}$$

where $\beta = 1/kT$ and F is the free energy. The coordinates $\boldsymbol{q}_{\nu\sigma}$ run over the volume V to which the plasma is confined. The symbol $\prod_{\boldsymbol{r}} \mathrm{d}\boldsymbol{A}^T(\boldsymbol{r})\, \mathrm{d}\boldsymbol{E}^T(\boldsymbol{r})/4\pi c$ is used to indicate a functional integration in which $\boldsymbol{A}^T(\boldsymbol{r})$ and $\boldsymbol{E}^T(\boldsymbol{r})$ each take all values consistent with the condition that they should be transverse vector fields.

The integration over the $\boldsymbol{p}_{\nu\sigma}$ can be carried out immediately and yields the factor

$$\prod_{\nu}\left(\frac{2\pi m_{\nu}}{\beta}\right)^{\frac{3}{2}N_{\nu}} .$$

The remaining integral decomposes into an integral over the $\boldsymbol{q}_{\nu\sigma}$ and an integral over the field variables, so that one has

$$F = -\tfrac{3}{2}kT\sum_{\nu} N_{\nu}\log\left(2\pi m_{\nu}kT\right) + F_{\text{pot}} + F_{\text{rad}} ,$$

$$\mathrm{e}^{-\beta F_{\text{pot}}} = \frac{1}{\prod_{\nu} N_{\nu}!}\int_{V}\ldots\int_{V} \mathrm{e}^{-\beta \mathcal{H}_{\text{pot}}}\prod_{\nu\sigma}\mathrm{d}\boldsymbol{q}_{\nu\sigma} . \tag{11}$$

F_{rad} is identical with the free energy of the free radiation field and is therefore irrelevant for the properties of the plasma. Thus the problem reduces to evaluating (11).

This cannot be done exactly, so that we have to look for an approximation method. The following method is based on the long range of the interacting force. It will turn out that the result is an expansion in the parameter $e^{3}n^{\frac{1}{2}}T^{-\frac{3}{2}}$, which is small for hot, dilute plasmas according to (1).

2.2. *The method of cells*

We subdivide the volume V in cells of volume Δ, large enough to contain many particles, but so small that φ_{ex} is practically constant in each cell. Let the position of cell λ be $\boldsymbol{R}^{\lambda}$, and let N_{ν}^{λ} be the number of particles of component ν in this cell. Then one has approximately

$$\mathcal{H}_{\text{pot}} = \tfrac{1}{2}\sum_{\nu\nu'\lambda\lambda'}\frac{e_{\nu}N_{\nu}^{\lambda}\, e_{\nu'}N_{\nu'}^{\lambda'}}{\left|\boldsymbol{R}^{\lambda}-\boldsymbol{R}^{\lambda'}\right|} + \sum_{\nu\lambda} e_{\nu}N_{\nu}^{\lambda}\,\varphi_{\text{ex}}(\boldsymbol{R}^{\lambda}) . \tag{12}$$

It is reasonable to replace the denominator for $\lambda = \lambda'$ by the diameter $\Delta^{\frac{1}{3}}$ of the cell; this term, however, will not effect the final result.

The integration in (12) can now be reduced to

$$\mathrm{e}^{-\beta F_{\text{pot}}} = \sum_{\{N_{\nu}^{\lambda}\}}{}' \left\{\prod_{\nu\lambda}\frac{\Delta^{N_{\nu}^{\lambda}}}{N_{\nu}^{\lambda}!}\right\} \mathrm{e}^{-\beta\mathcal{H}_{\text{pot}}\left(\{N_{\nu}^{\lambda}\}\right)} . \tag{13}$$

The summation extends over all non-negative values of the N_ν^λ subject to the conditions

$$\sum_\lambda N_\nu^\lambda = N_\nu .$$

The proper volume of the particles has been neglected for simplicity.

Using Stirling's formula we write (13) in the form

$$\mathrm{e}^{-\beta F_{\mathrm{pot}}} = \sum_{\{N_\nu^\lambda\}}{}' \mathrm{e}^{\Psi(\{N_\nu^\lambda\})} , \tag{14}$$

$$\Psi(\{N_\nu^\lambda\}) = -\frac{\beta}{2} \sum_{\nu\nu'\lambda\lambda'} \frac{e_\nu N_\nu^\lambda e_{\nu'} N_{\nu'}^{\lambda'}}{|R^\lambda - R^{\lambda'}|} - \sum_{\nu\lambda} N_\nu^\lambda \left\{ \beta e_\nu \varphi_{\mathrm{ex}}(R^\lambda) + \log \frac{N_\nu^\lambda}{\Delta} - 1 \right\} . \tag{15}$$

The maximum term in the sum (14) is found from

$$\frac{\partial \Psi}{\partial N_\nu^\lambda} \equiv -\beta e_\nu \left[\sum_{\nu'\lambda'} \frac{e_\nu N_\nu^{\lambda'}}{|\boldsymbol{R}^\lambda - \boldsymbol{R}^{\lambda'}|} + \varphi_{\mathrm{ex}}(\boldsymbol{R}^\lambda) \right] - \log \frac{N_\nu^\lambda}{\Delta} = \alpha_\nu , \tag{16}$$

where the α_ν are Lagrange multipliers. Clearly the quantity [ ] is the total potential $\varphi(\boldsymbol{R}^\lambda)$ at $\boldsymbol{R}^\lambda$, so that the N_ν^λ that maximize Ψ are given by

$$N_\nu^\lambda/\Delta = \mathrm{e}^{-\alpha_\nu} \, \mathrm{e}^{-\beta e_\nu \varphi(\boldsymbol{R}^\lambda)} .$$

This simply states that the most probable state obeys Boltzmann's distribution law. The left side of this equation is the density $n_\nu(\boldsymbol{R}^\lambda)$ of the ν-th component at the point R^λ. The potential φ obeys evidently

$$\nabla^2 \varphi = -4\pi \sum_\nu e_\nu n_\nu + \nabla^2 \varphi_{\mathrm{ex}} .$$

Now take $\varphi_{\mathrm{ex}} = 0$. Then it is readily verified that (16) is obeyed by

$$N_\nu^\lambda = N_\nu \Delta / V , \qquad \varphi = \mathrm{const} , \qquad \mathrm{e}^{-\alpha_\nu} = N_\nu / V .$$

Substitution in (15) yields

$$\Psi_{\max} = -\frac{\beta}{2}\varphi \sum_{\nu} e_\nu N_\nu - \sum_{\nu} N_\nu \left(\log \frac{N_\nu}{V} - 1\right) .$$

Supposing that the total plasma is neutral, one has $\Sigma\, e_\nu N_\nu = 0$ and therefore

$$\Psi_{\max} = -\sum_{\nu} N_\nu \left(\log \frac{N_\nu}{V} - 1\right) = -\sum_{\nu} \int_V n_\nu(r)\, \{\log n_\nu(\boldsymbol{r}) - 1\}\, \mathrm{d}\boldsymbol{r} .$$

If one substitutes this in (14) as a first approximation, *the resulting free energy is identical to the free energy of a mixture of ideal gases,* and is not affected by the presence of the Coulomb interaction.

2.3. *Next approximation and density fluctuations*

To find the next approximation, Ψ has to be expanded around its maximum to second order in $\delta N_\nu^\lambda = N_\nu^\lambda - (N_\nu^\lambda)_{\max}$. One finds (again for $\varphi_{\mathrm{ex}} = 0$)

$$\Psi - \Psi_{\max} = -\tfrac{1}{2} \sum_{\nu\lambda} \frac{(\delta N_\nu^\lambda)^2}{(N_\nu^\lambda)_{\max}} - \frac{\beta}{2} \sum_{\nu\nu'\lambda\lambda'} \frac{e_\nu\, \delta N_\nu^\lambda e_{\nu'} \delta N_{\nu'}^{\lambda'}}{|\boldsymbol{R}^\lambda - \boldsymbol{R}^{\lambda'}|} .$$

This is a negative definite quadratic form in the δN_ν^λ, which we shall abbreviate to

$$\Psi - \Psi_{\max} = -\tfrac{1}{2} \sum_{\nu\nu'\lambda\lambda'} Q_{\nu\nu'}^{\lambda\lambda'}\, \delta N_\nu^\lambda\, \delta N_{\nu'}^{\lambda'} . \tag{17}$$

Taking these second order terms into account one finds for the free energy (14)

$$\mathrm{e}^{-\beta F_{\mathrm{pot}}} = \mathrm{e}^{\Psi_{\max}} \int_{-\infty}^{\infty} \ldots \int \mathrm{e}^{-\frac{1}{2}\Sigma Q_{\nu\nu'}^{\lambda\lambda'} \delta N_\nu^\lambda \delta N_{\nu'}^{\lambda'}} \prod_{\nu\lambda} \delta N_\nu^\lambda$$

$$= \mathrm{e}^{\Psi_{\max}} \text{ const. } D^{-\frac{1}{2}} , \tag{18}$$

where D is the determinant of Q.

Incidentally, since in this approximation the probability distribution of the fluctuations δN_ν^λ is evidently Gaussian, one has, according to a general theorem,

$$\langle \delta N_\nu^\lambda \, \delta N_{\nu'}^{\lambda'} \rangle = (Q^{-1})_{\nu\nu'}^{\lambda\lambda'} \,. \tag{19}$$

Hence we shall find without additional labor the correlation function of the density fluctuations.

It is possible to transform Q to principal axes by a Fourier transform in space [12], and hence to find both D and Q^{-1}. From a mathematical point of view this is a rather circuitous method, since the determination of eigenvalues requires the solution of an algebraic equation, whereas D and Q^{-1} are rational functions of the elements of Q. We shall instead use an alternative method, which only involves rational operations.

To avoid irrelevant complications we now assume that there are only two components, viz., electrons and one kind of positive ions. Moreover, we assume that the positive ions are fixed in space and simply constitute a smeared out, homogeneous, positive background, which only serves to keep the plasma neutral. What is left is the electron gas, so that $e_\nu = -e$, and the subscript ν may be dropped everywhere. The matrix Q then reduces to

$$Q^{\lambda\lambda'} = \frac{\delta_{\lambda\lambda'}}{n_0 \Delta} + \frac{\beta e^2}{|\boldsymbol{R}^\lambda - \boldsymbol{R}^{\lambda'}|} \,,$$

where $n_0 = N/V$ is the equilibrium density of the electrons.

2.4. *Computation of free energy and density correlations*

We first compute Q^{-1}, which amounts to solving the vector X in terms of Y from

$$Y^\lambda = \sum_{\lambda'} Q^{\lambda\lambda'} X^{\lambda'} \equiv \frac{X^\lambda}{n_0 \Delta} + \beta e^2 \sum_{\lambda'} \frac{X^{\lambda'}}{|\boldsymbol{R}^\lambda - \boldsymbol{R}^{\lambda'}|} \,.$$

It is convenient to replace $\boldsymbol{R}^\lambda$, $\boldsymbol{R}^{\lambda'}$ by continuous variables $\boldsymbol{r}$, $\boldsymbol{r}'$. Accordingly we define $X(\boldsymbol{r})$, $Y(\boldsymbol{r})$ by setting $X^\lambda = \Delta \cdot X(\boldsymbol{R}^\lambda)$ and $Y^\lambda = \Delta \cdot Y(\boldsymbol{R}^\lambda)$ and obtain

$$\frac{X(\boldsymbol{r})}{n_0} + \beta e^2 \int \frac{X(\boldsymbol{r}')}{|\boldsymbol{r} - \boldsymbol{r}'|} \, \mathrm{d}\boldsymbol{r}' = \Delta \cdot Y(\boldsymbol{r}) \,.$$

Applying ∇^2 to each term

$$\nabla^2 X(\boldsymbol{r}) - 4\pi n_0 \beta e^2 X(\boldsymbol{r}) = n_0 \Delta \cdot \nabla^2 Y(\boldsymbol{r}) \,.$$

With the abbreviation $4\pi n_0 \beta e^2 = \kappa^2$ the solution is

$$X(\boldsymbol{r}) = -\frac{n_0\Delta}{4\pi}\int \frac{e^{-\kappa|\boldsymbol{r}-\boldsymbol{r}'|}}{|\boldsymbol{r}-\boldsymbol{r}'|}\nabla'^2\, Y(\boldsymbol{r}')\,d\boldsymbol{r}'$$

$$= n_0\Delta\, Y(\boldsymbol{r}) - n_0\Delta\frac{\kappa^2}{4\pi}\int \frac{e^{-\kappa|\boldsymbol{r}-\boldsymbol{r}'|}}{|\boldsymbol{r}-\boldsymbol{r}'|}\, Y(\boldsymbol{r}')\,d\boldsymbol{r}' .$$

Thus, returning to discrete variables,

$$(Q^{-1})^{\lambda\lambda'} = n_0\Delta\delta_{\lambda\lambda'} - \frac{n_0\kappa^2}{4\pi}\Delta^2\frac{e^{-\kappa|\boldsymbol{R}^{\lambda}-\boldsymbol{R}^{\lambda'}|}}{|\boldsymbol{R}^{\lambda}-\boldsymbol{R}^{\lambda'}|} .$$

According to (19) this represents the correlation function $\langle\delta N^{\lambda}\delta N^{\lambda'}\rangle$. Again it may be more conveniently written in continuous variables; δN^{λ} is then replaced with the density fluctuation $\delta n(\boldsymbol{R}^{\lambda}) = \delta N^{\lambda}/\Delta$,

$$\langle\delta n(\boldsymbol{r})\,\delta n(\boldsymbol{r}')\rangle = n_0\delta(\boldsymbol{r}-\boldsymbol{r}') - \frac{n_0\kappa^2}{4\pi}\frac{e^{-\kappa|\boldsymbol{r}-\boldsymbol{r}'|}}{|\boldsymbol{r}-\boldsymbol{r}'|} . \tag{20}$$

This result describes the existence of the *Debye sphere*: any local density increase at $\boldsymbol{r}$ causes a decrease of the average density at $\boldsymbol{r}' \neq \boldsymbol{r}$, which becomes negligible at a distance κ^{-1}, the *Debye length*.

In order to compute $D(\beta)$ we put

$$Q^{\lambda\lambda'}(\beta) = (n_0\Delta)^{-1}\,\delta_{\lambda\lambda'} + \beta P^{\lambda\lambda'}$$

and note that

$$\log D(0) = -(V/\Delta)\log(n_0\Delta) ,$$

$$\frac{d}{d\beta}\log D(\beta) = \sum_{\lambda\lambda'} P^{\lambda\lambda'}\frac{\partial}{\partial Q^{\lambda\lambda'}}\log D(\beta) .$$

On the other hand, for any matrix $A_{\lambda\mu}$ one has

$$(A^{-1})_{\lambda\mu} = \frac{\partial}{\partial A^{\mu\lambda}}\{\log \mathrm{Det}\, A\} .$$

With the use of this identity we find

$$\frac{\mathrm{d}}{\mathrm{d}\beta}\log D(\beta) = \sum_{\lambda\lambda'} P^{\lambda\lambda'}(Q^{-1})^{\lambda'\lambda} = \mathrm{Tr}\, PQ^{-1} \tag{21}$$

$$= e^2 n_0 \Delta \sum_{\lambda\lambda'} \frac{1}{|\boldsymbol{R}^\lambda - \boldsymbol{R}^{\lambda'}|}\left\{\delta_{\lambda\lambda'} - \frac{\kappa^2}{4\pi}\frac{\mathrm{e}^{-\kappa|\boldsymbol{R}^{\lambda'}-\boldsymbol{R}^\lambda|}}{|\boldsymbol{R}^{\lambda'}-\boldsymbol{R}^\lambda|}\right\} \tag{22}$$

$$= e^2 n_0 V\left\{\Delta^{-\frac{1}{3}} - \frac{\kappa^2}{4\pi}\int \frac{\mathrm{e}^{-\kappa r}}{r^2}\,\mathrm{d}\boldsymbol{r}\right\}$$

$$= e^2 n_0 V\{\Delta^{-\frac{1}{3}} - \kappa\}\ .$$

Integrating β from 0 to β,

$$\log D(\beta) = -\,(V/\Delta)\log(n_0\Delta) + e^2 n_0 V\,\Delta^{-\frac{1}{3}}\beta - \tfrac{4}{3}V\pi^{\frac{1}{2}}e^3 n_0^{\frac{3}{2}}\beta^{\frac{3}{2}}\ .$$

Substitution in (18) finally yields

$$F_{\mathrm{pot}} = -kT\Psi_{\mathrm{max}} - (kTV/2\Delta)\log(n_0\Delta) - \tfrac{1}{2}Ne^2\Delta^{-\frac{1}{3}}$$

$$-\tfrac{2}{3}\pi^{\frac{1}{2}}Ve^3 n_0^{\frac{3}{2}}(kT)^{-\frac{1}{2}}\ .$$

The first term is the ideal gas contribution found in the previous approximation. The second term is a correction to the ideal gas free energy. The third term represents the electrostatic self-energy of the electrons, but may be omitted since it is a constant. Hence the free energy due to the electron gas is

$$F_{\mathrm{el}} = -\tfrac{2}{3}\pi^{\frac{1}{2}}\,Ve^3 n_0^{\frac{3}{2}}(kT)^{-\frac{1}{2}}\ . \tag{23}$$

2.5. *Field fluctuations*

The density fluctuations (20) give rise to fluctuations in the Coulomb potential, whose correlation function can be readily calculated to be [13] *:

$$\langle\delta\varphi(\boldsymbol{r})\,\delta\varphi(\boldsymbol{r}')\rangle = kT\,\frac{1-\mathrm{e}^{-\kappa|\boldsymbol{r}-\boldsymbol{r}'|}}{|\boldsymbol{r}-\boldsymbol{r}'|}\ .$$

* The result is easily verified by applying to it the differential operators ∇^2 and ∇'^2 and comparing the outcome with (20).

The correlations between the components of the longitudinal field follow immediately,

$$\langle E_i^L(\boldsymbol{r})\, E_j^L(\boldsymbol{r}')\rangle = kT\nabla_i\nabla_j' \frac{1-\mathrm{e}^{-\kappa|\boldsymbol{r}-\boldsymbol{r}'|}}{|\boldsymbol{r}-\boldsymbol{r}'|} . \tag{24}$$

Note that, in addition to a term with Debye range, there is also a term with infinite range. However, we shall now show that in the fluctuation of the total field this long range term is cancelled by the contribution of the transverse field [14]. The apparent long range correlations are therefore only due to the artificial separation of the total field in a longitudinal and a transverse part.

From (10) one finds after integrating over the $\boldsymbol{p}_{\nu\sigma}$ that the probability distribution of the transverse field is proportional to

$$\exp[-\frac{\beta}{8\pi}\int\{(\boldsymbol{E}^T)^2 + (\mathrm{curl}\ \boldsymbol{A}^T)^2\}\,\mathrm{d}\boldsymbol{r}] .$$

It is essential to take into account that only transverse fields are included. This can be done by first taking a distribution over *all* fields with probability

$$\exp[-\frac{\beta}{8\pi}\int\{\boldsymbol{E}^2 + (\mathrm{curl}\ \boldsymbol{A})^2 + \gamma^2\,(\mathrm{div}\ \boldsymbol{E})^2 + \gamma^2\,(\mathrm{div}\ \boldsymbol{A})^2\}\,\mathrm{d}\boldsymbol{r}] \tag{25}$$

and putting afterwards $\gamma = \infty$. The exponent in (25) is a quadratic form in the quantities $E_i(\boldsymbol{r})$, whose matrix is

$$\langle i,\boldsymbol{r}|Q|j,\boldsymbol{r}'\rangle = \frac{\beta}{4\pi}[\delta_{ij}\,\delta(\boldsymbol{r}-\boldsymbol{r}') + \gamma^2\nabla_i\nabla_j'\ \delta(\boldsymbol{r}-\boldsymbol{r}')] .$$

Hence the correlations can be found by the same device as in the previous section.

The reciprocal is found by solving $X(\boldsymbol{r})$ from

$$\frac{\beta}{4\pi}[X_i(\boldsymbol{r}) - \gamma^2\nabla_i\ \mathrm{div}\ X(\boldsymbol{r})] = Y_i(\boldsymbol{r}) . \tag{26}$$

Taking the divergence of this equation one finds

$$(1-\gamma^2\nabla^2)\ \mathrm{div}\ X(\boldsymbol{r}) = (4\pi/\beta)\ \mathrm{div}\ Y(\boldsymbol{r}) ,$$

from which follows

$$\mathrm{div}\ X(\boldsymbol{r}) = \frac{1}{\beta\gamma^2}\int \frac{\mathrm{e}^{-|\boldsymbol{r}-\boldsymbol{r}'|/\gamma}}{|\boldsymbol{r}-\boldsymbol{r}'|}\ \mathrm{div}\ Y(\boldsymbol{r}')\,\mathrm{d}\boldsymbol{r}' .$$

Substitution in (26) yields

$$X_i(\boldsymbol{r}) = \frac{4\pi}{\beta} Y_i(\boldsymbol{r}) + \frac{1}{\beta}\nabla_i \int \frac{e^{-|\boldsymbol{r}-\boldsymbol{r}'|/\gamma}}{|\boldsymbol{r}-\boldsymbol{r}'|} \operatorname{div} Y(\boldsymbol{r}')\, d\boldsymbol{r}' ,$$

and the exponential factor reduces to unity on putting $\gamma=\infty$. Thus we find

$$\langle E_i^T(\boldsymbol{r})\, E_j^T(\boldsymbol{r}')\rangle = \frac{4\pi}{\beta}\delta_{ij}\,\delta(\boldsymbol{r}-\boldsymbol{r}') - \frac{1}{\beta}\nabla_i\nabla_j' \frac{1}{|\boldsymbol{r}-\boldsymbol{r}'|} .$$

Adding this result to (24) and realizing that there is no statistical dependence between $\boldsymbol{E}^L$ and $\boldsymbol{E}^T$, one obtains for the total field

$$\langle E_i(\boldsymbol{r})\, E_j(\boldsymbol{r}')\rangle = kT\left\{4\pi\delta_{ij}\delta(\boldsymbol{r}-\boldsymbol{r}') - \nabla_i\nabla_j' \frac{e^{-\kappa|\boldsymbol{r}-\boldsymbol{r}'|}}{|\boldsymbol{r}-\boldsymbol{r}'|}\right\} . \qquad (27)$$

Clearly the range of the correlation is the Debye length.

Other correlation functions may be computed in the same way. The correlations between the field components of the magnetic field, for example, are found to be

$$\langle B_i(\boldsymbol{r})\, B_j(\boldsymbol{r}')\rangle = kT\left\{4\pi\delta_{ij}\,\delta(\boldsymbol{r}-\boldsymbol{r}') - \nabla_i\nabla_j' \frac{1}{|\boldsymbol{r}-\boldsymbol{r}'|}\right\} .$$

They are not affected by the presence of the plasma, and are therefore not subject to Debye shielding. Note, however, that the long range disappears in the formula

$$\langle \boldsymbol{B}(\boldsymbol{r})\cdot \boldsymbol{B}(\boldsymbol{r}')\rangle = 8\pi kT\,\delta(\boldsymbol{r}-\boldsymbol{r}') .$$

It is not possible to find the mean magnetic energy density from this by putting $\boldsymbol{r}=\boldsymbol{r}'$, but if one first averages the field over a small region v one obtains an interesting formula

$$\langle \frac{1}{8\pi}\{\frac{1}{v}\int_v \boldsymbol{B}(\boldsymbol{r})\, d\boldsymbol{r}\}^2\rangle = \frac{kT}{v} .$$

Similarly one finds

$$\langle \frac{1}{8\pi}\{\frac{1}{v}\int_v \boldsymbol{E}(\boldsymbol{r})\, d\boldsymbol{r}\}^2\rangle = \frac{3}{2}\frac{kT}{v} ,$$

provided that v is large compared to a Debye sphere. The factor $\frac{3}{2}$ can be explained by the fact that $\boldsymbol{E}$ also has a longitudinal component. These formulas are of interest to astrophysicists, who feel that there ought to be some kind of equipartition between the field energy and the kinetic energy of the particles.

2.6. *Additional remarks*

a) The internal energy of the electron gas, on account of (18) and (21), is given by

$$U_{\mathrm{el}} = \frac{\partial}{\partial\beta}\,\beta F = \tfrac{1}{2}\frac{\partial}{\partial\beta}\log D = \tfrac{1}{2}\operatorname{Tr} PQ^{-1} \; . \tag{28}$$

The explicit value is given by (22). The meaning is clear: $P^{\lambda\lambda'}$ is the Coulomb energy of two particles at $\boldsymbol{R}^{\lambda}$ and $\boldsymbol{R}^{\lambda'}$ and Q^{-1} represents the probability for finding two particles at these points.

It is also possible to write (28) in the form

$$U_{\mathrm{el}} = \tfrac{1}{2}\operatorname{Tr} PQ^{-1} = \tfrac{1}{2}e^2 \operatorname{Tr}\frac{\partial Q}{\partial e^2}\,Q^{-1} = e^2\,\frac{\partial F^{\mathrm{el}}}{\partial e^2} \, ,$$

a known formula.

b) An alternative calculation of the free energy (11) starts from Mayer's cluster expansion for the partition function of a non-ideal gas [15]. This leads to an expansion of the free energy in the density, in which each term consists of a number of irreducible cluster integrals. If one now selects in each term only the so-called *ring diagrams,* and neglects all contributions of other diagrams, it is possible to evaluate each term and sum the series expansion. If one allows oneself another approximation in the evaluation of the terms, the result is again (23). This calculation is very neat, but the physical picture seems to me less clear than on the method of cells.

c) The density correlation function (20) may be interpreted in terms of the pair distribution function,

$$g(\boldsymbol{r},\boldsymbol{r}') = 1 - \frac{\kappa^2}{4\pi n_0}\,\frac{e^{-\kappa|\boldsymbol{r}-\boldsymbol{r}'|}}{|\boldsymbol{r}-\boldsymbol{r}'|} \; . \tag{29}$$

It states that if electron a is known to be at $\boldsymbol{r}_a$ the probability $g(\boldsymbol{r}_a,\boldsymbol{r}_b)/V$ for finding electron b at $\boldsymbol{r}_b$ is decreased in the region covered by the Debye sphere. The integrated *density decrease* of all electrons other than a in the Debye sphere is

$$\frac{N-1}{V}\int\frac{\kappa^2}{4\pi n_0}\frac{e^{-\kappa|\boldsymbol{r}_a-\boldsymbol{r}_b|}}{|\boldsymbol{r}_a-\boldsymbol{r}_b|}\mathrm{d}\boldsymbol{r}_b' = 1 \; ;$$

together with the electron a at $\boldsymbol{r}_a$ this makes up the average density.

On the other hand, the number of electrons in the Debye sphere is of the order

$$n_0\kappa^{-3} \sim (kT/e^2 n^{\frac{1}{3}}) \gg 1 \; , \tag{30}$$

on account of (1). The statistical fluctuations in the number of particles within the range κ^{-1} is therefore also much larger than 1. Hence the fact that the average is decreased by 1 has only a purely statistical significance: the Debye sphere has a rather ghost-like existence. In particular it is incorrect to treat plasmas by simply substituting a Debye interaction potential for the actual Coulomb interaction.

d) By confining ourselves to the electron gas we have avoided the difficulty that the partition function (11) does not converge when opposite charges occur, owing to the divergence of

$$\iint \exp\left[-\frac{\beta}{2}\frac{e_\sigma e_{\sigma'}}{|\boldsymbol{r}_\sigma-\boldsymbol{r}_{\sigma'}|}\right]\mathrm{d}\boldsymbol{r}_\sigma\,\mathrm{d}\boldsymbol{r}_{\sigma'}$$

for $\boldsymbol{r}_\sigma = \boldsymbol{r}_\sigma'$ when $e_\sigma e_{\sigma'} < 0$. This exhibits the fact that classically two oppositely charged particles cannot be in equilibrium since they gain an infinite amount of energy by coalescing. In the case of an electron and an ion it is reasonable to cut off the integration by a smallest distance of the order of the ion radius. In the case of an electron and a proton, however, this is the famous difficulty that was solved by Bohr's quantization of the orbits. Hence it will be roughly correct to cut off at the radius a_0 of the first orbit. A more detailed study investigates how the quantum orbits merge into classical orbits when the quantum number increases [16]. We shall here use a simpler approach, based on the same idea.

The quantummechanical partition function is

$$\begin{aligned} Z_{\mathrm{qu}} &= \mathrm{Tr}\,\exp\left[\frac{\beta\hbar^2}{2m}\nabla^2+\frac{\beta e^2}{r}\right] \\ &= \mathrm{Tr}\,\{e^{\beta e^2/r}\,e^{-\beta e^2/2r}\exp[\quad]\,e^{-\beta e^2/2r}\} \\ &= \sum_\alpha\int u_\alpha^*(\boldsymbol{r})\,\{\quad\}\,u_\alpha(\boldsymbol{r})\,\mathrm{d}\boldsymbol{r}\;, \end{aligned}$$

where $u_\alpha(\boldsymbol{r})$ is any complete orthonormal set. This may be written in the form

$$Z_{\mathrm{qu}} = \left(\frac{2\pi m}{\beta\hbar^2}\right)^{\frac{3}{2}} \int e^{\beta e^2/r}\, \chi_\beta(r)\, d\boldsymbol{r} \;,$$

$$\chi_\beta(r) = \left(\frac{\beta\hbar^2}{2\pi m}\right)^{\frac{3}{2}} \sum_\alpha u^*_\alpha(\boldsymbol{r})\, e^{-\beta e^2/2r} \exp\left[\frac{\beta\hbar^2}{2m}\nabla^2 + \frac{\beta e^2}{r}\right] e^{-\beta e^2/2r} u_\alpha(\boldsymbol{r}) \,.$$

Thus $\chi_\beta(\boldsymbol{r})$ is the correction factor that has to be inserted in the classical partition function to take into account quantum effects and hence to cancel the classical divergence at $r=0$.

We shall verify that $\chi_\beta(r) \approx 1$ when r is larger than the mean distance $d = n_0^{-\frac{1}{3}}$ between protons. This justifies the use of $\chi_\beta(r)$ as a cutting-off factor for each electron-proton pair, rather than treating the whole plasma as a quantummechanical many-body system. Using the operator identity

$$e^{-B/2} e^{A+B} e^{-B/2} = \exp\{A + \tfrac{1}{12}[[A,B],A] + \tfrac{1}{24}[[A,B],B] + \ldots\}$$

one finds

$$e^{-\beta e^2/2r} \exp\left[\frac{\beta\hbar^2}{2m}\nabla^2 + \frac{\beta e^2}{r}\right] e^{-\beta e^2/2r} =$$

$$\exp\left\{\frac{\beta\hbar^2}{2m}\nabla^2 + \frac{\beta^3 e^2\hbar^4}{48m^2}[[\nabla^2, \tfrac{1}{r}], \nabla^2] + \frac{\beta^3 e^4\hbar^2}{48m}[[\nabla^2, \tfrac{1}{r}], \tfrac{1}{r}] + \ldots\right\} . \quad (31)$$

The last of these three terms is smaller than the first one by a factor of order

$$\frac{\beta^2 e^4}{r^2} \sim (\kappa d)^4 \left(\frac{d}{r}\right)^2 . \quad (32)$$

Since $\kappa d \ll 1$ this term is small when the electron has a distance comparable to the distance to the next proton. The middle term in (31) is of the order (32) times the additional factor a_0/r, where a_0 is the radius of the first Bohr orbit. For $r \approx d$ this factor is less than 1, unless the plasma is a degenerate quantum gas, like the electrons in a metal.

The dots in (31) represent terms that involve higher powers of (32) and a_0/r. Hence for $r \gtrsim d$ one has approximately

$$\chi_\beta(r) = \left(\frac{\beta\hbar^2}{2\pi m}\right)^{\frac{3}{2}} \sum_\alpha u_\alpha^*(\boldsymbol{r}) \exp\left[\frac{\beta\hbar^2}{2m}\nabla^2\right] u_\alpha(\boldsymbol{r}) .$$

Take $u_\alpha(\boldsymbol{r}) = V^{-\frac{1}{2}} e^{i\boldsymbol{k}.\boldsymbol{r}}$, and one finds in the usual way

$$\chi_\beta(r) = \left(\frac{\beta\hbar^2}{2\pi m}\right)^{\frac{3}{2}} V^{-1} \sum_{\boldsymbol{k}} \exp\left(\frac{\beta\hbar^2}{2m}\boldsymbol{k}^2\right) = 1 .$$

The value of $\chi_\beta(r)$ for small r can of course only be found by more explicit computations involving the quantum orbits of the hydrogen atom.

3. KINETIC THEORY OF THE ELECTRON GAS

3.1. *The basic equations*

Again we shall confine ourselves to the classical electron gas with a constant positive background, confined to a volume V. We shall also omit the transverse part of the electromagnetic field. It is true that it does not drop out altogether, like in the equilibrium case, but for slow motions of the plasma it will be relatively unimportant. The Hamiltonian function of such a gas containing N particles is

$$H_N(\boldsymbol{r}_1, \boldsymbol{p}_1, \boldsymbol{r}_2, \boldsymbol{p}_2, \ldots \boldsymbol{r}_N, \boldsymbol{p}_N) =$$

$$\sum_{\sigma=1}^{N} \frac{\boldsymbol{p}_\sigma^2}{2m} + \sum_{\sigma<\sigma'} \Psi(\boldsymbol{r}_\sigma - \boldsymbol{r}_{\sigma'}) - n_0 \sum_{\sigma=1}^{N} \int_V \Psi(\boldsymbol{r}_\sigma - \boldsymbol{r}_0)\, d\boldsymbol{r}_0 .$$

Here $\Psi(\boldsymbol{r}) = e^2/r$ and n_0 is the density of the background. According to the standard procedure one introduces an ensemble described by a probability density

$$D(\boldsymbol{r}_1, \boldsymbol{p}_1, \boldsymbol{r}_2, \boldsymbol{p}_2, \ldots, \boldsymbol{r}_N, \boldsymbol{p}_N; t)$$

in phase space, which is symmetric in all particles. The time evolution of this function is determined by the Liouville equation

$$\frac{\partial D}{\partial t} = \sum_\sigma \left(\frac{\partial H_N}{\partial \boldsymbol{r}_\sigma}\cdot\frac{\partial D}{\partial \boldsymbol{p}_\sigma} - \frac{\partial H_N}{\partial \boldsymbol{p}_\sigma}\cdot\frac{\partial D}{\partial \boldsymbol{r}_\sigma}\right) = \{H_N, D\} . \tag{33}$$

Next one defines the *reduced distribution function for a subset of s electrons* by

$$f_s(\boldsymbol{r}_1, \boldsymbol{p}_1, \ldots, \boldsymbol{r}_s, \boldsymbol{p}_s; t) = V^s \int D \, \mathrm{d}\boldsymbol{r}_{s+1} \mathrm{d}\boldsymbol{p}_{s+1} \ldots \mathrm{d}\boldsymbol{r}_N \mathrm{d}\boldsymbol{p}_N .$$

The evolution of this function is found from (33) by integrating over all particles $s+1, \ldots, N$:

$$\frac{\partial f_s}{\partial t} = \{H_s, f_s\} + \frac{N-s}{V} \int \{ \sum_{\sigma=1}^{s} \Psi(\boldsymbol{r}_\sigma - \boldsymbol{r}_{s+1}), \; f_{s+1} \} \mathrm{d}\boldsymbol{r}_{s+1} \, \mathrm{d}\boldsymbol{p}_{s+1} \; . \quad (34)$$

The first term on the right is the Poisson bracket that describes the evolution of the s-particle distribution function due to the background and to their mutual interaction. The second term on the right takes into account the interaction of each electron $1, \ldots, s$ with any one of the remaining $N-s$. This set of equations expresses the rate of change of each f_s in terms of f_s itself and of f_{s+1} and is called the BBGKY hierarchy.

We rewrite this hierarchy more explicitly, replacing the momenta $\boldsymbol{p}_\sigma$ with the velocities $\boldsymbol{v}_\sigma = \boldsymbol{p}_\sigma / m$, and using the abbreviation

$$\boldsymbol{F}_{\sigma,\sigma'} = -\nabla_\sigma \frac{1}{|\boldsymbol{r}_\sigma - \boldsymbol{r}_{\sigma'}|} = \frac{\boldsymbol{r}_\sigma - \boldsymbol{r}_{\sigma'}}{|\boldsymbol{r}_\sigma - \boldsymbol{r}_{\sigma'}|^3} .$$

Moreover we go to the "thermodynamic limit", so that $(N-s)/V \to n_0$ for any fixed s. The result is

$$\begin{aligned} &\frac{\partial f_s}{\partial t} + \sum_{\sigma=1}^{s} \boldsymbol{v}_\sigma \cdot \frac{\partial f_s}{\partial \boldsymbol{r}_\sigma} \\ &- \frac{e^2}{m} \sum_{\sigma=1}^{s} \frac{\partial f_s}{\partial \boldsymbol{v}_\sigma} \cdot \Big(\sum_{\substack{\sigma'=1 \\ \sigma' \neq \sigma}}^{s} \boldsymbol{F}_{\sigma,\sigma'} - n_0 \int \boldsymbol{F}_{\sigma,0} \, \mathrm{d}\boldsymbol{r}_0 \Big) \qquad (35) \\ &- \frac{n_0 e^2}{m} \sum_{\sigma=1}^{s} \int \boldsymbol{F}_{\sigma,s+1} \cdot \frac{\partial f_{s+1}}{\partial \boldsymbol{v}_\sigma} \mathrm{d}\boldsymbol{r}_{s+1} \, \mathrm{d}\boldsymbol{v}_{s+1} \; . \end{aligned}$$

The equation contains two physical constants, n_0 and e^2/m. By an appropriate choice of the units of length and time they can both be reduced to unity. Thus *equation* (35) *contains no dimensionless quantity which could be used as an expansion parameter.* However, if one is interested in solutions near a thermal equilibrium state, an additional constant kT appears, so that a dimensionless parameter $e^2 n_0^{\frac{1}{3}}/kT$ can be formed, which is small according to (1). Accordingly we take as unit of length the Debye radius $\kappa^{-1} = (kT/4\pi n_0 e^2)^{\frac{1}{2}}$, and as unit of time the reciprocal plasma frequency $\omega_P^{-1} = (m/4\pi n_0 e^2)^{\frac{1}{2}}$. Equation (35) then becomes

$$\frac{\partial f_s}{\partial t} + \sum_{\sigma=1}^{s} \boldsymbol{v}_s \cdot \frac{\partial f_s}{\partial \boldsymbol{r}_s} = -\epsilon \sum_{\substack{\sigma=1 \\ \sigma\neq\sigma'}}^{s} \sum_{\sigma'=1}^{s} \boldsymbol{F}_{\sigma,\sigma'} \cdot \frac{\partial f_s}{\partial \boldsymbol{v}_\sigma}$$

$$+ \frac{1}{4\pi} \sum_{\sigma=1}^{s} \frac{\partial f_s}{\partial \boldsymbol{v}_\sigma} \cdot \int \boldsymbol{F}_{\sigma,0}\, \mathrm{d}\boldsymbol{r}_0 - \frac{1}{4\pi} \sum_{\sigma=1}^{s} \int \boldsymbol{F}_{\sigma,s+1} \cdot \frac{\partial f_{s+1}}{\partial \boldsymbol{v}_\sigma} \mathrm{d}\boldsymbol{r}_{s+1} \mathrm{d}\boldsymbol{v}_{s+1}, \tag{36}$$

in which the small parameter

$$\epsilon = \frac{1}{4\pi n_0} \left(\frac{4\pi n_0 e^2}{kT} \right)^{\frac{3}{2}}$$

occurs, which is roughly the reciprocal number of particles in a sphere whose radius is the Debye length. Note that ϵ multiplies the mutual interaction within the group of s particles. Hence for hot dilute plasmas the interaction between any small group of particles is small compared to the interaction with the remaining plasma. This exhibits the profound difference with a Boltzmann gas.

3.2. *The Vlasov equation*

The lowest approximation consists in putting $\epsilon = 0$. A solution of the set of equations may then be constructed [17] by setting

$$f_s(\boldsymbol{r}_1, \boldsymbol{v}_1, \boldsymbol{r}_2, \boldsymbol{v}_2, \ldots, \boldsymbol{r}_s, \boldsymbol{v}_s; t) = \prod_{\sigma=1}^{s} f_1(\boldsymbol{r}_\sigma, \boldsymbol{v}_\sigma, t), \tag{37}$$

where f_1 obeys (36) for $s=1$ and $\epsilon = 0$, i.e.,

$$\frac{\partial f_1(\boldsymbol{r}_1, \boldsymbol{v}_1, t)}{\partial t} + \boldsymbol{v}_1 \cdot \frac{\partial f_1}{\partial \boldsymbol{r}_1} + \frac{1}{4\pi} \frac{\partial f_1}{\partial \boldsymbol{v}_1} \cdot \int \boldsymbol{F}_{1,2} \left\{ \int f_1(\boldsymbol{r}_2, \boldsymbol{v}_2, t) \mathrm{d}\boldsymbol{v}_2 - 1 \right\} \mathrm{d}\boldsymbol{r}_2 = 0. \tag{38}$$

Equation (37) states that there are no statistical correlations between the electrons, in agreement with the fact that the mutual interaction among any group of s electrons has been omitted in this approximation. In equation (38) the factor $\{\quad\}$ on the right represents the average density of the electrons minus the density of the background. The multiplication by $\boldsymbol{F}_{1,2}$ and integration over $\boldsymbol{r}_2$ produce the electrostatic field due to both densities. Hence (38) may be written, in the original units,

$$\frac{\partial f_1(\boldsymbol{r},\boldsymbol{v},t)}{\partial t} + \boldsymbol{v}\cdot\frac{\partial f_1}{\partial \boldsymbol{r}} - \frac{e}{m}\boldsymbol{E}(\boldsymbol{r},t)\cdot\frac{\partial f_1}{\partial \boldsymbol{v}} = 0 \tag{39}$$

$$\operatorname{div}\boldsymbol{E}(\boldsymbol{r},t) = 4\pi e\{n_0 - n(\boldsymbol{r},t)\}\,, \qquad \operatorname{curl}\boldsymbol{E} = 0 \tag{40}$$

$$n(\boldsymbol{r},t) = \int f_1(\boldsymbol{r},\boldsymbol{v},t)\,\mathrm{d}\boldsymbol{v}\,. \tag{41}$$

(39) may be regarded as the Liouville equation for a one-particle distribution function in a given electrostatic field $\boldsymbol{E}$; it is called the *Vlasov equation,* cf. eq. (9). Eq. (40) represents Coulomb's law and (41) defines the electron density.

Note that this derivation proves that the field acting on an electron is simply the over all smeared-out field. In contrast to Lorentz' theory of a dielectric medium, no correction for the internal or local field is necessary. The reason is that in this approximation there is no microscopic correlation between the positions of the electrons. This point has been the subject of some discussion [18].

As mentioned in section 1, for small deviations from equilibrium a linear approximation can be used, which consists in replacing f_1 in the last term of (39) by the equilibrium distribution. The resulting linear set can be solved and gives rise to *plasma waves,* subject to *Landau damping* [6], which has been confirmed experimentally [19].

The appearance of an irreversible effect like damping is of some interest in view of the fact that the set (39)-(41), and of course also its linearized version, is manifestly invariant for time reversal. Attempts have been made to explain the physical mechanism by which the plasma waves dissipate their energy into heat motion of the electrons. However, these explanations, inasmuch as they invoke nonlinear effects, are inadequate, because Landau damping was found as a rigorous consequence of the linearized Vlasov equation. The correct explanation is based on "phase mixing", which means the following.

In a longitudinal plasma wave, both $n(\boldsymbol{r},t)$ and $\boldsymbol{E}(\boldsymbol{r},t)$ vary sinusoidally in space. However, such a wave is not a pure normal mode of the linearized Vlasov equation. With a given space dependence of n and $\boldsymbol{E}$ there is still an infinity of normal modes, owing to the fact that not only $\boldsymbol{r}$, but also $\boldsymbol{v}$ occurs as a variable in the equation. Thus a normal mode is a function in the six-dimensional $\boldsymbol{r}, \boldsymbol{v}$-space. Hence a plasma wave is in general a superposition of many normal modes. These normal modes propagate with slightly different phase velocities, and as a result the initial variations of n and $\boldsymbol{E}$ in space will gradually smear out and ultimately disappear [20]. This is Landau damping.

The fact that the density n tends to its equilibrium value n_0 does not imply that the full distribution function f_1 also tends to its equilibrium. That this cannot be the case follows from the invariance for time reversal. For, knowing f_1 at any future time, one can always calculate backwards and find the initial f_1. What happens during the motion of the plasma is that the variation of f_1 with $\boldsymbol{r}$ is gradually transformed into a variation of f_1 with $\boldsymbol{v}$, which does not show after the integration (41) is performed.

3.3. *Collisions*

To first order in ϵ the hierarchy can no longer be solved by functions of the form (37), since the interaction of the particles will create correlations between their positions and velocities. In order to solve (36) to first order in ϵ we set (in abbreviated notation)

$$f_2(1,2) = f_1(1)\, f_1(2) + \epsilon g_2(1,2)\,, \tag{42}$$

$$f_3(1,2,3) = f_1(1)\, f_1(2)\, f_1(3) + \epsilon\{f_1(1)\, g_2(2,3) + f_1(2)\, g_2(3,1) + f_1(3)\, g_2(1,2)\}\,. \tag{43}$$

In general

$$f_s(1,2\ldots,s) = \prod_{\sigma=1}^{s} f_1(\sigma) + \epsilon \prod_{\substack{\text{all}\\ \text{pairs}}} g_2(\sigma',\sigma'') \prod_{\substack{\sigma=1\\ \sigma\neq\sigma',\sigma''}}^{s} f_1(\sigma)\,. \tag{44}$$

This implies the assumption that correlations between three and more electrons will be of higher order in ϵ. The result will show that solutions of this type actually exist.

Substitution in the hierarchy (36) first leads to an equation for f_1 involving g_2,

$$\frac{\partial f_1}{\partial t} + \boldsymbol{v}_1 \cdot \frac{\partial f_1}{\partial \boldsymbol{r}_1} + \frac{1}{4\pi} \frac{\partial f_1(1)}{\partial \boldsymbol{v}_1} \cdot \int \boldsymbol{F}_{1,2} \{ \int f_1(2) \, \mathrm{d}\boldsymbol{v}_2 - 1 \} \, \mathrm{d}\boldsymbol{r}_2$$

$$= - \frac{\epsilon}{4\pi} \int \boldsymbol{F}_{1,2} \cdot \frac{\partial g_2(1,2)}{\partial \boldsymbol{v}_1} \, \mathrm{d}\boldsymbol{r}_2 \, \mathrm{d}\boldsymbol{v}_2 \, . \qquad (45)$$

The left-hand side is identical with the Vlasov expression (38). The right-hand side represents the direct interaction with a second particle at $\boldsymbol{r}_2$, taking into account the probability for having a particle at $\boldsymbol{r}_2$. This is the desired collision term added to the Vlasov equation.

Next one obtains an equation for g_2 involving f_1,

$$\frac{\partial g_2(1,2)}{\partial t} + \boldsymbol{v}_1 \cdot \frac{\partial g_2}{\partial \boldsymbol{r}_1} + \boldsymbol{v}_2 \cdot \frac{\partial g_2}{\partial \boldsymbol{r}_2} =$$

$$- \frac{1}{4\pi} \frac{\partial g_2}{\partial \boldsymbol{v}_1} \cdot \int \boldsymbol{F}_{1,3} \{ \int f_1(3) \, \mathrm{d}\boldsymbol{v}_3 - 1 \} \, \mathrm{d}\boldsymbol{r}_3$$

$$- \frac{1}{4\pi} \frac{\partial g_2}{\partial \boldsymbol{v}_2} \cdot \int \boldsymbol{F}_{2,3} \{ \int f_1(3) \, \mathrm{d}\boldsymbol{v}_3 - 1 \} \, \mathrm{d}\boldsymbol{r}_3$$

$$- \frac{1}{4\pi} \frac{\partial f_1(1)}{\partial \boldsymbol{v}_1} \cdot \int \boldsymbol{F}_{1,3} \, g_2(2,3) \, \mathrm{d}\boldsymbol{r}_3 \, \mathrm{d}\boldsymbol{v}_3$$

$$- \frac{1}{4\pi} \frac{\partial f_1(2)}{\partial \boldsymbol{v}_2} \cdot \int \boldsymbol{F}_{2,3} \, g_2(1,3) \, \mathrm{d}\boldsymbol{r}_3 \, \mathrm{d}\boldsymbol{v}_3$$

$$- \frac{\partial f_1(1)}{\partial \boldsymbol{v}_1} \cdot \boldsymbol{F}_{1,2} \, f_1(2) - \frac{\partial f_1(2)}{\partial \boldsymbol{v}_2} \cdot \boldsymbol{F}_{2,1} \, f_1(1) \, . \qquad (46)$$

The last line contains the interaction between particles 1 and 2, while all other terms describe the rate of change of g_2 due to the surrounding cloud of particles.

This is merely the result of substituting (42) and (43) in the first two equations of the hierarchy (36). However, it may subsequently be verified by direct substitution that if f_1 and g_2 obey (45) and (46), the whole hierarchy is satisfied by (44), to order ϵ. Hence the problem of solving the hierarchy reduces to that of solving the two coupled nonlinear equations (45) and (46). This will be studied in the next section, but first we verify that the equilibrium distribution obeys (45) and (46).

Equilibrium is described by a Maxwell distribution for f_1, in the present units is simply

$$f_M(\boldsymbol{v}) = (2\pi)^{-\frac{3}{2}} \, e^{-\frac{1}{2}v^2} \, .$$

Moreover one expects that g_2 has the form

$$g_2(1,2) = f_M(v_1) f_M(v_2) \, \gamma(|\boldsymbol{r}_1 - \boldsymbol{r}_2|) \, ,$$

where γ will turn out to be the Debye function. Substitution in (45) makes all terms on the left zero, and the right-hand side also vanishes owing to the spherical symmetry of g_2. Substitution in (46) yields

$$(\boldsymbol{v}_1 - \boldsymbol{v}_2) \cdot \frac{\partial \gamma(\boldsymbol{r}_{12})}{\partial \boldsymbol{r}_{12}} = (\boldsymbol{v}_1 - \boldsymbol{v}_2) \cdot \left\{ \boldsymbol{F}(\boldsymbol{r}_{12}) + \frac{1}{4\pi} \boldsymbol{G}(\boldsymbol{r}_{12}) \right\} \, , \qquad (47)$$

where

$$\boldsymbol{r}_{12} = \boldsymbol{r}_1 - \boldsymbol{r}_2 \, , \qquad \boldsymbol{F}(\boldsymbol{r}_{12}) = \boldsymbol{F}_{12} \, ,$$

and

$$\boldsymbol{G}(\boldsymbol{r}_{12}) = \int \boldsymbol{F}(\boldsymbol{r}_1 - \boldsymbol{r}_3) \, \gamma(\boldsymbol{r}_2 - \boldsymbol{r}_3) \, d\boldsymbol{r}_3 = - \boldsymbol{G}(\boldsymbol{r}_{21}) \, .$$

Eq. (47) yields for $\gamma(\boldsymbol{r})$ the integral equation

$$\nabla \gamma(\boldsymbol{r}) = \boldsymbol{F}(\boldsymbol{r}) + \frac{1}{4\pi} \int \boldsymbol{F}(\boldsymbol{r} - \boldsymbol{r}') \, \gamma(\boldsymbol{r}') \, d\boldsymbol{r}' \, .$$

On taking the divergence of this vector equation one finds

$$\nabla^2 \gamma(\boldsymbol{r}) = 4\pi \delta(\boldsymbol{r}) + \gamma(\boldsymbol{r}) \, ,$$

which is the familiar Debye equation *.

If one substitutes for f_1 an arbitrary function of $|\boldsymbol{v}|$ alone, it is easily found that no solution is obtained, unless this function is

* The connection between γ and the pair distribution function g in (29) is: $\gamma = n_0(g-1)$.

Maxwellian. The collisionless Vlasov equation, on the other hand, admits any function of $|\boldsymbol{v}|$ as a stationary solution. This is entirely analogous to the familiar case of the dilute and the ideal gas.

3.4. *Time-dependent solutions of the equations* (45) *and* (46)

The equations (45) and (46) constitute a set of two coupled non-linear integro-differential equations for the two functions f_1 and g_2. They are of first order in $\partial/\partial t$ and therefore determine all future values of f_1 and g_2 when their initial values are given *. However, the future values of f_1 are not determined by the initial f_1 alone, but also depend on the initial g_2, as is evident from (45). Hence there is no "*kinetic equation*", i.e., an equation governing the evolution of f_1 by itself.

One might hope that it is still possible to find an approximate kinetic equation [21]. Suppose that there exists a time period τ such that

(i) f_1 does not vary appreciably during any time interval of length τ;

(ii) for fixed f_1 the function g_2 approaches an asymptotic form in a time less than τ.

In that case it is possible to compute this asymptotic form of g_2 from (46) for any given fixed f_1, the result being a functional $\tilde{g}(1,2|f_1)$ of f_1 which does not contain t. Substituting this functional in (45) one would get an approximately correct equation for f_1 alone.

Unfortunately, as has been mentioned before [22], even in the case of dilute gases there is no convincing reason why such a separation of time scales should be possible and in fact this assumption has led to profound difficulties [23]. In the case of plasmas it seems rather obvious that this assumption cannot be true in general. The rate at which f_1 varies is given by (45), and the terms on the left of that equation do not contain any parameter that could cause them to be small compared to the terms in (46). Explicit calculations for plasma waves have verified that g_2 does *not* approach a limiting form in a time short compared to the period of the waves [24].

The situation is different, however, in the special case of a *homogeneous* plasma, i.e.,

$$f_1 = f_1(\boldsymbol{v}, t)\,, \qquad g_2 = g_2(\boldsymbol{r}_1 - \boldsymbol{r}_2; \boldsymbol{v}_1, \boldsymbol{v}_2; t)\,.$$

* The past is also determined by their values at any one time. The equations are still invariant for time reversal, since nothing has been done that could break the symmetry in time.

Owing to the homogeneity the integral of f_1 cancels against the background, so that the terms { } in (45) and (46) are zero. Thus (45) reduces to

$$\frac{\partial f_1(\boldsymbol{v},t)}{\partial t} = -\frac{\epsilon}{4\pi} \int \boldsymbol{F}_{1,2} \cdot \frac{\partial g_2(1,2)}{\partial \boldsymbol{v}_1} \, \mathrm{d}\boldsymbol{r}_2 \mathrm{d}\boldsymbol{v}_2 \,,$$

which shows that f_1 does indeed vary slowly owing to the small parameter ϵ. In the next two sections we carry out the program mentioned above for such a homogeneous plasma.

An alternative way of dealing with (45) and (46) is due to Guernsey [25]. He did not assume a homogeneous plasma and did not rely on Bogolyubov's assumption. Instead, he considered functions f_1 and g_2 which deviate little from the equilibrium solution, and linearized the equations in these deviations. The result has not, of course, the form of a kinetic equation for f_1 alone, since the initial value of g_2 also enters. We shall not reproduce his calculation here.

3.5. *Bogolyubov's method for the homogeneous plasma*

The asymptotic form $\tilde{g}(\boldsymbol{r}_1 - \boldsymbol{r}_2; \boldsymbol{v}_1, \boldsymbol{v}_2 | f_1)$ of g_2 is a time-independent solution of (46) for an arbitrary but fixed $f_1(\boldsymbol{v})$. Hence $\tilde{g}(\boldsymbol{r}; \boldsymbol{v}_1, \boldsymbol{v}_2)$ obeys

$$\begin{aligned}(\boldsymbol{v}_1 - \boldsymbol{v}_2) \cdot \frac{\partial \tilde{g}}{\partial \boldsymbol{r}} = &-\frac{1}{4\pi} \frac{\partial f_1(\boldsymbol{v}_1)}{\partial \boldsymbol{v}_1} \cdot \int \boldsymbol{F}(\boldsymbol{r} - \boldsymbol{r}') \tilde{g}(-\boldsymbol{r}'; \boldsymbol{v}_2, \boldsymbol{v}_3) \, \mathrm{d}\boldsymbol{r}' \mathrm{d}\boldsymbol{v}_3 \\ &+ \frac{1}{4\pi} \frac{\partial f_1(\boldsymbol{v}_2)}{\partial \boldsymbol{v}_2} \cdot \int \boldsymbol{F}(\boldsymbol{r} - \boldsymbol{r}') \tilde{g}(\boldsymbol{r}'; v_2, v_3) \, \mathrm{d}\boldsymbol{r}' \mathrm{d}\boldsymbol{v}_3 \\ &- \boldsymbol{F}(\boldsymbol{r}) \cdot \left\{ \frac{\partial f_1(\boldsymbol{v}_1)}{\partial \boldsymbol{v}_1} f_1(\boldsymbol{v}_2) - \frac{\partial f_1(\boldsymbol{v}_2)}{\partial \boldsymbol{v}_2} f_1(\boldsymbol{v}_1) \right\} .\end{aligned}$$

The homogeneity in space suggests that one take the Fourier transforms

$$g_{\boldsymbol{k}}(\boldsymbol{v}_1, \boldsymbol{v}_2) = \int \tilde{g}(\boldsymbol{r}; \boldsymbol{v}_1, \boldsymbol{v}_2) \, \mathrm{e}^{-i\boldsymbol{k}\cdot\boldsymbol{r}} \, \mathrm{d}\boldsymbol{r}$$

$$\boldsymbol{F}_{\boldsymbol{k}} = \int \boldsymbol{F}(\boldsymbol{r}) \, \mathrm{e}^{-i\boldsymbol{k}\cdot\boldsymbol{r}} \, \mathrm{d}\boldsymbol{r} = -4\pi i \boldsymbol{k}/k^2 \,.$$

The equation then becomes

$$\boldsymbol{k}\cdot(\boldsymbol{v}_1-\boldsymbol{v}_2)\,g_{\boldsymbol{k}}=\frac{\boldsymbol{k}}{k^2}\cdot\left\{\frac{\partial f_1(\boldsymbol{v}_1)}{\partial \boldsymbol{v}_1}\,h^*_{\boldsymbol{k}}(\boldsymbol{v}_2)-\frac{\partial f_1(\boldsymbol{v}_2)}{\partial \boldsymbol{v}_2}\,h_{\boldsymbol{k}}(\boldsymbol{v}_1)\right\}$$

$$+\frac{4\pi\boldsymbol{k}}{k^2}\left\{\frac{\partial f_1(\boldsymbol{v}_1)}{\partial \boldsymbol{v}_1}\,f_1(\boldsymbol{v}_2)-\frac{\partial f_1(\boldsymbol{v}_2)}{\partial \boldsymbol{v}_2}\,f_1(\boldsymbol{v}_1)\right\},\quad (48)$$

where we have put

$$h_{\boldsymbol{k}}(\boldsymbol{v})=\int g_{\boldsymbol{k}}(\boldsymbol{v},\boldsymbol{v}')\,\mathrm{d}\boldsymbol{v}'\,.$$

From (48) one can extract an equation for $h_{\boldsymbol{k}}(\boldsymbol{v})$ alone by dividing by $\boldsymbol{k}\cdot(\boldsymbol{v}_1-\boldsymbol{v}_2)$ and integrating over $\boldsymbol{v}_2$. For this purpose we decompose $\boldsymbol{v}$ in a component u parallel to $\boldsymbol{k}$ and a component $\boldsymbol{w}$ perpendicular to $\boldsymbol{k}$. One then finds from (48)

$$h_{\boldsymbol{k}}(\boldsymbol{v})=\frac{1}{k^2}\,\frac{\partial f_1(\boldsymbol{v})}{\partial u}\int\frac{h^*(\boldsymbol{v}')}{u-u'}\,\mathrm{d}\boldsymbol{v}'-\frac{h(\boldsymbol{v})}{k^2}\int\frac{\partial f_1(\boldsymbol{v}')/\partial u}{u-u'}\,\mathrm{d}\boldsymbol{v}'$$

$$+\frac{4\pi}{k^2}\,\frac{\partial f_1(\boldsymbol{v})}{\partial u}\int\frac{f_1(\boldsymbol{v}')}{u-u'}\,\mathrm{d}\boldsymbol{v}'-\frac{4\pi f_1(\boldsymbol{v})}{k^2}\int\frac{\partial f_1(\boldsymbol{v}')/\partial u'}{u-u'}\,\mathrm{d}\boldsymbol{v}'\,.$$

The integration over $\boldsymbol{w}'$ can be carried out. We abbreviate

$$\int h_{\boldsymbol{k}}(u,\boldsymbol{w})\,\mathrm{d}^2\boldsymbol{w}=H(u)$$

$$\int f_1(u,\boldsymbol{w})\,\mathrm{d}^2\boldsymbol{w}=F(u)$$

and note that this definition of $F(u)$ may also be written

$$F(u)=k\int f_1(\boldsymbol{v})\,\delta(\boldsymbol{k}\cdot\boldsymbol{v}-ku)\,\mathrm{d}\boldsymbol{v}\,.\quad (49)$$

Using these notations one finds

$$h_{\boldsymbol{k}}(\boldsymbol{v})=\frac{1}{k^2}\int_{-\infty}^{\infty}\frac{\mathrm{d}u'}{u-u'}\left\{\frac{\partial f_1(\boldsymbol{v})}{\partial u}H^*(u')-h(\boldsymbol{v})\,\frac{\mathrm{d}F(u')}{\mathrm{d}u'}\right\}$$

$$+\frac{4\pi}{k^2}\int_{-\infty}^{\infty}\frac{\mathrm{d}u'}{u-u'}\left\{\frac{\partial f_1(\boldsymbol{v})}{\partial u}\,F(u')-f_1(\boldsymbol{v})\,\frac{\mathrm{d}F(u')}{\mathrm{d}u'}\right\}.\quad (50)$$

Integrating this equation over $\boldsymbol{w}$ one is left with an equation for H alone

$$H(u) = \frac{1}{k^2} \int_{-\infty}^{\infty} \frac{\mathrm{d}u'}{u-u'} [F'(u)\, H^*(u') - H(u)\, F'(u') + 4\pi\, F'(u)\, F(u') - 4\pi\, F(u)\, F'(u')] \, . \tag{51}$$

We still have to decide how to integrate across the pole at $u' = u$. This pole originated from dividing (48) by $\boldsymbol{k}\cdot(\boldsymbol{v}_1 - \boldsymbol{v}_2)$,

$$g_{\boldsymbol{k}}(\boldsymbol{v}_1, \boldsymbol{v}_2) = \frac{R_{\boldsymbol{k}}(\boldsymbol{v}_1, \boldsymbol{v}_2)}{\boldsymbol{k}\cdot(\boldsymbol{v}_1 - \boldsymbol{v}_2)} ,$$

where $R_{\boldsymbol{k}}(\boldsymbol{v}_1, \boldsymbol{v}_2)$ is short for the right-hand side of (48). The corresponding $\boldsymbol{r}$-representation is

$$\tilde{g}(\boldsymbol{r}; \boldsymbol{v}_1, \boldsymbol{v}_2) = \frac{1}{8\pi^3} \frac{R_{\boldsymbol{k}}(\boldsymbol{v}_1, \boldsymbol{v}_2)}{\boldsymbol{k}\cdot(\boldsymbol{v}_1 - \boldsymbol{v}_2)} \mathrm{e}^{i\boldsymbol{k}\cdot\boldsymbol{r}}\, \mathrm{d}\boldsymbol{k} \, .$$

The asymptotic behavior of $\tilde{g}$ for large r is determined by the pole alone. Writing $k_{\parallel}$ for the component of $\boldsymbol{k}$ parallel to $\boldsymbol{v}_1 - \boldsymbol{v}_2$, and $\boldsymbol{k}_{\perp}$ for the perpendicular component, we have therefore asymptotically

$$\tilde{g}(\boldsymbol{r}; \boldsymbol{v}_1, \boldsymbol{v}_2) \simeq \frac{1}{8\pi^3} \frac{1}{|\boldsymbol{v}_1 - \boldsymbol{v}_2|} \int_{-\infty}^{\infty} \frac{\mathrm{e}^{ik_{\parallel} r_{\parallel}}}{k_{\parallel}} \mathrm{d}k_{\parallel} \times \int \mathrm{e}^{i\boldsymbol{k}_{\perp}\cdot \boldsymbol{r}_{\perp}} R_{\boldsymbol{k}_{\perp}, 0}(\boldsymbol{v}_1, \boldsymbol{v}_2)\, \mathrm{d}^2\boldsymbol{k}_{\perp} \, .$$

At this point enters the physical assumption that there should be *no statistical correlation between particles before the collision* ("Stosszahlansatz"). This assumption is responsible for destroying the invariance for time reversal and introducing irreversibility. In the present formalism the assumption amounts to

$$\tilde{g}_2(\boldsymbol{r}_1 - \boldsymbol{r}_2; \boldsymbol{v}_1, \boldsymbol{v}_2) \to 0 \text{ as } (\boldsymbol{r}_1 - \boldsymbol{r}_2)_{\parallel} \to -\infty \, .$$

This condition is met by choosing the integration path *below* the pole. Alternatively one may replace $1/k_{\parallel}$ in the denominator by $(k_{\parallel} - i\epsilon)^{-1}$, or by $2\pi i \delta_{-}(k_{\parallel})$. Hence

$$\frac{1}{u-u'} = 2\pi i \delta_-(u-u') = \mathrm{P}\frac{1}{u-u'} + \pi i \delta(u-u') , \tag{52}$$

where P indicates that on integrating the principal value should be taken.

3.6. *The Balescu-Lenard equation*

We are now in a position to apply the ingenious argument of Lenard [26]. *The first step* is to conclude from (51) that $H(u)$ must be real. For, taking the imaginary part of (51), using (52), one obtains an equation for $\mathrm{Im} H(u) \equiv I(u)$ alone,

$$I(u) = -\frac{1}{k^2}\,\mathrm{P}\int_{-\infty}^{\infty} \frac{du'}{u-u'}\{F'(u)I(u') + I(u)F'(u')\} . \tag{53}$$

Any reasonable * function $\varphi(u)$ on the real axis can be uniquely decomposed in a "positive-frequency part" $\varphi_+(u)$, which is holomorphic in the upper half of the complex u-plane and vanishes at infinity, and a "negative frequency" part $\varphi_-(u)$, which is holomorphic and zero at infinity in the lower half plane. Moreover

$$\mathrm{P}\int_{-\infty}^{\infty} \frac{\varphi(u')}{u-u'}du' = -\pi i\{\varphi_+(u) - \varphi_-(u)\} .$$

With the aid of this fact (53) may be written

$$I_+ + I_- = (2\pi i/k^2)(F'_+ I_+ - F'_- I_-) .$$

The uniqueness of the decomposition implies that the positive and negative frequency parts cannot cancel but must vanish separately,

$$(k^2 - 2\pi i F'_+)\, I_+ = (k^2 + 2\pi i F'_-)\, I_- = 0 .$$

Hence $I_+ = I_- = 0$, so that $H(u)$ is real.

The second step is to solve Im $h_{\boldsymbol{k}}(v)$ from (50). Take the real and imaginary parts of (50), using the fact that $H(u)$ is real, and eliminate Re $h_{\boldsymbol{k}}(v)$ from both equations; by means of some long-winded but elementary algebra and using (51) one finds

* That is, belonging to the Lebesgue class L_2. See E.C.Titchmarsh, *Introduction to the Theory of Fourier integrals* (Clarendon Press, Oxford, 1937) ch. V.

$$\operatorname{Im} h_{\boldsymbol{k}}(\boldsymbol{v}) = \frac{4\pi^2}{k^2} \frac{\dfrac{\partial f(\boldsymbol{v})}{\partial u} F(u) - f(\boldsymbol{v}) \dfrac{\mathrm{d}F(u)}{\mathrm{d}u}}{|1+(2\pi i/k^2)F'_-(u)|^2} = \frac{4\pi^2/k^2}{|1+(2\pi i/k^2)F'_-(\boldsymbol{k}\cdot\boldsymbol{v}/k)|^2}$$

$$\times\, \boldsymbol{k}\cdot \int \delta(\boldsymbol{k}\cdot\boldsymbol{v} - \boldsymbol{k}\cdot\boldsymbol{v}') \left\{ \frac{\partial f_1(\boldsymbol{v})}{\partial \boldsymbol{v}} f_1(\boldsymbol{v}') - f_1(\boldsymbol{v}) \frac{\partial f_1(\boldsymbol{v}')}{\partial \boldsymbol{v}'} \right\} \mathrm{d}\boldsymbol{v}' . \tag{54}$$

The remarkable feature is that $H(u)$ has dropped out, so that it will not be necessary to solve the integral equation (51), which only served to show that $H(u)$ is real.

The third step consists in expressing the collision term in (45) in terms of $h_{\boldsymbol{k}}(v)$ and to note that, owing to $h_{-\boldsymbol{k}}(\boldsymbol{v}) = h^*_{\boldsymbol{k}}(\boldsymbol{v})$, only the imaginary part of $h_{\boldsymbol{k}}(\boldsymbol{v})$ contributes:

$$= -\frac{\epsilon}{4\pi} \frac{\partial}{\partial \boldsymbol{v}_1} \cdot \int \boldsymbol{F}(\boldsymbol{r})\, \tilde{g}(\boldsymbol{r}; \boldsymbol{v}_1, \boldsymbol{v}_2)\, \mathrm{d}\boldsymbol{r}\, \mathrm{d}\boldsymbol{v}_2$$

$$= -\frac{\epsilon}{4\pi} \frac{\partial}{\partial \boldsymbol{v}_1} \cdot \int \frac{\mathrm{d}\boldsymbol{k}}{8\pi^3} \boldsymbol{F}_{\boldsymbol{k}} \int g^*_{\boldsymbol{k}}(\boldsymbol{v}_1, \boldsymbol{v}_2)\, \mathrm{d}\boldsymbol{v}_2$$

$$= -\frac{\epsilon}{4\pi} \frac{\partial}{\partial \boldsymbol{v}_1} \cdot \int \frac{\mathrm{d}\boldsymbol{k}}{8\pi^3} \frac{-4\pi i \boldsymbol{k}}{k^2} h^*_{\boldsymbol{k}}(\boldsymbol{v}_1)$$

$$= \frac{\epsilon}{8\pi^3} \frac{\partial}{\partial \boldsymbol{v}_1} \cdot \int \frac{\boldsymbol{k}}{k^2} \{\operatorname{Im} h_{\boldsymbol{k}}(\boldsymbol{v}_1)\}\, \mathrm{d}\boldsymbol{k} .$$

Substitution of the result (54) finally gives for (45)

$$\frac{\partial f_1(\boldsymbol{v})}{\partial t} = \frac{\partial}{\partial \boldsymbol{v}_1} \cdot \int Q(\boldsymbol{v}_1, \boldsymbol{v}_2) \cdot \left\{ \frac{\partial f_1(\boldsymbol{v}_1)}{\partial \boldsymbol{v}_1} f_1(\boldsymbol{v}_2) - f_1(\boldsymbol{v}_1) \frac{\partial f_1(\boldsymbol{v}_2)}{\partial \boldsymbol{v}_2} \right\} \mathrm{d}\boldsymbol{v}_2 ,$$

where the tensor $Q_{ij}(\boldsymbol{v}_1, \boldsymbol{v}_2)$ is an abbreviation for

$$Q_{ij} = \frac{2e^4 n_0}{m^2} \int \frac{k_i k_j}{k^4} \frac{\delta(\boldsymbol{k}\cdot\boldsymbol{v}_1 - \boldsymbol{k}\cdot\boldsymbol{v}_2)}{|Z(\boldsymbol{k}, \boldsymbol{v}_1)|^2}\, \mathrm{d}\boldsymbol{k}$$

in the original units, and

$$Z(\boldsymbol{k},\boldsymbol{v}) = 1 + \frac{4\pi n_0 e^2}{mk^2} \int \frac{\mathrm{d}\boldsymbol{v}'}{\boldsymbol{k}\cdot(\boldsymbol{v}-\boldsymbol{v}')-i\epsilon} \left(\boldsymbol{k}\cdot\frac{\partial f_1(\boldsymbol{v}')}{\partial \boldsymbol{v}'}\right) .$$

Summarizing: Application of Bogolyubov's assumption to a homogeneous plasma has led to a kinetic equation for $f_1(\boldsymbol{v},t)$ alone, which is called the *Balescu-Lenard equation.*

Without the factor $|Z|^{-2}$ it is identical with the equation that Landau [8] derived from the Boltzmann equation for dilute gases by substituting in it a Coulomb force for the interaction. It is easily seen that in his case the integral for Q_{ij} diverges at $k = 0$, that is, for large distance between the interacting particles. He remedied this divergence by cutting off at the Debye length, but in the Balescu-Lenard equation this is automatically taken care of by the factor $|Z|^{-2}$.

On the other hand, the integral for Q_{ij} still diverges at $k = \infty$, that is, for close encounters. This is due to the fact that the dynamics of the collisions have not been treated correctly but replaced with the "impulse approximation", which breaks down for close collisions. The remedy is to cut off the small values of the impact parameter, corresponding to "hard" collisions, i.e., collisions which scatter the particles by an angle of the order of 90°. This corresponds to a cut-off at $k_{\max} = 3kT/e^2$. Landau used the same impulse approximation with the same cut-off, but only for convenience, because in his treatment it is easy to avoid this divergence by inserting the exact dynamics of the collision.

REFERENCES

[1] M. Kac, G. E. Uhlenbeck and P. C. Hemmer, J. Math. Phys. 4 (1963) 216 and 229; 5 (1964) 60 and 75;
N. G. van Kampen, Phys. Rev. 135 (1964) 362;
J. L. Lebowitz and O. Penrose, J. Math. Phys. 7 (1966) 98.
[2] J. A. Shercliff, *A textbook of magnetohydrodynamics* (Pergamon, Oxford, 1965).
[3] L. Spitzer, *Physics of fully ionized gases* (Interscience, New York, 2nd ed. 1962).
[4] A. Vlasov, J. Phys. (USSR) 9 (1945) 25.
[5] I. B. Bernstein, J. Greene and M. D. Kruskal, Phys. Rev. 108 (1957) 546.
[6] L. D. Landau, J. Phys. (USSR) 10 (1946) 25;
N. G. van Kampen, Physica 21 (1955) 949;
K. M. Case, Ann. Phys. (N.Y.) 7 (1959) 349.
See also e.g. T. H. Stix, *The theory of plasma waves* (McGraw-Hill, New York, 1962).
[7] T. Pradhan, Phys. Rev. 107 (1957) 1222;
B. U. Felderhof, Physica 29 (1963) 293.
[8] L. D. Landau, Phys. Z. Sowjetunion 10 (1936) 156.
[9] M. Rosenbluth, W. MacDonald and D. Judd, Phys. Rev. 107 (1957) 1.

[10] R.Balescu, Phys. Fluids 3 (1960) 52.
[11] A.Lenard, Ann. Phys. (N.Y.) 10 (1960) 390.
[12] N.G. van Kampen, Phys. Rev. 135 (1964) 362.
[13] I.Oppenheim and N.G. van Kampen, Phys. Fluids 7 (1964) 813.
[14] B.U.Felderhof, Physica 31 (1965) 295.
[15] J.E.Mayer, J. Chem. Phys. 18 (1950) 1426.
[16] R.Dekeyser, Physica 32 (1965) 1405.
[17] N.Rostoker and M.N.Rosenbluth, Phys. Fluids 3 (1960) 1.
[18] V.L.Ginzburg, *The propagation of electromagnetic waves in plasmas* (Pergamon, Oxford, 1964).
[19] H.Derfler and T.C. Simonen, Phys. Rev. Letters 17 (1966) 172; J.H.Malmberg and C.B.W.Wharton, Phys. Rev. Letters 17 (1966) 175.
[20] N.G. van Kampen, Physica 21 (1955) 949.
[21] N.N.Bogolyubov, in: Studies in Statistical Mechanics, vol. 1, eds. J.De Boer and G.E.Uhlenbeck (North-Holland, Amsterdam, 1962).
[22] N.G. van Kampen, in: *Fundamental Problems in Statistical Mechanics*, ed. E.G.D.Cohen (North-Holland, Amsterdam, 1962).
[23] E.G.D.Cohen, these proceedings.
[24] B.D.Fried and H.W.Wyld, Phys. Rev. 122 (1961) 1.
[25] R.Guernsey, Phys. Fluids 5 (1962) 322.
[26] A.Lenard, Ann. Phys. (N.Y.) 10 (1960) 390.

GENERAL REFERENCES

W.B.Thompson, *An Introduction to Plasma Physics* (Pergamon, Oxford 1962).

R.Balescu, *Statistical Mechanics of Charged Particles* (Interscience, New York 1963).

D.C.Montgomery and D.A.Tidman, *Plasma Kinetic Theory* (McGraw-Hill, New York 1964).

N.G. van Kampen and B.U.Felderhof, *Theoretical Methods in Plasma Physics* (North-Holland Publishing Co., Amsterdam 1967).